THE WAY
SCIENCE
WORKS

THE WAY SCIENCE WORKS

MACMILLAN • USA

The Way Science Works

MACMILLAN
A Simon & Schuster Macmillan Company
1633 Broadway
New York, NY 10019

Created, edited and designed by
Duncan Baird Publishers,
Castle House, 75-76 Wells Street,
London W1P 3RE

Library of Congress Cataloging-in-Publication Data

The Way science works : an illustrated exploration of technology in
    action / foreword by John Durant
        p. cm.
    Includes bibliographical references and index.
    ISBN 0-02-860822-4
    1. Technology—Encyclopedias. 2. Science—Encyclopedias.
T9.W32 1995
603—dc20   95–19702
                                                                CIP

10 9 8 7 6 5 4 3 2 1

Managing Editor: Marek Walisiewicz
Editor: Tom Ruppel
Researcher: Melanie Porte
Art Editor: Ted McCausland
Design Manager/Senior Designer: Iona McGlashan
Designers: Jill Mumford, Roger Hutchins
Picture Researcher: Jan Croot
U.S. Editor: Mary Ann Lynch

Typeset in Times NR MT, Times Ten, Trade Gothic and Goudy
Reproduction by Colourscan Overseas
Manufactured in China by Imago

Contributors
Scott Beagrie
Clifford Bishop
David Brodie
Tobias Chapman
Chris Cooper
Paul Doughty
Nigel Dudley
Professor John Durant
Roger Ford
Dr. Ian Killilea
Peter Lafferty
Eleanor Lawrence
Dr. Steve Matcher
Dr. Michael Mooney
Guy Norris
Paulette Pratt
Dr. Rebecca Renner
Graham Ridhout
Len Sanford
Giles Sparrow
David Tymm
Dr. Wendy Waddington

Illustrators
Main contributors:
Mike Badrocke
Hugh Dixon
Eugene Fleury
Ron Haywood
Trevor Hill
Ed Stuart
David Russell
Leslie D. Smith
Others:
Roy Flooks
Mick Gillah
Pavel Kostel
Roddy Murray
Jim Robins
Colin Rose
Peter Sarson
Colin Salmon
Tony Townsend

Consultants
Martina Blum
Dr. Ralf Bodemann
Dr. Hans-Liudger Dienel
Dr. Christine Gündisch
Dr. Birte Hantke
Dr. Walter Hauser
Dr. Manfred Hoffmeister
Stefan Ittner
Dr. Cornelia Kemp
Dr. Matthias Knopp
Peter Leitmeyr
Ulrich Marsch
Luitgard Marschall
Dr. Elke Müller
Peter Schimkat
Dr. Stefan Stein
Dr. Gudrun Wolfschmidt

Foreword

I once heard a story about an anthropologist who was studying one of the Indian communities of the South American rain forests. As anthropologists do, he spent many months living in his chosen community, and of course he got to know the people quite well. One day, he explained to his new friends that he had to go back to his own tribe for a few weeks; among other things, he needed a fresh supply of paper and pencils with which to work. This explanation caused the anthropologist's friends some puzzlement. They asked what paper and pencils were made of; and on discovering that the answer was wood, they said, "Well, there are plenty of trees here; why don't you make some more paper and pencils for yourself?"

Science and technology have transformed our lives, but not quite in the ways we often think. Obviously, they have given us a great deal of new knowledge and power. From automobile-making to computer-networking, and from hydro-electric power generation to heart transplantation, we can do things that our pre-scientific ancestors never dreamed of; but this is not because each of us is cleverer or more capable than they were. The fact is that we live in an age of specialists, and we depend on others for most of what we need in our daily lives. Where pre-scientific cultures are astonishingly self-sufficient, we are in the same position as the anthropologist: we look to others to provide the paper and pencils.

There are many advantages in specialization, but there is at least one huge disadvantage. In our complicated world, there is a real risk that we may lose touch altogether with the scientific and technological dimensions of our everyday lives. With the accelerating pace of scientific and technological change, this risk increases all the time. It took millennia to move from pencil and paper technology to printing, but only centuries to move from printing to broadcasting, and just a few decades to move from broadcasting to interactive electronic networking. Today, we are faced with an apparently endless cascade of new communications technologies, and it is a brave person indeed who will predict what lies ahead.

Why is it important for us to keep in touch with science and technology? Well, for one thing we need to know about them. We need to comprehend our world – to understand why, for example, we've been able to put men on the moon but not (as yet) to find a cure for cancer, and why we can see back in time using powerful telescopes, but cannot predict for certain the fate of the universe. We need to cope with our world – to make sensible personal decisions about, for instance, lifestyle and health care. And we need to contribute effectively to our world – to make informed judgments about political issues involving science and technology, such as energy policy, the information superhighway, and environmental pollution. In each case, we can live far more effectively if we understand some science and technology.

Foreword

Science and technology are not merely useful; they are also interesting. We human beings are curious creatures, and we want to know what's going on. Children are perhaps the most curious of all, and it's often been said that scientists and technologists are really grown-ups who've never given up their childhood curiosity about how the world works, or might be made to work. How is it possible for machines to defy gravity, to play games, or to create accurate three-dimensional images of the inside of someone's head without causing so much as a scratch on the skin? You don't need to know these things in order to fly, to play computer chess, or to receive medical treatment; but if you have even an ounce of curiosity, it is impossible to do any of them without at least wondering.

The Way Science Works is intended for anyone who has a sense of wonder about science and technology. It provides an ideal introduction to the inventions with which we work, run our homes, and entertain ourselves. It deals both with traditional, "harder" technologies involved in, for example, transportation, power generation, and manufacturing, and with modern, "softer" technologies such as information technology and biotechnology; and in addition, it introduces many of the key scientific principles that underlie these technologies. For each topic, a clear description accompanies imaginative illustrations that reveal step-by-step how things really work.

It is impossible to predict the future. The only thing we can say with confidence is that, partly because of scientific discovery and technological innovation, the future will be different from the past. If we are to cope with a rapidly changing world, we shall need to be scientifically and technologically literate; and *The Way Science Works* is an ideal way to start.

Professor John Durant
The Science Museum, London

About this book

The end products of technological innovation are the machines and processes that shape our everyday lives. *The Way Science Works* is organized into seven dynamic themes – Transportation, Information and Entertainment, Domestic Technology, Power and Manufacturing, Research and Medicine, Space, and Molecules and Materials – that reflect the way these products are put to use. These main sections are divided into double-page spreads, each of which addresses a particular aspect of technology. On each spread, examples of products or key manufacturing processes are anatomized in a main full-color illustration, and fully explained by the accompanying captions. The supporting illustrations, photographs, and main text provide a broader context, and give an introduction to the scientific principles drawn upon by that branch of technology.

Each double-page spread of the seven main sections can be read as a self-contained article that leads to a rapid understanding of an isolated topic. But to make full use of the book, the *Connections* listed at the bottom of every spread should be consulted. Here, the subject in question is cross-referenced to related technologies. This reflects the fact that we live in a complex society, in which the development of one technology depends on the existence of another, and manufacturing processes are highly interdependent. Oil rigs, for example, are used to supply refineries, which in turn provide fuel oil for power stations and feedstock for the chemicals industry: and oil rigs are serviced by giant tankers, and maintained by divers or remotely-operated submersibles. By tracing the relationships laid out by the *Connections,* it is possible not only to build up a more complete picture of how an oil rig functions, but to gain an appreciation of how different technological disciplines and industries are related.

The *Connections* serve another very important purpose, referring to spreads in the 64-page *Principles* section at the end of the book. Here, short, self-contained brief essays illustrated with simple diagrams, explain the basic science that is drawn upon in technological design. The treatment of these fundamental principles is entirely non-mathematical and illuminated with everyday examples, making it immediately comprehensible to the layperson.

Contents

Car production lines operate 24 hours a day. Repetitive operations, such as the spot welding of body panels, are carried out with unerring precision by robots.

1
Transportation

Internal Combustion Engine

Cars: Systems

Cars: Safety

Cars: New Developments

Trains

Railroad Systems

Jet Engines

Aircraft: Aerodynamics

Aircraft: Systems

Helicopters

Aircraft: New Developments

Hovercraft and Hydrofoils

Ships

Underwater Technology

Navigation Aids

The Seat of Power

How an internal combustion engine works

Every year, industry produces more than 300 million vehicles, almost all of them powered by a version of the internal combustion engine. These motors can range from tiny single-cylinder models that propel model aircraft, to the 20-liter turbocharged diesel engines of construction trucks that produce the output of more than fifteen family car engines. The basic design of the gasoline engine has changed little in over a century since its invention by Benz and Daimler, but today's engines are hundreds of times more powerful and efficient.

A car traveling at speed is being pushed along by over a hundred tiny explosions taking place inside its engine each second. In a typical *Otto-cycle* motor, a highly combustible mixture of fuel and air is sucked into a cylinder above a moving piston, and compressed to about one-eighth of its original volume. A spark plug ignites the mixture, driving the piston downward. The up-and-down movement of the piston is converted by a crank and connecting rod into a rotation in the crankshaft. In an automobile this is used to turn the wheels, but the rotation can equally be coupled to the propeller of an aircraft or an electricity generator.

Combustion is a highly *exothermic* chemical reaction, releasing a large amount of heat that increases the pressure of the exhaust gases. In addition, the gas products of the reaction have a greater volume than the fuel and air mixture they are produced from, magnifying the effect. The difference in gas pressures means that more force is produced by the exhaust gases pushing the piston downward than is needed to compress the original air and fuel mixture. The excess energy is the useful output of the engine.

Very few engines run constantly at the same speed, or need to provide a constant power output. As an automobile starts to climb a hill, its engine will be called upon to provide more power while turning at the same rate. The driver's response is to press down on the accelerator pedal, which has the effect of letting a greater volume of air and fuel into each cylinder. More fuel means a more powerful explosion, and thus a greater turning force on the crankshaft. In the past, a carburetor was used to control fuel/air flow, but today this is usually done by an electronic fuel injection and engine management system. This controls the air/fuel mixture, together with the timing of the ignition spark, so that the engine constantly works in the most efficient – and the least polluting – way possible.

Engineering alternatives

In a four-stroke cycle each cylinder only produces power for a quarter of the time. This explains why most production motors have four cylinders, so that one of them is always producing force to keep the crankshaft turning. Using even more cylinders means that the power strokes of the pistons overlap, and output is smoother, so 6-, 8-, or 12-cylinder engines are popular on luxury cars.

Keeping cool
Internal combustion engines produce more waste heat than useful power. Thus a cooling system [C] is needed to prevent overheating. A "jacket" of pipes surrounds each cylinder. Water pumps through these and is cooled by air flowing past the fine pipes that make up the radiator. The whole system is pressurized so the water can heat up to 250°F without boiling.
The smooth running of the engine is ensured by the oil supply. A pump sucks oil from the sump, filters it, and then feeds it at high pressure to the crankshaft and camshaft bearings.

At the fourth stroke
Most automobile engines use the four-stroke Otto cycle [A]. In the induction stroke [1], the turning crankshaft moves the piston downward, sucking a mixture of fuel and air through the open inlet valve. This closes as the piston rises in the compression stroke [2], squeezing the fuel and air. At the top of the piston's travel, the spark plug ignites the mixture. The heat produced rapidly expands the gas, forcing the piston down in the power stroke [3]. When the piston begins to rise through the exhaust stroke [4], the exhaust valve opens and lets the spent gases out.

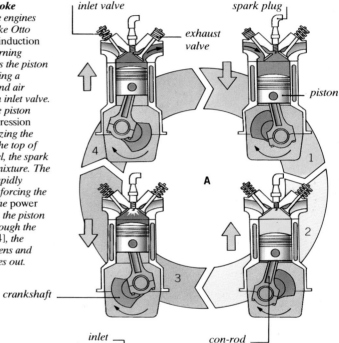

Oil burner
A diesel engine [B] has a similar configuration to an Otto-cycle engine. Air is sucked in [1] and compressed into a space 22 times smaller than its original volume [2], a compression ratio far greater than that of a spark-ignition engine. This great pressure heats the air to high temperature. When fuel oil is injected at the top of the stroke [3], it spontaneously ignites, the explosion driving the piston down again [4]. When the motion of the crankshaft pushes the piston back upward once more, the exhaust valve opens, allowing the spent products of combustion to escape [5].

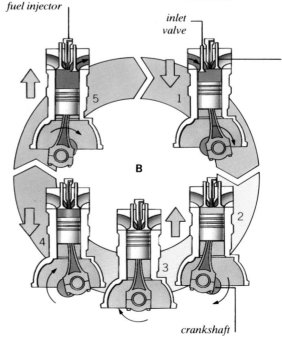

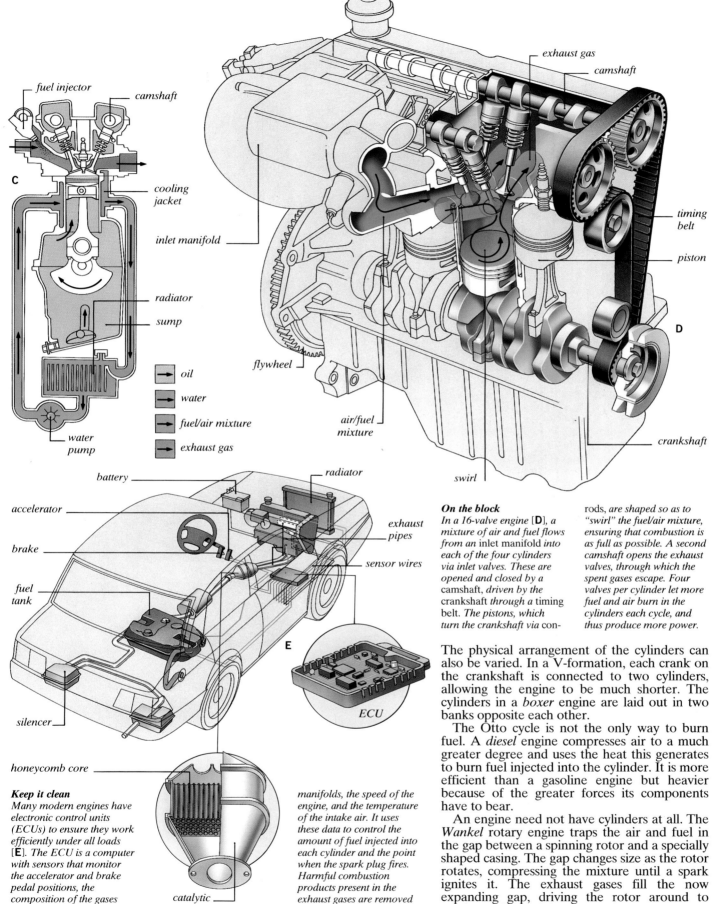

fuel injector

camshaft

C

cooling jacket

inlet manifold

radiator

sump

→ oil

→ water

→ fuel/air mixture

→ exhaust gas

water pump

exhaust gas

camshaft

timing belt

piston

D

crankshaft

flywheel

air/fuel mixture

swirl

battery

radiator

accelerator

brake

exhaust pipes

fuel tank

sensor wires

silencer

E

ECU

honeycomb core

catalytic converter

Keep it clean

Many modern engines have electronic control units (ECUs) to ensure they work efficiently under all loads [**E**]. *The ECU is a computer with sensors that monitor the accelerator and brake pedal positions, the composition of the gases in the inlet and exhaust*

manifolds, the speed of the engine, and the temperature of the intake air. It uses these data to control the amount of fuel injected into each cylinder and the point when the spark plug fires. Harmful combustion products present in the exhaust gases are removed by a catalytic converter.

On the block

In a 16-valve engine [**D**], *a mixture of air and fuel flows from an inlet manifold into each of the four cylinders via inlet valves. These are opened and closed by a camshaft, driven by the crankshaft through a timing belt. The pistons, which turn the crankshaft via con-*

rods, are shaped so as to "swirl" the fuel/air mixture, ensuring that combustion is as full as possible. A second camshaft opens the exhaust valves, through which the spent gases escape. Four valves per cylinder let more fuel and air burn in the cylinders each cycle, and thus produce more power.

The physical arrangement of the cylinders can also be varied. In a V-formation, each crank on the crankshaft is connected to two cylinders, allowing the engine to be much shorter. The cylinders in a *boxer* engine are laid out in two banks opposite each other.

The Otto cycle is not the only way to burn fuel. A *diesel* engine compresses air to a much greater degree and uses the heat this generates to burn fuel injected into the cylinder. It is more efficient than a gasoline engine but heavier because of the greater forces its components have to bear.

An engine need not have cylinders at all. The *Wankel* rotary engine traps the air and fuel in the gap between a spinning rotor and a specially shaped casing. The gap changes size as the rotor rotates, compressing the mixture until a spark ignites it. The exhaust gases fill the now expanding gap, driving the rotor around to produce a lightweight source of power.

On the Move

How car transmissions work

In the course of a single century, the automobile has evolved from an expensive plaything of the rich to an everyday means of transportation for the masses. The huge demand for automobiles – around 40 million are made each year – leads manufacturers to strive constantly to improve comfort, safety and economy. These considerations, together with the constraints of mass-production engineering mean that most automobiles follow the same basic design. Nevertheless, hand-built vehicles capable of speeds in excess of 220 miles per hour remain popular with the wealthy.

A modern automobile is a combination of several complex systems, each designed to control one aspect of its behavior. Beneath their widely differing bodies, most automobiles are similar. In almost every case there is an *internal combustion engine*, driving the front wheels of small automobiles, or the rear ones of larger models, via a *clutch* and *transmission*. The engine, though likely to have electronic systems to maximize power while minimizing pollution, is essentially the same type that was fitted to automobiles since they were first manufactured.

Modern automobiles look very different, however. Their flowing and rounded shapes are perfected using computer-aided design programs to ensure that the minimum amount of air is disturbed as they move forward. This reduces the amount of *drag* – wind resistance – and so helps to reduce fuel consumption, as well as allowing higher speeds to be achieved.

In gear

An automobile engine typically produces its useful power when it spins at between 3000 and 5000 rpm. But the road wheels that push the automobile along need to turn around at the most only 1000 times each minute. So an automobile needs a transmission, or *gear train*, operated either manually or automatically. Both types work by bringing down the fast-spinning input from the engine to the small number of revolutions per minute needed by the wheels. Engines only develop useful power over a relatively narrow band of speeds, so several gear ratios are needed to provide for the wide range of speeds an automobile travels at.

A manual transmission is driven from the engine through a clutch, which enables the drive to be broken when a new ratio is being selected. A plate on the engine flywheel is coated with a high-friction material. A powerful spring pushes this against a similarly coated plate connected to the input of the gear train, transmitting power. Depressing the clutch pedal makes the spring arch away from the driven wheel, freeing it and cutting off the power.

Automatic transmissions do not have a clutch. Instead the engine's power flows through a *torque converter*. This has no direct connection between input and ouput. Instead the engine turns an *impeller* which throws oil out, turning a *reactor* turbine attached to the transmission. The fluid coupling allows slippage – when the parts move at different speeds – but can be

doorpillar

box sections

A

spring/damper

wishbone

B

track rod

rack

pinion

collapsible section

spring/damper unit

MacPherson strut

front subframe

anti-roll bar

Shaped for safety
A modern automobile body [**A**] *is a* monocoque *– its strength comes from the shape of its precisely formed and welded sheet-metal panels. Areas where the stresses are highest, such as doorpillars, are made of even stronger* box-sections *and the engine and suspension are carried on separate subframes. At the front and rear the panels and floorpan are designed to crumple in a collision, so that the energy of an impact is absorbed progressively, reducing the forces felt by the passengers. Side-impact protection bars* strengthen *the more vulnerable flanks of the vehicle.*

Steering group
One of the most popular types of front suspension is the MacPherson strut *system* [**B**]. *The spring/damper unit allows the wheel to move up and down, while the lower* wishbone *stops any back and forth movement. An anti-roll bar* connects *the* *two sides of the suspension to keep the automobile level when cornering. The steering wheel is connected, via a collapsible section for safety, to a small pinion. This moves a toothed* rack *from side to side. Track rods* connected *to its ends transmit this movement to the wheels.*

Connections: Internal Combustion Engine 12 Cars 14 16 18 Oil Refining 186 Principles 220 222 224 228 230 236 238 240 242

made to lock up at higher speeds for maximum efficiency. The gears are shifted either hydraulically or, on more sophisticated automobiles, by a computer. This monitors the road speed, accelerator position, and engine load, and decides the best time to shift ratios.

Keeping the wheels on the road

The suspension is one of the most difficult parts of an automobile to design. It has to perform the task of keeping the wheels in contact with the road, whether the vehicle is braking, accelerating, or cornering. A variety of different linkages ensures that the wheels can move up and down, but also that they do not move from front to rear under braking and acceleration. Springs provide the amount of vertical movement needed. However, like all sprung things, automobiles would tend to bounce after passing a bump if they were not fitted with shock absorbers or *dampers*. These are telescoping cylinders filled with thick oil that resist rapid movement, stopping up-and-down oscillation.

The driven wheels of most small, front-wheel drive automobiles are held in a *MacPherson strut* suspension. This consists of a spring mounted outside a long telescopic damper, forming a vertical link and a lower wishbone. The top of the damper is bolted to the suspension tower, a specially strengthened area of bodywork, and its lower end is prevented from moving back and forth by a wishbone assembly.

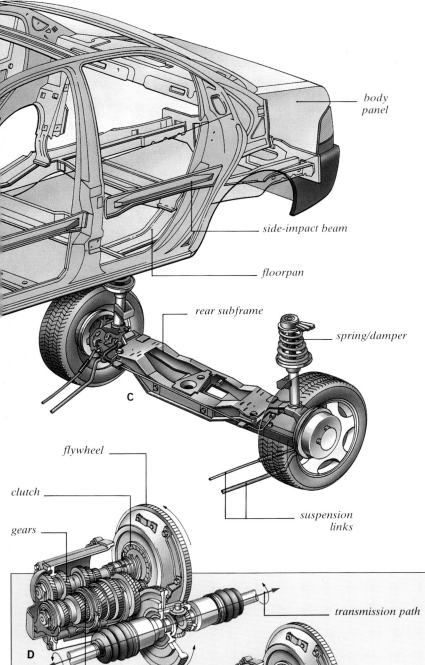

body panel

side-impact beam

floorpan

rear subframe

spring/damper

C

flywheel

clutch

gears

suspension links

Holding the line
An automobile's rear suspension [C] performs a comparatively complex job. To keep the moving wheels perpendicular to the road, many different links to the bodywork are needed. Advanced designs actually make the wheel point inward slightly when the

automobile is cornering, to give more balanced handling. Coil springs absorb the vehicle's movement, but on their own would make the car bounce long after going over a bump. Shock absorbers or dampers stop this happening by opposing any rapid movement.

transmission path

D

dog-clutch

E

driveshaft

differential

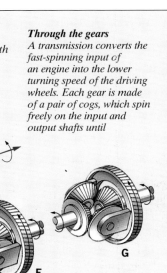

F

G

Through the gears
A transmission converts the fast-spinning input of an engine into the lower turning speed of the driving wheels. Each gear is made of a pair of cogs, which spin freely on the input and output shafts until

connected rigidly by dog-clutches. First gear [D] uses a small gear on the input shaft to turn a larger cog on the output shaft. In fifth gear [E] the cogs on each shaft are a similar size so the ouput shaft spins much more quickly. From the transmission, power travels to the wheels via a differential, which allows the driving wheels to turn at different rates when cornering. When the vehicle is traveling straight ahead [F], both driveshafts turn at the same speed. If one wheel is held stationary [G], the conical gears turn so all power goes to the free wheel.

Driving Safely

How cars help drivers avoid and survive accidents

Although the car has revolutionized modern life, it is also the cause of many thousands of fatalities each year. A fast-moving car is a two-ton projectile that can cause immense damage if it goes out of control. The structure of a vehicle is immensely strong and can survive huge forces, but the human body is not so robust, and ways must be found to reduce the forces that act on a body in an accident. Safety belts are a simple but effective example: their use has been shown to reduce certain types of injury by more than seventy percent.

A family car weighs roughly two tons and can travel at 90 mph. This means that in a collision a lot of kinetic energy must be dissipated if the passengers are to survive. A vehicle's safety systems work in two ways: firstly, they make the car's structure absorb the worst of the impact; and secondly, they directly restrain and protect the passenger's body.

The car's steel body is designed to crumple progressively when hit from the front or rear. If it were rigid, the car and its passengers would be brought to a halt almost instantaneously in a collision. Crumpling allows for slower deceleration, reducing the forces on the passengers. Engine mountings are designed to push the powerplant safely beneath the passenger compartment in an impact. And the fuel tank is situated in a rigid area between the wheels where it is less likely to burst.

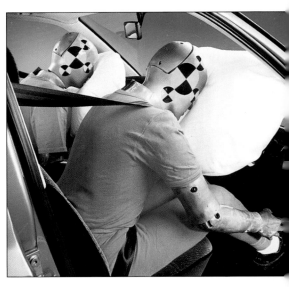

Air cushions
Although a safety belt should be worn at all times when driving, airbags (right) give further protection in a collision. Microswitches detect the sudden deceleration of an impact and trigger the inflation of the bags in a 100th of a second.

Calling a halt
*A car's brakes are under hydraulic control [**A**]. Pushing down on the brake pedal causes a master cylinder to move. Hydraulic fluid transmits this movement to slave cylinders in the wheel hubs. Because the master cylinder has a smaller area than the slave cylinders, the force exerted by the foot is greater by the time it reaches the wheels.*
The rear wheels are fitted with drum brakes. Hydraulic pressure in the slave cylinder pushes two pistons outward, pressing brake shoes against the inside of a drum, which turns with the wheel. Friction slows the drum's rotation. The rear brakes can also be operated by a system of cables (green) connected to the handbrake.
A car's weight is thrown forward as it decelerates, so more powerful brakes are needed on the front wheels. These are usually disk brakes of a different design to the rear drum brakes. Hydraulic pressure against a

piston in a calliper pushes two high-friction pads against the sides of a metal disk, slowing its rotation. In cars fitted with antilock brakes (ABS), sensors on each wheel send signals (red) to a computerized control box. This adjusts the hydraulic pressure applied to each brake (blue).

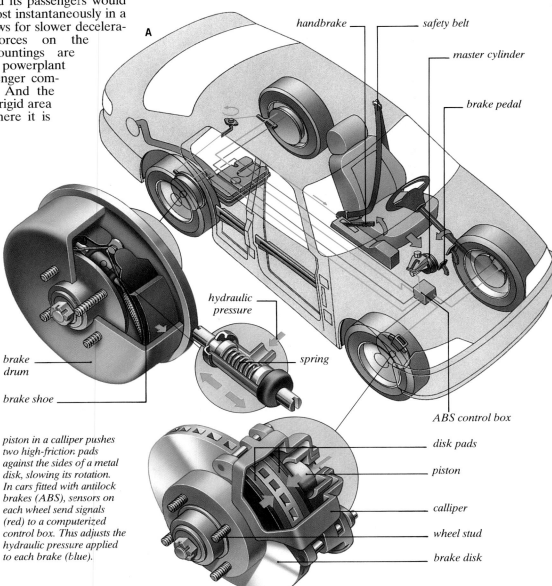

handbrake

safety belt

master cylinder

brake pedal

A

hydraulic pressure

spring

brake drum

brake shoe

ABS control box

disk pads

piston

calliper

wheel stud

brake disk

Tightening grip
*A safety belt only works effectively if it is held tight against the wearer's body. Many cars now have pre-tensioners to ensure that this happens and the body cannot slip forward under the lap belt [**B**]. Normally the belt can be pulled out from its spool freely [3]. But the violent deceleration of an accident causes a mechanism to swing forward, firmly clamping the webbing of the belt so that no more can be played out [4]. At the same time a second mechanism pulls the buckle assembly sharply downward, and presses the wearer tightly against the seat [1,2].*

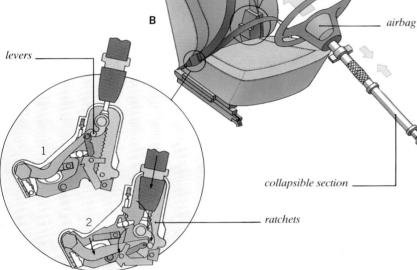

belt locked

spool

swings left

B

levers

airbag

collapsible section

ratchets

Inside the passenger cabin are many other safety features. Dashboards and other exposed surfaces are free of sharp edges that could cause injury. Safety belts prevent passengers being thrown through the windshield, and head restraints guard against whiplash. However, a safety belt works only if it holds the occupant's body tightly against the seat. When a belt is put on it is difficult to ensure that it has the right tension and that the fit is correct. Some new automobiles are fitted with belt pre-tensioners that tighten the belt as soon as it is put on, and then tighten it even more in the sudden deceleration of a collision.

Airbags, inflated by mechanisms that sense the jolt of a collision, deflate slowly as a body hits them, making its deceleration more gradual and less harmful.

Accident prevention

A car also has mechanical features that prevent accidents from ever happening. Safe roadholding is maintained only if the tires keep a constant grip on the road as the car moves along. In icy or wet conditions hard braking can make the wheels lock up so that the car skids and is impossible to control. Equally dangerously, if too much power is transmitted to a wheel, it is liable to break its grip and start spinning. Antilock brake and traction control systems monitor and control these dangerously unstable occurrences.

Planned weakness
*Impact with the steering column in an accident [**B**] is cushioned by an airbag, which emerges from the steering wheel boss 1/100th of a second after impact. At the same time a cagework section in the steering column allows it to collapse safely.*

Control at all times
*ABS [**C**] works by pulsing the hydraulic pressure to an automobile's brakes several times a second, quickly releasing and reapplying the brakes so that a skid cannot develop. TCS (traction control system) works in a similar way to avoid dangerous wheelspin.*

Keeping control
*In icy conditions, a car's wheels may lock completely under heavy braking [**C**], causing the driver to lose control of the vehicle. This is prevented by the ABS (anti-lock braking system). When braking begins, the wheels start to turn more slowly [1]. If a wheel locks [2], a sensor detects its lack of movement and sends a signal to the ABS control unit. This releases the hydraulic pressure in the pipe leading to the brake, allowing the wheel to turn once again [3]. The brake is then reapplied, and the process begins again. ABS allows a driver to steer the car safely when braking.*

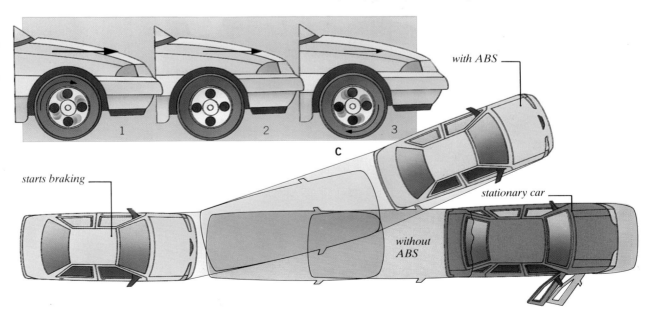

with ABS

starts braking

stationary car

without ABS

C

The Road Ahead

How the vehicles of the future may look

The state of California has set a worldwide trend with laws decreeing that by 1998 two percent of all new cars should be zero emission vehicles – automobiles that make no direct contribution to the high levels of pollution choking modern cities. In 2003 this figure will rise to ten percent. These clean vehicles will most likely be powered by electricity, produced by a variety of new batteries made from exotic and highly reactive metals. Another possible source of power may be highly efficient fuel cells, a product of the American space program.

At present, most vehicles are powered by some form of internal combustion engine. Despite the development of devices, such as catalytic converters and lean-burn engines, designed to reduce emissions, pollution from conventional engines is still a growing problem. And it will become increasingly difficult to refine the existing types of engine so that they comply with the stricter regulations of the future. For these reasons it seems likely that the vehicles of the future will be electrically powered.

Electric cars and trucks already exist, but their performance is severely restricted by the lead-acid batteries – normal car batteries – that they use. These powerpacks are heavy and can store only relatively small amounts of power, enough to give the vehicle a range of roughly 60 miles between recharges.

A *hybrid* vehicle has a much greater range. It is powered by electric motors, which draw current from normal batteries when driving around town. But on the open road, the driver can switch to a high-efficiency gasoline engine, which drives a generator to recharge the batteries and also power the electric motors. The batteries themselves will be new designs. *Sodium-sulfur* cells can match the performance of much heavier lead-acid batteries, but only operate at a dangerous 570°F. An alternative is the aluminum-air cell. This battery is not strictly rechargeable, but can be "refueled" by simply replacing its aluminum anode.

Fastest under the sun
The solar-powered car Sunraycer (right) triumphed in a 1900 mile race across Australia, powered by electricity generated in the solar cells covering its aerodynamic upper surface. The lightweight vehicle completed the race at speeds in excess of 50mph.

Future fuel
One possible design for a car of the future [A] uses a hydrogen-oxygen fuel cell to produce the electricity needed by its electric motors. The fuel cell is a highly efficient way of extracting energy from a fuel (see box) that was first developed for use in the manned missions of the American space program. Hydrogen and oxygen are potentially explosive substances, so a working car would probably obtain its hydrogen by the chemical breakdown of methanol, a much safer substance to keep on board. The oxygen needed would be extracted from the air.

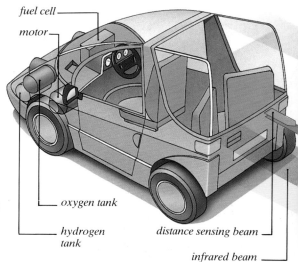

fuel cell

motor

oxygen tank

hydrogen tank

distance sensing beam

infrared beam

Fuel Cells

Car designers are always searching for ways of increasing fuel efficiency – the percentage of the chemical energy in a fuel converted into the kinetic energy of movement. A diesel engine has an efficiency of around 35 percent. An aluminum-air battery, charged directly from the domestic electricity supply, has an efficiency of no more than 50 percent.

But the highest efficiency – more than 70 percent – is provided by *fuel cells*. These, like batteries, harness a chemical reaction to produce electricity. But, as in an engine, the reactants in a fuel cell have to be constantly replenished. Most of these devices utilize the reaction between hydrogen and oxygen, which in the open air can be dangerously explosive. In a fuel cell, however, the process takes place more controllably, between ions dissolved in a liquid electrolyte.

Hydrogen power
A fuel cell consists of a carbon anode and cathode immersed in an electrolyte. Both electrodes contain platinum, which acts as a catalyst, breaking down oxygen molecules into atoms as the gas bubbles past the cathode. Here oxygen absorbs electrons while combining with water to form hydroxide ions. At the same time hydrogen gives up electrons at the anode as it combines with hydroxide ions to form water. The current that flows between the electrodes is used to drive a motor.

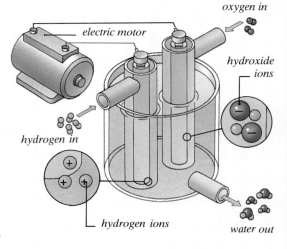

oxygen in

electric motor

hydroxide ions

hydrogen in

hydrogen ions

water out

Electronic guidance

Almost all of the systems in a future vehicle will be under electronic control. Computers will monitor the use of power to ensure maximum efficiency. Steering, braking, and acceleration may all be controlled by a single joystick, connected to motors and steering gear through fiber-optic data links. Active suspension systems, in which a vehicle's attitude is constantly adjusted by computer-controlled hydraulic rams, will ensure the smoothest possible ride and the safest roadholding.

Computers may even take over the job of a driver. Infrared sensors would constantly monitor a vehicle's distance from one in front. Using these sensors, a computer would control each vehicle, communicating with similarly equipped cars to form "trains" of vehicles traveling at high speed along busy highways. The short distances separating the cars would make optimum use of existing road space.

High efficiency engines
*Two-stroke engines waste oil and are noisy. But their high fuel efficiency makes them an attractive power unit for the future. An improved design [**D**] has a cycle that can be split into 4 steps. Air is pumped into the cylinder as the piston rises, and a dose of fuel is added through an injector [1]. The air/fuel mixture is further compressed [2] until the piston reaches the top of its stroke, when a spark ignites the mixture [3]. Hot exhaust gases drive the piston downward, turning the crankshaft, before being scavenged [4], blown out of the exhaust valve by air pumped through the inlet.*

New power generation
*A car of the future [**A**] may use hybrid power, drawing electricity from a generator driven by an efficient 2-stroke motor or from a bank of sodium-sulfur batteries. Each battery contains concentric layers of sodium, aluminum, and sulfur [**C**]. As sodium atoms [1] diffuse into the layer of aluminum [4], they give up electrons [2] that flow to the anode. Sulfur atoms [3] absorb electrons from the metal cathode and combine with sodium ions to form sodium sulfide.*

*The voltage that this process creates between anode and cathode can be used to power compact electric motors in each front wheel hub. The car can be slowed by regenerative braking: the motors act as generators, converting the energy of motion back into electricity, which is stored again in the batteries [**B**].*

The ride is smoothed by active suspension, controlled hydraulically by computer. Control over braking, acceleration, and steering is combined in a single joystick. Next to its headlights, the car has infrared lights: a night camera picks up reflected infrared, and projects a night-vision image in front of the driver on a head-up display. A second infrared beam measures the distance to the vehicle in front.

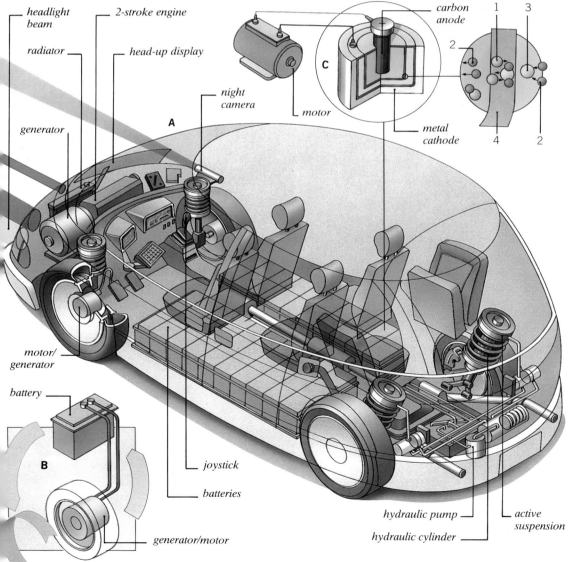

exhaust

scavenging fan

inlet

headlight beam

2-stroke engine

radiator

head-up display

night camera

motor

carbon anode

metal cathode

generator

joystick

batteries

motor/ generator

battery

generator/motor

hydraulic pump

hydraulic cylinder

active suspension

On the Fast Track

How high-speed trains are making continents smaller

Every day, *Eurostar* trains depart from London, England, each carrying 800 passengers to the French and Belgian capitals of Paris and Brussels in just three hours. Running on three separate rail networks at speeds approaching 185 miles per hour, these are among the fastest and most advanced trains ever built. Eurostar's twelve electric motors generate more than 12,000 kilowatts of tractive power needed to pull the 1310-foot-long, 900-ton train; and its sophisticated control and safety systems ensure that one driver can operate the train single-handedly between cities.

The latest generation of high-speed passenger trains are pulled by advanced electric locomotives that pick up high-voltage power from overhead lines or electrified rails. In areas where it is not practical or economic to electrify lines, diesel-electric locomotives are used. Their primary power supply is a diesel engine, coupled to an electric generator. The current this produces is fed to motors mounted in the *power bogies*. (The wheels of a locomotive or car are not directly connected to the bodywork, but are held in separately sprung chassis or *bogies*. These isolate the passenger compartments from the vibration of the wheels and pivot to go around corners.)

Crossing borders

Eurostar is an electric train derived from the 185 mph French *Train à Grande Vitesse* (TGV), which has been in service since 1981. A specially adapted TGV set the world speed record for a passenger train of 320 mph. The train is capable of such high speeds because of its streamlined shape and its lightweight, articulated structure. Normal trains have two bogies per car, one at each end. TGVs instead have one bogie between each pair of neighboring cars (apart from the power cars), halving the number needed per train. This saves around five tons of weight per car.

Major changes to the bodywork and power systems of the standard TGV had to be made for the Eurostar train. First, it was slimmed down to fit the narrower British *loading gauge* – the maximum permitted height and width of the rolling stock. Second, safety features were fitted for the journey through the Channel Tunnel; the most important of these is the capacity to uncouple the train in the middle so that the two halves can be pulled out of the tunnel separately in an emergency. Third, the train's power systems were modified. Although all the lines that the train runs on are electrified, power supply varies from country to country. In France and through the tunnel a 25 kV alternating current (a.c.) is carried through overhead wires. This is picked up by a folding contact called a *pantograph*. Belgium also has overhead lines, but these carry a 3 kV direct current (d.c.). In England there is a 750 kV direct current supply running in a third rail. To function on all lines, Eurostar is therefore equipped with two pantographs and, next to the wheels, a set of sliding shoe contacts for the third rail.

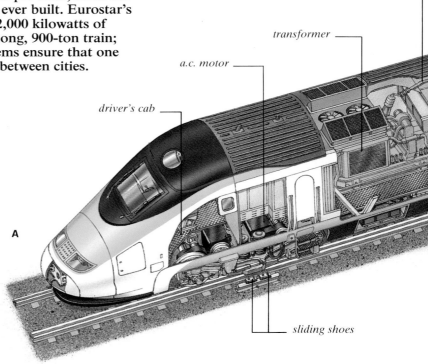

power line

transformer

a.c. motor

driver's cab

A

sliding shoes

Uniting Europe
*The electricity that drives the Eurostar train [**A**] undergoes several conversions before finally reaching its motors. Most of the time, a* pantograph *picks up 25 kV a.c. power from an* overhead catenary. *Silicon* diodes *and* thyristors *inside an* inverter *convert this to an 1800 V d.c. supply that is passed to the three* motor blocks *(one for each power bogie – two on each driving car and a third at the front of the first car). These are boxes of power electronics that reconvert the d.c. to a.c., but at a frequency matching the turning speed of the motors. The power output is altered by changing the voltage of the supply sent to each motor.*

A second pantograph is needed for the Belgian 3 kV d.c. supply, which passes directly to the motor blocks, as does the 750 V d.c. electricity picked up by sliding shoes from the third rail in the UK.

All Eurostar trains are reversible – either of the two locomotives at each end can be used to lead the train of eighteen cars. All the systems, including the air conditioning, power doors and brakes in each car, are centrally controlled via an optical *data-bus* – a digital data highway running the length of the train.

Calling a halt

Even more important than the way the train is made to move is how it is stopped. Eurostar has two complementary braking systems. *Rheostatic braking* uses the train's motors as generators that convert the moving train's *kinetic energy* into an electric current. This then flows through large resistors, or *rheostats*, where it is converted into heat energy. At lower speeds these are backed up by high-powered disk brakes, with four disks on each axle. The computers on board each car automatically blend together the two types of stopping power.

Power station on wheels
*A diesel-electric locomotive, such as the British Inter-City 125 [**B**], is used where electrification of a railway is not feasible. Inside, a twelve cylinder turbocharged diesel engine burns fuel oil to turn a high-capacity a.c. generator. Unlike Eurostar, the 125 has d.c. motors, so a rectifier is needed to convert the a.c. to a d.c. supply. The motors and the gearboxes connecting them to the wheels are mounted inside the locomotive's bogies. The diesel-electric arrangement enables the engines to run at their most efficient and least polluting speed when most power is needed – when a train pulls out of a station.*

Connections: Internal Combustion Engine 12 Railroad Systems 22 Electricity Transmission 116 Principles 236 238 240 246 248 250

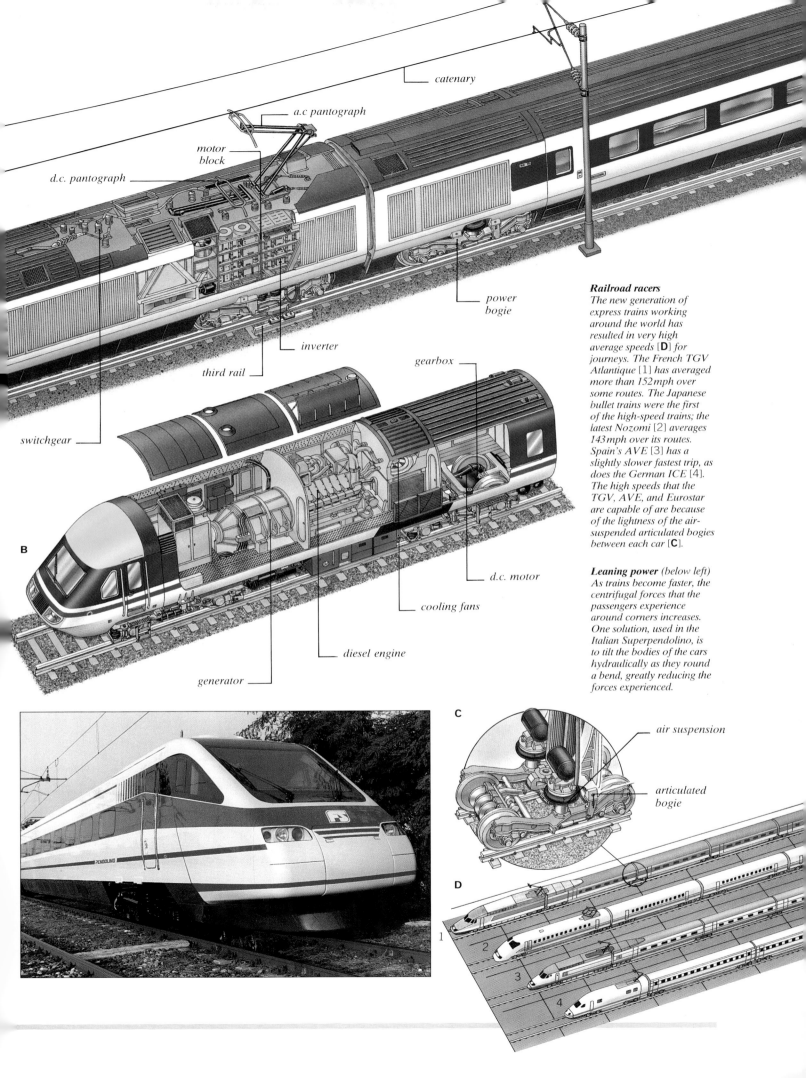

catenary

a.c pantograph

motor block

d.c. pantograph

power bogie

inverter

third rail

gearbox

switchgear

B

d.c. motor

cooling fans

diesel engine

generator

Railroad racers
The new generation of express trains working around the world has resulted in very high average speeds [D] for journeys. The French TGV Atlantique [1] has averaged more than 152mph over some routes. The Japanese bullet trains were the first of the high-speed trains; the latest Nozomi [2] averages 143mph over its routes. Spain's AVE [3] has a slightly slower fastest trip, as does the German ICE [4]. The high speeds that the TGV, AVE, and Eurostar are capable of are because of the lightness of the air-suspended articulated bogies between each car [C].

Leaning power *(below left)*
As trains become faster, the centrifugal forces that the passengers experience around corners increases. One solution, used in the Italian Superpendolino, is to tilt the bodies of the cars hydraulically as they round a bend, greatly reducing the forces experienced.

C

air suspension

articulated bogie

D

1
2
3
4

Signaling the Way
How train movements are controlled

The huge volume of passengers that a modern railroad must handle necessitates frequent, fast-moving trains. On the French TGV Nord line, for example, trains running at more than 185 miles per hour are separated by a time interval of just three-minutes. Only a complex signaling system can ensure that the trains, which may take several miles to brake to a halt, are a safe distance apart. Almost all such systems rely on a constant traffic of electric signals passing along the rails, sensing a train's presence and keeping the driver informed of dangers ahead.

A signaling system's task is to maintain a safe distance between trains. In the first railroad networks this was done by merely holding a train in a station until a fixed time had passed since the last train left. Thankfully, modern systems are not so risky. The track is divided into *blocks* – sections that a train cannot enter if occupied by another train. Signals, like traffic lights at the trackside, tell a driver how to proceed. Red means stop, green means track clear, and single and double yellow lights warn the driver to brake to a lower speed.

The French *Train à Grande Vitesse* (TGV) travels so quickly that a driver would find it easy to miss trackside signals. Instead, the maximum permitted speeds in the current and next blocks are displayed inside the cabin. The trains pick up instructions either from *transpon-ders* – radio transmitters buried in the track – or from electric currents in the tracks themselves.

Standard gauge
The two rails that form a track have to be a fixed distance (the gauge) apart, usually 56.5 in [C]. Each axle of a train presses down with many tons of force, so the welded track needs to have some "give," or resilience. Traditionally this came from gravel ballast,
but modern track is bolted to concrete sleepers, themselves set in a bed of even more concrete. A porous mat between the sleepers and the rails provides resilience and, together with plastic clips, ensures that the rails are electrically isolated.

Moving freight
The cars in a freight train may share a common origin but their destinations may all be different. They are reordered in a line yard [**B**]. *A locomotive pushes a set of uncoupled cars up to the crest of a hill. Here, a barcode reader scans a code the side of each car to determine its destination. The car rolls down the hill and is automatically directed, via sets of switches, into one of many sidings where a train for one destination is built up. Because each car rolls at a different rate down the slope, its speed is monitored and checked by* retarders *on the tracks.*

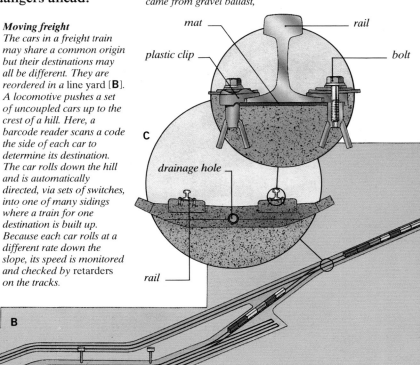

frog
swing rails
sidings
mat — rail
plastic clip
bolt
C
drainage hole
rail
A
B
uncoupled wagons
barcode reader

Turning out
*A train changes track at a set of switches [**A**]. Two swing rails* move sideways *to direct the train. The highly stressed* frog *at the crossing point is made of strong manganese steel.*

Magnetic Levitation

Aircraft, because they are freed from surface friction, are able to travel much faster than land-based forms of transportation. Similarly, if electromagnets are used to "levitate" a train, holding it slightly above a track, it can travel at high speeds almost without friction.

Such magnetic levitation, or *maglev*, trains work on a very simple principle – the weight of the train is supported by the repulsive forces between the like poles of two sets of magnets, one in the train itself and the other in the track. The force increases as the poles are pushed together, so that variations in weight are automatically accounted for.

The *maglev* has no wheels to provide the tractive force to push it along. Instead, it is powered by a *linear motor*. Functioning like a normal electric motor that has been cut open and laid out flat on the track, this pulls the train along at speeds in excess of 300 mph.

Connections: Trains 20 Computers: Networks 82 Electricity Transmission 116

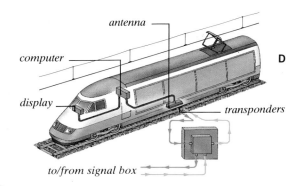

antenna

computer

display

D

transponders

to/from signal box

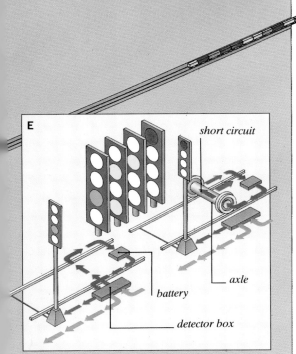

E

short circuit

axle

battery

detector box

Radio signals
On some lines, signaling information is passed to the train driver by transponders *on, or buried beneath, the track* [**D**]. *As a train passes overhead, it emits a radio signal that "interrogates" the transponders. They reply by sending digitally coded information about the speed limit on the present stretch of track, the condition of the track ahead, and the distance to the next transponder. These messages are decoded by a computer onboard the train and displayed in the cab. In some metro systems, transponders are used to control the trains fully automatically.*

Circuit breakers
The presence of a train in a track sector is detected by a track circuit [**E**]. *At its simplest this is a battery connected between the two rails. Current flows from one rail to the other via a detector box, connected to the signaling system. When a train enters the sector, the current is short-circuited through its metal wheels and axles. The detector box notes the missing current and sets the signal at the beginning of the sector to red. Preceding signals are set to yellow, double yellow and green, to warn of danger ahead.*

Automatic safety
On an automated railroad, computers do three signaling jobs – protection, operation, and control. Automatic Train Protection (ATP) keeps trains a safe distance apart. Using *track circuits*, small electric currents used to tell when a train has entered a block, the signaling computers can set the signal immediately behind the train to red.

Points and signals, which route and control the trains, are operated through an *interlocking*. In early signal boxes this was an arrangement of bolts that prevented a signalman from setting the points in a way that let a train take a dangerous route. Today, the same job is done by microprocessors in a *solid-state interlocking* (SSI). A computer is programmed with all the possible combinations of signals and points positions in a stretch of track, and allows only safe ones to be chosen. For safety, the SSI uses three microprocessors at the same time. If one microprocessor disagrees with the other two, it is assumed to have failed and is shut down.

The electronic messages sent through the rails or transponders can be used to control the train's speed. A simple version makes a sound in the driver's cab every time a yellow signal is passed. If this is not acknowledged by pressure on the appropriate button, the brakes are automatically applied. Braking occurs automatically if the train passes a signal at red.

More complicated signaling systems completely control a train's movement. As a train enters a track circuit section, it receives two messages giving the maximum speeds in the section it is in and the next one that it will enter. If the track is clear ahead, the train is made to run at a constant speed. But if there is a slower train in front, a lower speed is set and the train brakes automatically. At a station, the system brings a train smoothly to a halt.

Low flying
In a simplified maglev train, unlike magnetic poles (here pink and yellow) on the track and train attract one another, pulling the train forward [1]. *Like poles (yellow-yellow; and red-red) repel one another, augmenting this effect. After the train has moved by a small distance, the polarity of its electromagnets is suddenly reversed, providing a repulsive force that keeps the train aloft* [2]. *After the train's inertia has carried it farther along, forward attractive and repulsive forces come into play again.* [3]. *The huge magnetic fields needed are generated by electromagnets containing superconducting wires, chilled to* 10 K.
A real maglev train, such as the German Transrapid, has a slightly different design, with electromagnets held in pods slung underneath the train [4,5].

forward motion

magnets

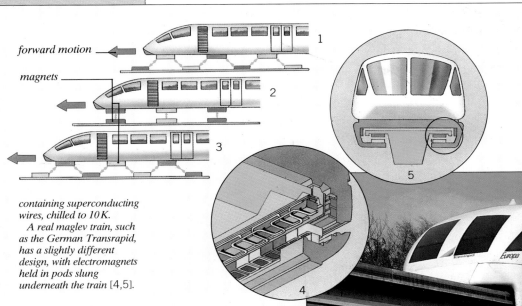

1

2

3

5

4

Air Power

How the jet engine works

Modern jet engines are the most efficient power units ever developed. Weighing around 5 tons, they can deliver a thrust of up to 500,000 newtons – enough to lift a 55-ton deadweight vertically. The jet engine has revolutionized air travel, allowing aircraft to fly faster, higher, and more economically than previously possible. But jet engines, also known as gas turbines, have other, less obvious applications: they are used to turn the propellers of warships, to power trains and hovercraft, and to drive industrial machines, such as high-speed drills and pumps.

Since the first flight of a jet aircraft – a German Heinkel HE-178 – in 1939, the jet engine has been in constant development. Today, nearly all commercial and military aircraft are powered by advanced derivatives of the jet engine.

Like the automobile engine, the jet is an internal combustion engine in which a mixture of air and vaporized fuel is burned. As the temperature rises, the mixture of gases produced in combustion increases rapidly in volume. This hot, expanding mixture is then put to useful work. In the automobile, it drives a set of pistons coupled to a crankshaft, which in turn drives the wheels. However, the jet engine harnesses the energy of combustion more directly. The expanding gas is forced through a nozzle, where it accelerates before being expelled from the rear of the engine at high speed (typically more than 1240mph).

As the molecules of pressurized gas are accelerated backwards through the exhaust nozzle, they exert an equal and opposite force on the inner walls of the engine. This force, known as the *thrust*, is transferred through the engine casing to the fuselage and wings, and pushes the aircraft forward.

Thrust for flight

Thrust is equivalent to the mass of gas expelled from the engine multiplied by its acceleration. This means that the same thrust is produced by a large mass of gas accelerated to modest speed, as by a smaller mass of gas accelerated to higher speed. Engines that produce high gas acceleration run at high temperatures and produce a lot of noise. Early jet aircraft used this type of engine, known as a *turbojet*, but most modern commercial airliners are powered by *turbofans*, which accelerate greater volumes of air to lower speeds, and are consequently quieter and more fuel-efficient.

Jet engineering

A simple turbojet has three main parts – a *compressor* at the front, a *combustion chamber* in the middle, and a *turbine* at the rear. The compressor, which is made up of a series of rotating airfoil blades, works like a giant high-speed fan. It sucks air into the engine and forces it through a progressively narrowing duct, compressing it to less than one-tenth of its original volume, before pushing it into the combustion chamber. Because the air is under pressure, fuel burns more quickly, resulting in greater power and

Turbojet design
*Air drawn into a turbojet [**A**] is compressed to a pressure of 10 atmospheres by rows of fast-spinning blades. It is forced into a combustion chamber into which fuel is injected and burns continuously. The hot gas formed rushes past sets of turbine blades linked by an axle to the compressors at the front. The assembly of compressors, axle, and turbines is known as a spool. As air flows past each set of compressor or turbine blades, it is made to swirl. Stationary blades or* stators *[**B**] between each set of moving blades "straighten" the airflow, minimizing eddy currents, which would reduce engine efficiency.*

efficiency. In the combustion chamber, air is mixed with vaporized jet fuel (kerosene) and the fuel–air mixture is ignited.

The temperature of the gas in the chamber rises to more than 2550°F. The hot mixture expands, increasing in speed as it is forced out of the engine. Before the exhaust gas is expelled a fraction of its kinetic energy is used to drive a bladed turbine. A shaft couples the turbine to the compressor, causing the latter to turn at the high speeds needed to pressurize incoming air.

In a turbojet, the remaining kinetic energy of the gas mixture provides the thrust: but in a variant of the turbojet, known as a *turboprop*, an additional set of turbines extracts most of this energy and uses it to drive a shaft connected via a reduction gear to a propeller. Turboprops are particularly efficient at lower speeds and commonly power short-haul aircraft. In industrial gas turbines, the driveshaft turns an electrical generator, drill bit, or pump.

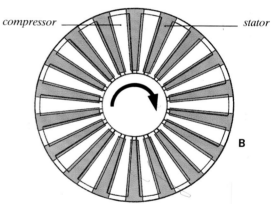

compressor — stator

B

A

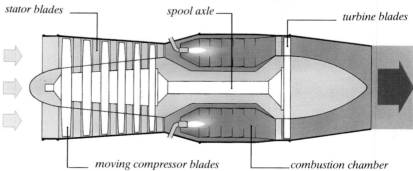

stator blades — spool axle — turbine blades

moving compressor blades — combustion chamber

B

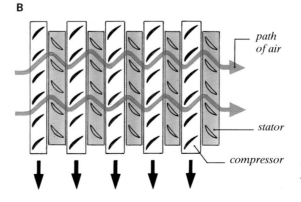

path of air

stator

compressor

Vertical takeoff (right) The Harrier fighter utilizes the jet's high power-to-weight ratio for direct lift, eliminating the need for a runway. Thrust is delivered via four rotatable nozzles, which point downward for vertical takeoff and landing, and rotate fully aft for forward flight.

Connections: Internal Combustion Engine 12 Aircraft: Systems 22 Helicopters 30 Ships 36 Hovercraft and Hydrofoils 34 Shaping Metals 190

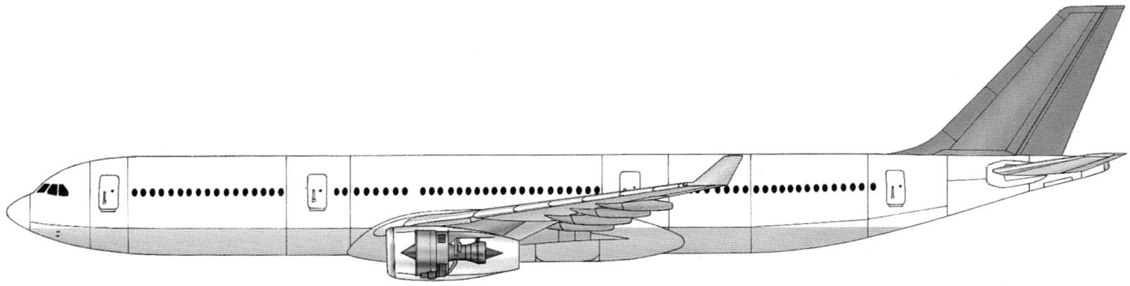
C

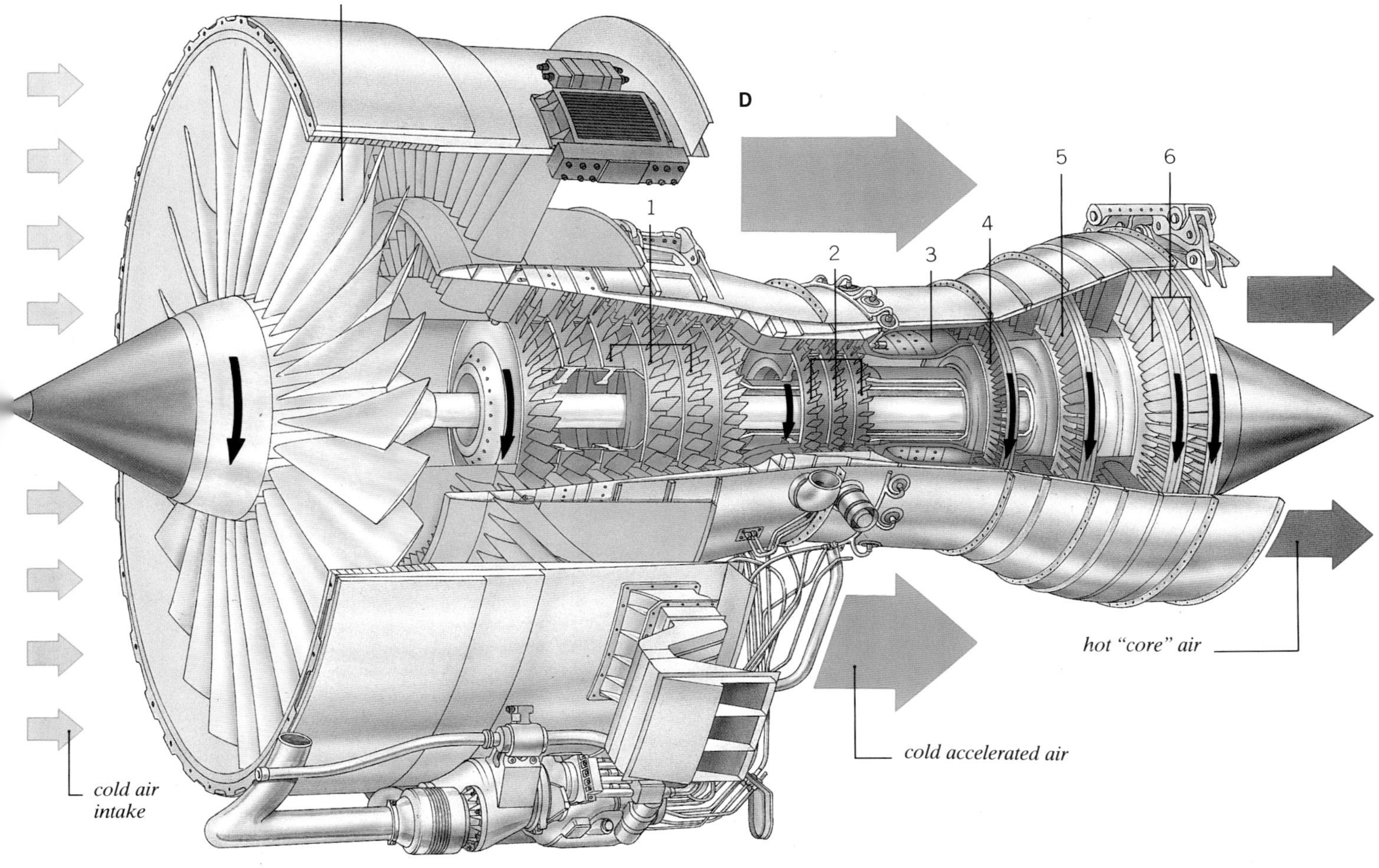
D
7
1
2
3
4
5
6
hot "core" air
cold accelerated air
cold air intake

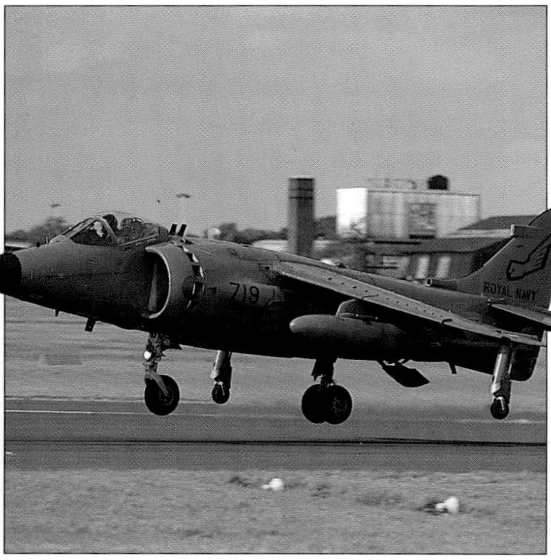

Turbofan technology

*Most modern airliners are powered by two, three, or four turbofan engines, usually mounted beneath the wings [**C**].*

*In essence, a turbofan [**D**] is a conventional turbojet fitted with an extra set of turbine blades: the rotation of the additional turbine is used to turn a huge fan.*

As in the turbojet, air is drawn into the engine core. Two rotating sets of compressor blades [1,2] squeeze the air to a pressure of 30 atmospheres before it enters the combustion chamber [3].

A hot, expanding mixture of burnt gas rushes out of the combustion chamber and past three sets of turbine blades [4,5,6] before being expelled through nozzles at the rear of the engine.

The first set of turbines (orange) is turned by exhaust gas at extremely high temperature and pressure. The small-bladed turbines are made to rotate extremely quickly. An axle couples their movement to small-bladed high-pressure compressors just in front of the combustion chamber. A second set of larger-bladed turbines (brown) is made to turn more slowly. Its movement is coupled to lower-pressure compressors. The third set of turbines (yellow) has still larger blades. These extract energy from the (by now) lower-pressure airstream, and transmit it via an axle to the huge fan at the front of the engine [7].

The fan rotates within a broad duct. The air "pushed" backward by the fan generates more than three-quarters of the engine's overall thrust. The remainder comes from the hot air pushed through the engine core. Because most of the engine's thrust results from a large volume of cold air accelerated by the fan, rather than from a small volume of hot gas accelerated through the core, the turbofan is quieter and more efficient than a standard turbojet.

Advanced Materials 194 Principles 224 234 238 240

On the Wing

How an aircraft flies

The idea of powered flight has always exercised a hold over the human imagination. But not until the early 20th century were the dreams of aviation's pioneers matched by their technical abilities. The Wright brothers brought together an understanding of aerodynamics and a new, lightweight power plant – the gasoline engine – to build the *Flyer*. The principles of aeronautics mapped out in its 12-second maiden flight on December 17, 1903 apply equally to today's giant airliners, which can carry more than 400 passengers across the Atlantic in a single hop.

To remain airborne an aircraft must experience an upward force, or *lift*, equal to or greater than its own weight. All heavier-than-air craft get their lift from aerodynamic forces produced when air passes over lifting surfaces, or *airfoils*. An aircraft's wings and tailplanes are airfoils, as are the rotors of a helicopter.

An airfoil generates lift by directing the flow of air around it so as to create an area of lower pressure on its top surface relative to its underside. This causes the wing to be simultaneously pushed and "sucked" upward. Although many geometrical shapes can produce lift in this way, the airfoil is specially designed to maximize lift while minimizing *drag* – the forces that slow an aircraft's passage through the air.

In flight, air meets the leading edge of an airliner's wing at a speed of several hundred miles per hour. Part of the airstream passes over the wing's steeply cambered upper surface, while the remainder is diverted around its flatter

Yaw, pitch and roll
A pilot steers an aircraft by moving flaps that change the shape of its airfoils – its wings, tailplanes, and rudder – and therefore change the amount of lift that each airfoil produces.

Left-to-right motion, or yaw [**A**], is controlled by a rudder (*mauve*) on the tail fin. When the rudder is moved to the left, the aircraft turns to the left.

Pitch [**B**] is altered by moving elevators (*green*) attached to the tailplanes. Moving the elevator up reduces lift generated by

the tail: the tail drops and the aircraft's nose is forced upward. Conversely, when the elevators are lowered, the aircraft's nose rises.

The aircraft is rolled [**C**] around its long axis by moving wing sections called ailerons (*orange*). Raising the aileron on the right wing and lowering the aileron on the left produces more lift on the left- than on the right-hand side, "tipping" the aircraft over to the right.

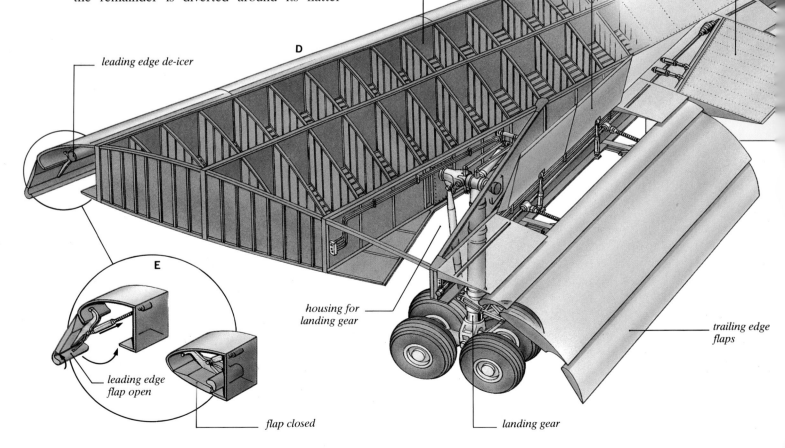

high-speed aileron

spoilers

fuel tanks

leading edge de-icer

D

E

leading edge flap open

flap closed

housing for landing gear

landing gear

trailing edge flaps

Connections: Aircraft: Systems 22 Helicopters 30 Hovercraft and Hydrofoils 34 Rockets 162 Space Shuttle: In Flight 164 Principles 230 236 238

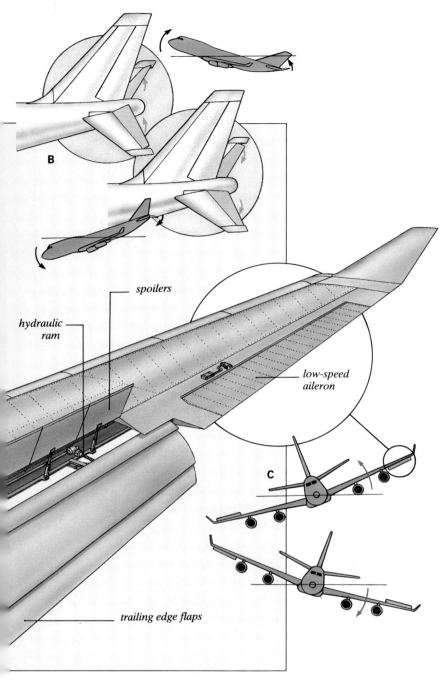

spoilers

hydraulic ram

low-speed aileron

trailing edge flaps

C

B

Shaped for flight
An airliner wing [D] is swept back at an angle of around 35° to minimize turbulence (and therefore drag) at high speed. A honeycomb structure of spars and ribs gives the wing strength while keeping it light and flexible. The "cells" between the ribs and spars hold fuel. Other sections house the landing gear assembly and hydraulic control lines. A duct on the leading edge carries hot air, which prevents ice build-up from spoiling the wing's aerodynamic profile.

The wing has an array of ailerons, flaps, and spoilers activated by hydraulic rams.

These let the pilot change the shape of the wing at different stages of a flight.
For example, at takeoff and landing, flaps are extended to increase the area and curvature of the wing, thus producing high lift at a low airspeed. Spoilers on top of the wing can be raised to induce drag: they are deployed when the pilot wishes to lose height rapidly, or after landing to slow the plane on the runway. During level cruising flight, the flaps and spoilers are retracted. Flaps are located on the trailing and leading edges [E] of both wings, extending down and forward to give extra lift for low-speed flight.

underside. Air passing over the top has a greater distance to travel before it rejoins the airstream behind the wing, and must accordingly move at a higher speed than the air beneath. According to Bernoulli's principle, a fast-moving fluid (in this case air) exerts less pressure than a slow-moving stream. So as air pressure above the wing falls below that on the underside, the wing experiences lift.

The pressure imbalance across the wing, and therefore the amount of lift, increases with greater airspeed: for this reason, aircraft must reach a high speed on the ground before take-off. Lift also increases with wing area and camber: so low-speed planes typically have long, steeply cambered wings, whereas the wings of high-speed craft tend to be shorter, and thinner and flatter in cross section.

Aerial acrobatics

A pilot can gain lift by tilting the aircraft's nose upward. This increases the angle that the wings present to the airstream – the *angle of attack* – which effectively increases the distance that air must travel over the top of the airfoil. At angles of attack of up to about 15°, extra lift is produced at the cost of some airspeed (unless engine power is increased to compensate). But at higher angles the smooth flow of air over the wings breaks up into eddy currents. This causes an abrupt drop in lift, known as *wing stall*, and the aircraft plummets downward, spinning out of control. Today's airliners are fitted with devices to warn of impending wing stall.

A pilot can change a wing's lifting characteristics by moving a series of flaps on its trailing edge. Lowering or extending the wing flaps exaggerates the airfoil section, providing additional lift at low speed – essential during takeoff and landing. Other movable airfoils on the wings and tail section – the elevators, ailerons, and rudder – allow for maneuvers such as *yaw, pitch,* and *roll.*

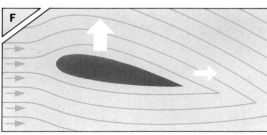

F

[F] airflow is smooth: lift is high and drag is low. At low speed, the pilot increases the angle of the wings into the wind to increase lift. Below a certain wind speed, air flow becomes irregular [G], forming eddies. At this stall *speed, lift is lost and the pilot must react quickly to retain control of the aircraft.*

Smooth flow
Drag – the force that slows an aircraft in flight – has a number of causes. It is due in part to friction, as the aircraft plows through the air. Drag is also induced when the smooth flow of air around the wings breaks down, and airflow becomes irregular. In fast, level flight

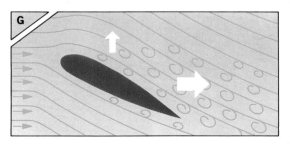

G

Fly by Wire

How computers are taking over the work of an airliner pilot

The pilot of a modern airliner can calmly pull back on the control stick sharply enough to put the aircraft into a deadly stall. But disaster is averted because there is no rigid link between the stick and the control surfaces; instead, computers monitor the pilot's requests and perform only what is safe. The cabin that the pilot sits in has changed: 20 years ago the Concorde boasted more than 130 different gauges for the aircrew to watch. Today, this jumble of individual dials has been replaced by computerized displays that alert the pilot only when something goes wrong.

Rather than a cumbersome collection of pulleys and wires, or bulky hydraulic tubes and cylinders, the latest airliners have no mechanical link between the pilot's controls and the aerodynamic surfaces. Instead there is a *fly-by-wire* system – several computers that monitor the pilot's requests on a joystick and pedals, converting them into electronic orders for *actuators*, motors that move the flight control surfaces. At the same time, the computers monitor the craft's airspeed and direction via an array of sensors.

The computers run software that describes the safe limits of the airliner's performance, and prevents these from being exceeded. Although this cuts down the responsibility that a pilot has for guiding the plane, it does have many benefits. The fast reaction time of the computers allows for the constant adjustment of ailerons, elevator, and rudder to correct for quirks in the aircraft's handling or a sudden updraft.

Connecting the computers, actuators and the various position sensors together is a *data-bus* – a single wire or fiber-optic cable that runs the entire length of the plane. The signals that it passes are digital. Each individual component receives every signal, but reacts only to those that carry its own code, vastly reducing the amount of wiring needed.

Signals from sensors can be fed to all the components that need them at the same time. The information from a *pitot tube*, which measures air pressure, is sent in digital form not only to the pilot's display, but also to the air conditioning and cabin pressurization systems.

Safety in numbers

Precautions have to be taken to ensure that the pilot can still control the plane if a computer develops a fault or the power supply fails. A combination of three computers control the plane's movements. Each is responsible for a different set of surfaces, but checks the orders sent out by the others for possible faults.

In the unlikely event that all three computers fail, there are a further two, much less powerful processors for emergencies. Some planes even have a simple set of mechanical links to be used in the last resort.

Connected to the data-bus is the flight management computer (FMC). Taking inputs from the various speed and position sensors, together with navigational data from satellites and ground beacons, this glorified autopilot can take

Under the skin
*The body of an airliner [**A**] combines great strength with low weight. Its fuselage is built from a skeleton of ribs and spars, made, like the skin covering them, from light aluminum alloy.*

More and more of the structure is now made up of composite materials – high-strength carbon or glass fibers woven like cloth and held rigid with a coating of resin. Composite parts are typically 20 percent lighter than the equivalent aluminum structure.

On the Airbus A-330, the most widely used alternative material is CFRP (Carbon-Fiber-Reinforced Plastic). This is strong and extremely light, and is ideal for highly stressed areas such as the ailerons and flaps. Toughened CFRP (TCFRP) forms larger structures such as the tail and floor beams, and a hybrid, or mixture of CFRP with another material, Nomex, is used on the rudder control surfaces.

The nose of the airliner houses a weather-radar dish, covered by a fiberglass radome. Its signals, along with those from a variety of

sensing instruments, are fed into the plane's data system.

The areas between the spars in each wing are sealed and used as fuel tanks. The airliner is balanced by pumping fuel along a network of pipes from tanks in the wings to tanks in the tail.

At the extreme rear of the craft is the APU (Auxiliary Power Unit), a small gas turbine which meets the electrical needs of the aircraft when the main engines are not running. Just in front of it is a domed bulkhead, which contains the pressurized atmosphere of the cabin.

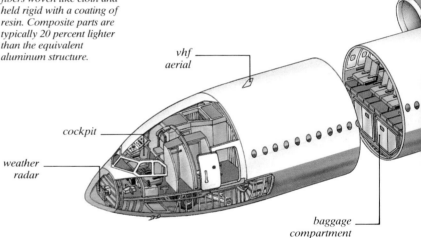

vhf aerial

cockpit

weather radar

baggage compartment

On-line information
Six computer monitors present the pilot of an Airbus A-330 (right) with flight data as and when it is needed. The simple cockpit design that results forms a stark contrast with the myriad dials crowded into the Concorde's 25-year-old cabin design (inset).

Connections: Jet Engines 24 Navigation Aids 40 Cellular Phones and Fiber Optics 48

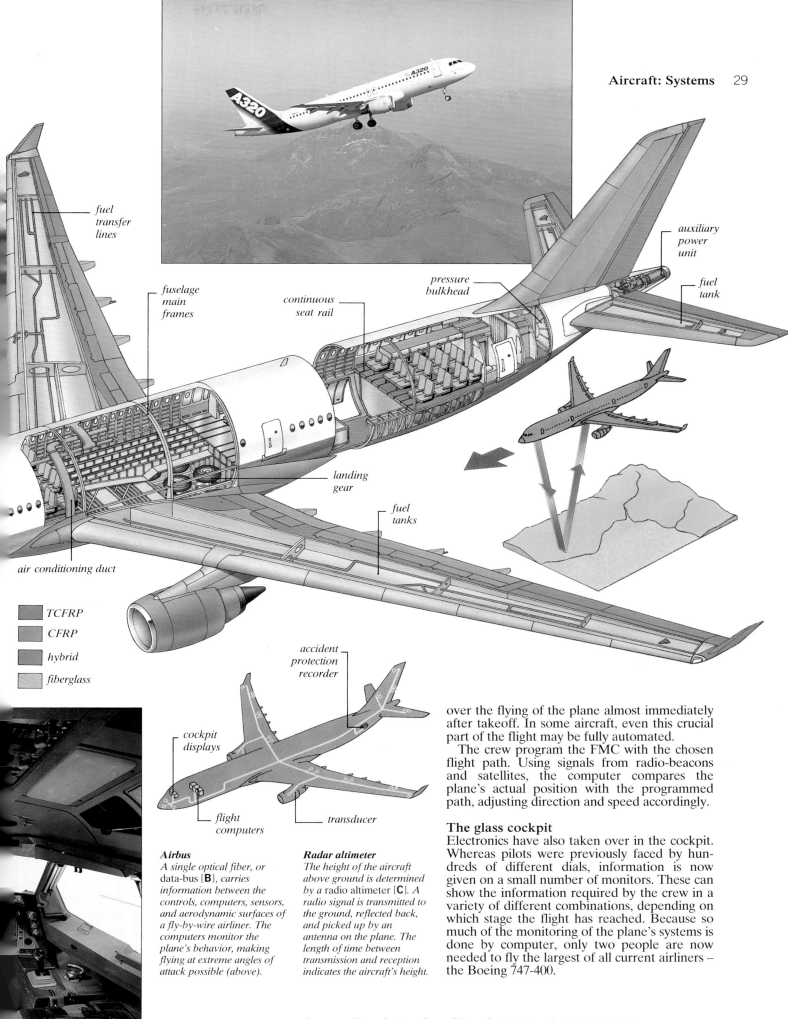

fuel
transfer
lines

fuselage
main
frames

continuous
seat rail

pressure
bulkhead

auxiliary
power
unit

fuel
tank

landing
gear

fuel
tanks

air conditioning duct

TCFRP
CFRP
hybrid
fiberglass

accident
protection
recorder

cockpit
displays

flight
computers

transducer

Airbus
*A single optical fiber, or
data-bus [B], carries
information between the
controls, computers, sensors,
and aerodynamic surfaces of
a fly-by-wire airliner. The
computers monitor the
plane's behavior, making
flying at extreme angles of
attack possible (above).*

Radar altimeter
*The height of the aircraft
above ground is determined
by a radio altimeter [C]. A
radio signal is transmitted to
the ground, reflected back,
and picked up by an
antenna on the plane. The
length of time between
transmission and reception
indicates the aircraft's height.*

over the flying of the plane almost immediately
after takeoff. In some aircraft, even this crucial
part of the flight may be fully automated.

The crew program the FMC with the chosen
flight path. Using signals from radio-beacons
and satellites, the computer compares the
plane's actual position with the programmed
path, adjusting direction and speed accordingly.

The glass cockpit
Electronics have also taken over in the cockpit.
Whereas pilots were previously faced by hundreds of different dials, information is now
given on a small number of monitors. These can
show the information required by the crew in a
variety of different combinations, depending on
which stage the flight has reached. Because so
much of the monitoring of the plane's systems is
done by computer, only two people are now
needed to fly the largest of all current airliners –
the Boeing 747-400.

Computers: Networks 82 Advanced Materials 194 Principles 250 254 258

Hanging in the Air

How helicopters are able to hover

Helicopters can reach places inaccessible to any other form of transportation. Able to hover motionless in the air, take off and land vertically, and maneuver in any direction, they have become essential equipment for a huge variety of tasks. The biggest helicopters are used as airborne cranes, lifting loads of up to 55 tons to the tops of buildings; smaller machines can be used for air–sea rescue or to dust crops and inspect pipelines. With no need for a runway, helicopters play a vital role in supplying ships, oilrigs, and remote settlements.

A conventional airplane flies because its wings generate upward force, or *lift*, as they slice through the air. A helicopter is held aloft by the same aerodynamic forces. It possesses between two and six rotor blades attached to a central hub. Each blade is essentially a long, thin wing, spinning at 300 revolutions per minute above the helicopter cabin. Whereas a fixed-wing aircraft must move at high speed to stay aloft, the "wings" of a helicopter are already moving, enabling it to hover, or climb and descend vertically, as well as fly along in any direction.

The angle at which a blade meets the air – its *pitch* – determines how much lift it generates. A helicopter is steered by adjusting the pitch of each rotor blade as it moves around its circular path, thereby controlling the amount of lift generated on different sides of the craft.

Increasing the pitch of all the rotor blades at the same time makes the helicopter climb vertically. Forward flight is achieved by increasing the pitch of each blade at the back of its journey around the central hub, and reducing its pitch as it passes around the front. This produces more lift at the back than at the front, pushing the helicopter's nose down and propelling the whole craft forward.

Flying ahead

As a helicopter flies forward, air rushes more quickly over the blades on one side of the rotor (the *advancing blades*) than those on the other side (the *retreating blades*). Air flowing faster over an airfoil increases lift, so the advancing blades create much more upward force than the retreating ones. This imbalance would tip the helicopter over if it was not corrected by the use of *flapping hinges*. These let the advancing blade pivot (or flap) upward, decreasing its angle of attack and reducing the lift it produces. The weight of the retreating blade forces it downward, increasing its lift.

Above a certain forward speed the retreating blade *stalls*, as the smooth airflow over its surfaces breaks up into swirling vortices. Lift is lost making the helicopter lose height and spin out of control. Because of this problem, the fastest helicopters have maximum speeds of only 250 mph – far slower than conventional aircraft.

A helicopter needs three to ten times more thrust to stay aloft than an airplane of similar size. Sufficiently powerful piston engines are too bulky and produce too much vibration to be used for anything other than the smallest craft, so most helicopters are powered by one or more *turboshaft* engines. This is a type of jet engine, in which the rotor blades are powered via a gearbox directly from the turbine shaft.

In a spin

The turning force, or *torque*, needed to turn the rotors "pushes" against the helicopter fuselage. By Newton's third law – for every force there is an equal and opposite reaction – this force tends to turn the fuselage in the opposite direction to the blades. In most helicopters this is counteracted by the *tail rotor*, a small propeller mounted on the tail boom that pushes air to one side, stopping the fuselage from spinning. Using foot pedals, the pilot can alter the pitch of the tail rotor blades, thereby varying the thrust they produce. The imbalance in the forces that results permits the pilot to point the helicopter to the left or the right.

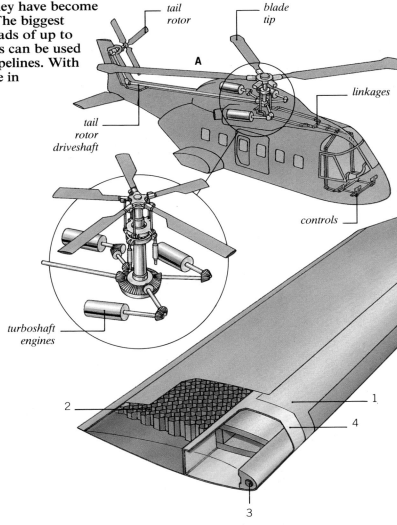

tail rotor — *blade tip*

A

linkages

tail rotor driveshaft

controls

turboshaft engines

2

1

4

3

Drive and direction
*Attaching the rotor blades to their driveshaft, the rotor head [**A**] is a mass of complex machinery. The drive from three turboshaft engines powers the shaft via bevel gears. A fourth shaft runs rearward to turn the tail rotor. Rods run up from the controls in the cockpit to the swash plate coupling on the main rotor and back to the tail rotor. Rubber elastomeric bearings allow the blades to flap up or down as their pitch is changed, and also let the blades lag as the drag increases. A damper between the hub and each blade prevents dangerous vibrations from building up.*

Connections: Jet Engines 24 Aircraft: Aerodynamics 26 Shaping Metals 190 Polymers 192 Advanced Materials 194 Principles 230 238 240 242

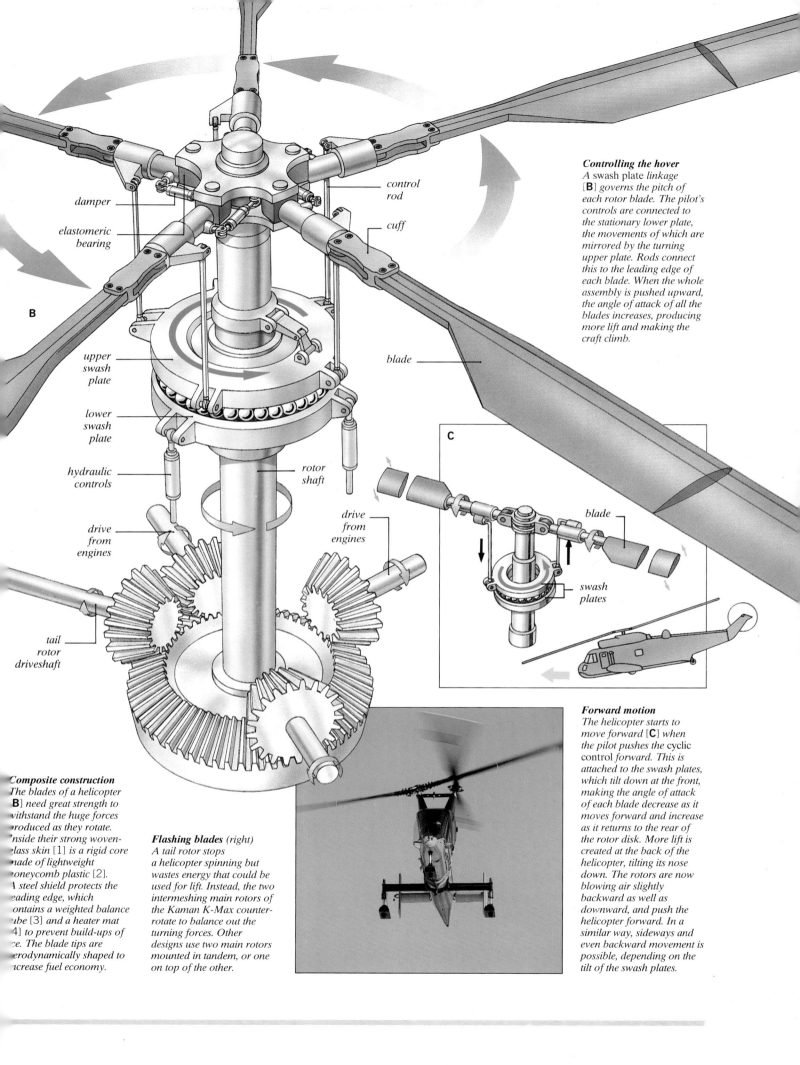

damper

elastomeric
bearing

B

upper
swash
plate

lower
swash
plate

hydraulic
controls

drive
from
engines

tail
rotor
driveshaft

control
rod

cuff

blade

rotor
shaft

drive
from
engines

C

blade

swash
plates

Controlling the hover
*A swash plate linkage
[**B**] governs the pitch of
each rotor blade. The pilot's
controls are connected to
the stationary lower plate,
the movements of which are
mirrored by the turning
upper plate. Rods connect
this to the leading edge of
each blade. When the whole
assembly is pushed upward,
the angle of attack of all the
blades increases, producing
more lift and making the
craft climb.*

Composite construction
*The blades of a helicopter
[**B**] need great strength to
withstand the huge forces
produced as they rotate.
Inside their strong woven-
glass skin [1] is a rigid core
made of lightweight
honeycomb plastic [2].
A steel shield protects the
leading edge, which
contains a weighted balance
tube [3] and a heater mat
[4] to prevent build-ups of
ice. The blade tips are
aerodynamically shaped to
increase fuel economy.*

Flashing blades (right)
*A tail rotor stops
a helicopter spinning but
wastes energy that could be
used for lift. Instead, the two
intermeshing main rotors of
the Kaman K-Max counter-
rotate to balance out the
turning forces. Other
designs use two main rotors
mounted in tandem, or one
on top of the other.*

Forward motion
*The helicopter starts to
move forward [**C**] when
the pilot pushes the cyclic
control forward. This is
attached to the swash plates,
which tilt down at the front,
making the angle of attack
of each blade decrease as it
moves forward and increase
as it returns to the rear of
the rotor disk. More lift is
created at the back of the
helicopter, tilting its nose
down. The rotors are now
blowing air slightly
backward as well as
downward, and push the
helicopter forward. In a
similar way, sideways and
even backward movement is
possible, depending on the
tilt of the swash plates.*

Flying into the Future
How aircraft will look in the 21st century

Increasing demand for air travel is pushing airliner development in four directions: bigger, faster, safer, and cleaner. Bigger means megatransports carrying up to 1000 passengers, double that of today's largest airliner, the Boeing 747. Faster means a supersonic airliner with twice the range and three times the capacity of the Concorde. New developments are also aimed at improving the safety and environmental acceptability of today's generation of airliners, to produce craft that are quieter and less damaging to the sensitive areas of the atmosphere through which they fly.

Airliners of the future may not have windshields at all. Instead, large flat-panel screens will display a computer-generated view of the outside. Using the data provided by an array of sensors, from video cameras to radar and lasers, all possible dangers will be superimposed on a detailed map of the area being flown through. Such virtual reality systems will be so accurate that planes will be able to land in practically blind weather conditions.

In the near future, commercial pilots will still have windows to look through, but will also be kept constantly informed of the plane's speed and heading via *head-up displays*. These use lasers to project the information onto the windshield, so that the pilot need never take his eyes off the scene outside.

Double decks

Every year, ever-greater numbers of people are taking to the air, making existing routes increasingly crowded. Airlines and manufacturers are addressing this problem by designing planes capable of carrying up to 1000 people over distances as great as 7500 miles. These new *superjumbos* will weigh around 500 tons when fully loaded, stretching the properties of current airframe materials to their limits.

Rather than simply building a longer fuselage, aircraft manufacturers are considering double- and even triple-passenger decks. Existing airports will not be able to accommodate

Megatransport
*The largest of existing airliners, the Boeing 747, can carry at most around 500 passengers. Its successor is now being planned, a giant capable of carrying around 1000 people [**B**]. To fit in existing airport spaces, the dimensions of the aircraft will be about the*

same as a 747, so a double-deck arrangement seems to be the best way to contain so many passengers. The simple ovoid shape [1] is not the only one being considered. Other possible concepts are a "clover-leaf" design [2] and a "horizontal double-bubble" [3], like two fuselages enclosed together.

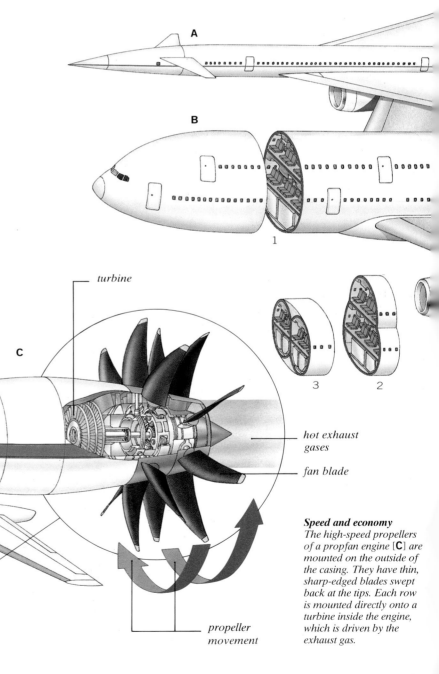

New blades
*The propfan [**C**] is a successor to two other types of jet engine. Turbofans contain turbines that turn huge fans mounted on the inside of their casings. The turbine in a slower but more economical turboprop, drives a propeller mounted via a transmission.*

turbine

hot exhaust gases

fan blade

Speed and economy
*The high-speed propellers of a propfan engine [**C**] are mounted on the outside of the casing. They have thin, sharp-edged blades swept back at the tips. Each row is mounted directly onto a turbine inside the engine, which is driven by the exhaust gas.*

propeller movement

Connections: Jet Engines 24 Aircraft 26 28 Helicopters 30 Space Flight: New Developments 180 Principles 236 240

All aboard
For the new superliners to be economic, passengers will have to be unloaded and loaded, together with their luggage, in as little as as twenty minutes. The empty plane will also have to be cleaned in this time, posing problems for already overcrowded airports.

Supersonic
At present there is only one supersonic airliner in service, the Concorde, but it is now an old design and in danger of becoming too polluting and noisy to be acceptable. So its successor is being designed.

The new aircraft [A] will be both larger – capable of carrying around 250 passengers compared with Concorde's 100 – and have a much longer range of 5000 miles. New, efficient engines with variable intake ducts and the extensive use of lightweight materials will enable the airliner to fly at twice the speed of sound – more than 1240 mph.

wingspans much greater than those of the present Jumbo, the Boeing 747-400, so new designs may have to incorporate folding wingtips. In other, more radical designs, known as *lifting bodies*, the increased upward force needed to keep the craft aloft is provided jointly by the wings and the fuselage, which is itself shaped like a giant airfoil.

A new Concorde

Supersonic flight is a difficult technical feat, full of compromises. Only one existing airliner, the *Concorde*, achieves this – at the expense of being cramped, greedy for fuel, and extremely noisy. International teams are working on its successor, which will have to be able to carry 250-300 people at least 5000 miles at altitudes above 65,000 ft. It will have to do all this but still keep within strict environmental rules governing noise levels and emissions that could further damage the sensitive ozone layer.

The engines of such a plane have to work in two sets of very different conditions. At takeoff and landing the plane moves slowly, so that its engines need large intakes like those of a 747 to gulp in air. Conversely, at supersonic speeds the motion of the plane rams air into the engines at high pressure, so that a much smaller intake is needed. Variable-size intakes, coupled with systems that let different amounts of air bypass the combustion chambers, make for complicated engines that have yet to be perfected.

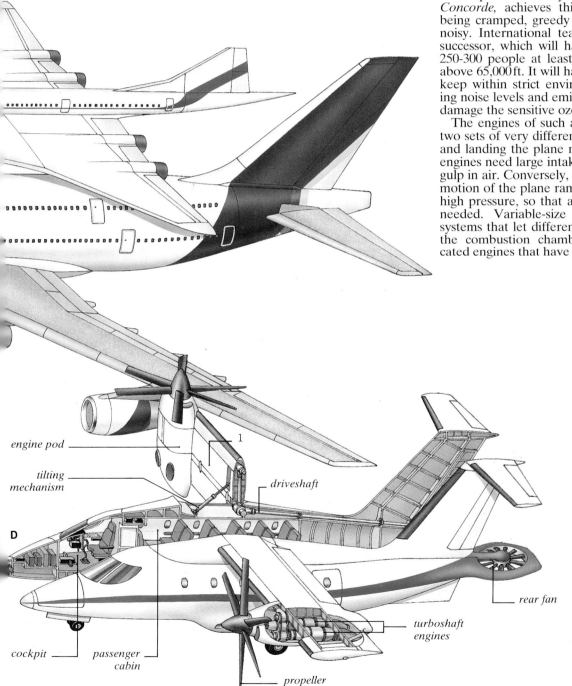

engine pod

tilting mechanism

D

cockpit

passenger cabin

driveshaft

propeller

turboshaft engines

rear fan

Transformer
One design that combines the virtues of helicopters and business jets – vertical landing and high cruising speeds – is a tilt-wing aircraft [D]. *It will be able to take off and land vertically or from short runways, yet cruise at around 280 mph.*

This design will carry 14 passengers and two crew. The aircraft will span only 42.5 ft between propeller tips, allowing it to operate from the same pad space as a medium-sized helicopter. It will be used for business and commuter flights, and by the emergency services.

The wing tilts to a vertical position [1] to act as a two-rotor helicopter for takeoff and landing, and moves to the horizontal [2] for conventional forward flight. The propellers are driven by four turboshaft *engines mounted in twin-engine pods on the wing. For safety in the case of an engine failure, driveshafts interlink the engines and also drive rear fans, necessary for vertical flight and hovering.*

Air Cushions and Water Wings

How hovercraft and hydrofoils skim the waves

Amid a cloud of sand- and salt-spray, a 330-ton SR-N4 hovercraft glides easily from the sea toward its terminal on the beach. The 400 passengers and 60 automobiles inside are held aloft, eight feet above the surface, by a 100,000-cubic-foot cushion of compressed air. Hydrofoils also free themselves from slowing effects of water, standing tall on legs supported by underwater wings. A Boeing Jetfoil can reach a speed of nearly 60 miles per hour, pushed along by the frothing fountains of sea water from its gas-turbine-powered waterjets.

Compared to air, water is a difficult substance to move through. As a ship sails forward, it uses up a lot of energy pushing water out of the way to form its wake. One way of reducing this waste of energy and making a boat travel faster is to lessen the proportion of the craft that is under water. Air is 800 times less dense than water, so a craft traveling through it meets much less resistance to its motion.

Hovercraft are not really boats at all – they are lifted completely clear of the surface by huge fans under their hulls. This makes them virtual airplanes, effectively "flying" close to the waves. However, one major problem is to contain the supporting air, which naturally seeks to escape through the gaps around the side of the craft. Most commercial vehicles have rubber skirts to trap the air cushion. On their lower edge are *fingers* – horseshoe-shaped extensions that bend, conforming to the shape of waves and minimizing leaks.

There are other ways to contain the air cushion. Small competition vehicles do not have skirts but instead blow an invisible curtain of air inward from the edge of the craft to maintain high pressure. Other passenger craft have rigid sides that dip into the water like a catamaran, with rubber skirts only at the front and the back. These, like hovercraft, are fast because they are lifted by a cushion of air that reduces drag, but cannot travel over land because part of the hull is always in contact with the water.

Given the push
A normal ship's propeller is no use to a craft entirely out of the water, so hovercraft instead use huge air fans to push them along. Those on the SR-N4 are the biggest propellers of their type ever manufactured. They can be pointed in different directions for steering; some craft also lean into corners by deflating the air cushion slightly on one side.

Hovercraft need extremely powerful but light engines to provide the enormous thrust to keep several hundred tons in the air. The largest use gas turbines, jet engines adapted to turn the propellers and centrifugal fans. However, these are extremely sensitive to corrosion from the dense spray of salt water thrown up by escaping air, so smaller craft use diesel engines, which need much less maintenance.

The hovering principle works not only on water. Hovercraft are truly amphibious – because the air cushion spreads the vehicle's weight over the widest possible area, they can travel over sensitive terrain (such as the tundra of northern Canada) without causing damage. Air is also a very good lubricant – a train has been designed that glides along on a thin layer at speeds of more than 200 mph.

Winged wonders
Just as an aircraft is held aloft by the rapid flow of air over its airfoil-shaped wing, an underwater wing, or *hydrofoil,* can lift a vessel clear of the water, thereby greatly reducing the drag experienced. Because lift is maintained only by the fast flow of water over the submerged foils, anything that slows the vessel down – such as a wave that hits the hull – makes for an uncomfortable journey. To avoid this, some hydrofoils are attached in a V-shaped surface-piercing formation. A large wave breaking over the vessel covers more of the wing with water, producing more lift to push the hull clear of the crest.

Floating on air
A hovercraft [**A**] is much faster than an ordinary ship, and some types are capable of speeds up to 75 mph. Because they ride along on a cushion of air, these craft are amphibious – able to travel over land and sea – and are suitable for a variety of jobs from passenger services to search-and-rescue missions.

The model depicted is powered by four 12-cylinder diesel engines. Two of these power large centrifugal fans that suck air in and blow it underneath the craft. Its main propulsive power is through two 8.9 ft-diameter propellers, powered via drive belts by their own diesel engines. Rudders fitted behind these propellers are used for steering. Other fans blow air through the bow thrusters, ducts which can be rotated to aid steering.

The aircushion underneath the hovercraft is contained by a rubber skirt, a flexible curtain that enables the hovercraft to ride high over obstacles and waves. The upper half of the skirt, known as the bag, is a double-walled structure that directs the incoming air into the plenum chamber – the gap below the hovercraft. Attached to the bag is a series of fingers [**B**] that touch the surface the craft is traveling along to form a seal around the air cushion. The fingers also greatly improve the ride the passengers experience, their up-and-down movement working as a giant shock absorber. Although the skirt is very tough, in the event of a rip or engine failure, a buoyancy tank keeps the hovercraft afloat.

radar aerial

A

Seaplane (right)
Hydrofoils skim along just above the surface of the sea, held aloft by underwater wings. Any contact between the hull and the wave tops would result in a sudden slowing of the craft and a dangerous loss of lift. Surface-piercing hydrofoils (right) avoid this problem through the action of their V-shaped wings. When the boat cuts through a wave, a longer length of wing becomes immersed, and more upward lift is produced. This pushes the underside of the craft clear of the wave crest and danger, and allows the hydrofoil to follow the contours of the sea.

Connections: Internal Combustion Engine 12 Jet Engines 24 Aircraft: Aerodynamics 26 Ships 35 Navigation Aids 40 Principles 228 230 236 238 240

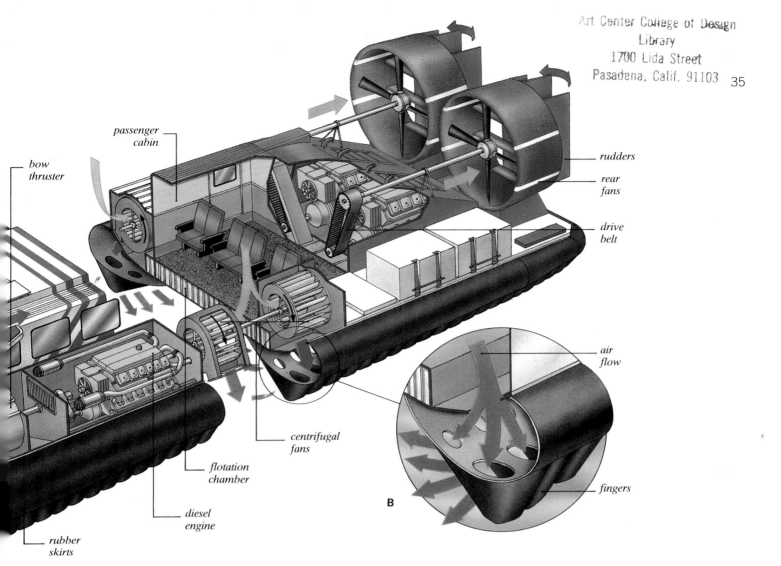

passenger
cabin

bow
thruster

rudders

rear
fans

drive
belt

air
flow

B

fingers

centrifugal
fans

flotation
chamber

diesel
engine

rubber
skirts

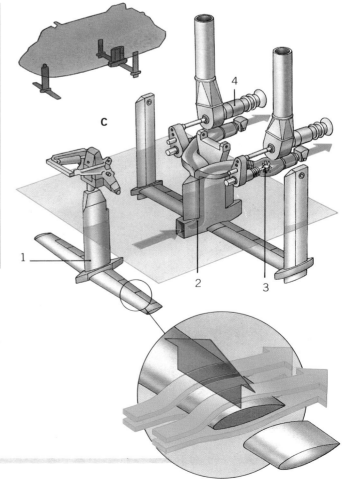

C

4

1

2

3

Riding high

The Boeing Jetfoil [C] is one of the fastest hydrofoil designs. Because its hull is held clear of the drag-inducing water, it can cruise comfortably at 46 mph.

The jetfoil is maneuvered by the steerable front strut [1] and flaps at the trailing edges of the foils. The flaps are also controlled by an automatic system that makes sure that the hull always stays a safe height above the waves.

A water jet pushes the craft along: a scoop between the rear foils [2] takes up water, which high-pressure pumps [3] force out of the rear of the craft at high speed. Gas turbines [4] power these pumps, their air intakes reaching upward to avoid sucking in sea-spray.

Water travels more quickly over the curved upper surface of its wing-like foils than below them, creating an area of low pressure that "sucks" the whole craft upward.

Naval Architecture

How ships are designed and built

At almost a third of a mile long and two hundred feet wide, the *Jahre Viking* is the largest ship, and, in fact, moving object, ever built. This lumbering 620,000-ton giant takes many tens of miles to be brought to a halt from its cruising speed and can only dock at the largest of deepwater ports. In the design of all vessels, function dictates form: the unwieldiness of an oil tanker contrasts starkly with the agility of a passenger ferry, which is equipped with extra propellers and even rudders at the bow to maneuver in the close confines of a harbor.

The shape of a boat's hull is a compromise among a number of different demands. First, the hull needs to slip through the water with the minimum of resistance – extreme examples of this are rowboats and canoes, which are long and thin with sharply pointed bows. Another way of reducing drag is to lift as much of the hull as possible clear of the water. Speedboats have undersides designed to *aquaplane* (produce a lifting force) at high speed, taking part of the hull out of the water.

The second consideration is stability. Sail boats are equipped with heavy keels, usually weighted with lead, to counteract the turning force of the wind on their sails. The keel's shape also helps to push the boat forward by providing extra resistance to sideways movement.

All types of hull are shaped to resist capsizing. The designer must ensure that the upward (buoyant) forces on the boat combine with the downward forces (the boat's weight) to produce an overall turning force that tends to right the ship when it is leaning over. A catamaran design is specially stable because of the wide spacing between its twin hulls, although it is harder to make a multihulled vessel as rigid as a similar-sized single-hulled boat.

The third consideration that has to be taken into account is the function and carrying capacity of the vessel. Load-carrying boats and ships, from barges to oil tankers, tend to be wider than the ideal streamlined shape in order to allow a greater volume in which to store cargo. Some hulls have particularly specialized forms: trawlers have high bows that enable them to ride through the largest ocean swell, but a much lower stern so that fishing nets can easily be let out and brought in. Icebreakers have spoon-shaped bows that gradually ride upward on pack ice. As the ship's weight bears down, the frozen sheet cracks and a path is cleared. Advanced icebreakers are equipped with air bubbling systems that create currents of water and air between the hull and ice, reducing the drag that the ship experiences.

Ocean power

Powering the variety of hull shapes are many different engine types. Until recently the largest cargo vessels and passenger liners were powered by gas turbines, adapted versions of jet engines used in airliners. To run efficiently, these engines need to spin at many thousands of revolutions per minute, far faster than the ideal

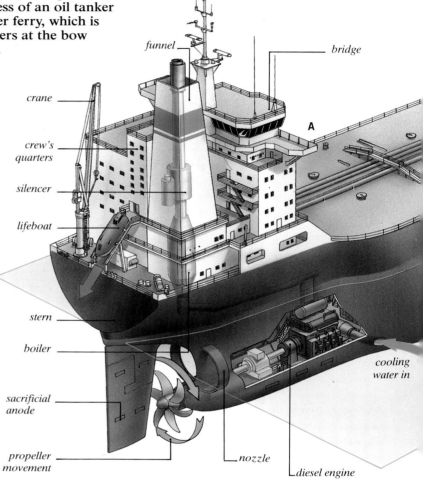

funnel — bridge
crane
A
crew's quarters
silencer
lifeboat
stern
boiler
cooling water in
sacrificial anode
propeller movement
nozzle
diesel engine

Sea giant
*A modern oil tanker [**A**] is designed to carry the largest possible quantity of oil at the smallest expense. The oil itself is contained in large tanks, split into smaller sections by baffles and bulkheads. These prevent the oil from slopping from side to side and upsetting the stability of the ship. The bow and stern narrow to a waist at the waterline before bulging out again under water. This ensures that the length of the waterline, where drag is at its greatest, is kept to a minimum, while keeping the displacement – and hence the buoyancy – high. The tanker is powered by a large low-speed diesel*

engine. It is cooled by sea water, sucked in through a scoop on the starboard side and ejected through a similar vent to port. As the hot exhaust gases pass out through the funnel, they are used to heat hot water in a boiler.

The engine turns slowly enough to be directly coupled to the propeller, a bronze casting more than 30 ft across. Nozzles concentrate the flow of water over the propeller to increase efficiency. The rudder steers the ship at sea by directing the wake from the propeller. In the tighter confines of a harbor maneuvering is made easier by bow thrusters, enclosed

propellers driven by electric motors that can push the front of the boat from side to side.

The rudder is studded with sacrificial anodes – pieces of zinc that are attacked by sea water in preference to the steel hull. The reaction produces an electric current through the water that further protects the hull from corrosion.

At the rear of the ship is the lifeboat. In an emergency, the crew can rapidly enter the boat and strap themselves tightly in. Then a cable is released and the boat free-falls into the water, allowing a much swifter escape than by traditional lifeboats.

Connections: Internal Combustion Engine 12 Jet Engines 24 Hovercraft and Hydrofoils 34 Oil: Extraction 104 Principles 228 236 238 240

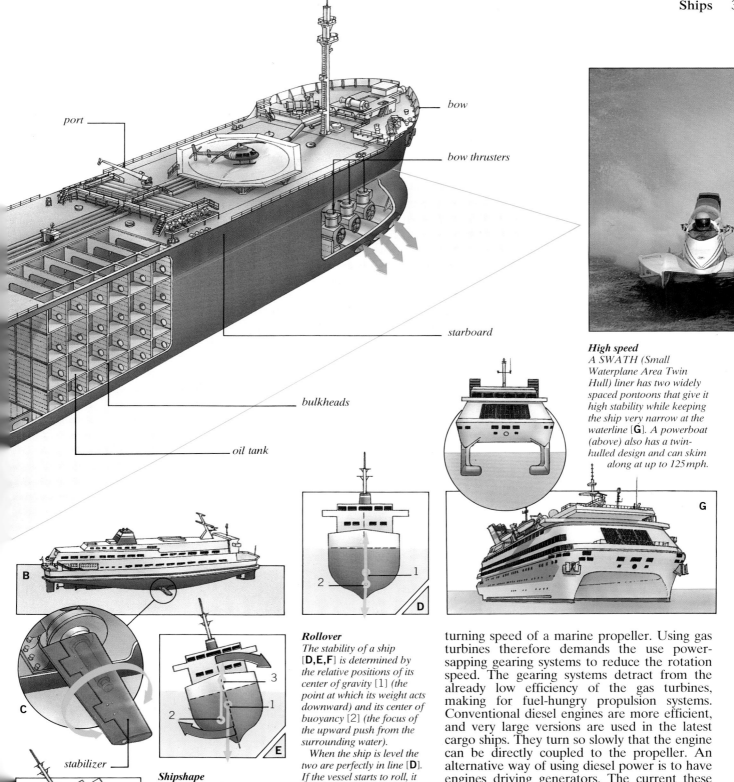

port

bow

bow thrusters

starboard

bulkheads

oil tank

High speed
A SWATH (Small Waterplane Area Twin Hull) liner has two widely spaced pontoons that give it high stability while keeping the ship very narrow at the waterline [**G**]. *A powerboat (above) also has a twin-hulled design and can skim along at up to 125 mph.*

B

C

stabilizer

F

E

2   1

D

3

2   1

G

Rollover
The stability of a ship [**D,E,F**] *is determined by the relative positions of its center of gravity* [1] *(the point at which its weight acts downward) and its center of buoyancy* [2] *(the focus of the upward push from the surrounding water).*

When the ship is level the two are perfectly in line [**D**]. *If the vessel starts to roll, it will right itself* [**E**] *if the point at which the upward (buoyant) thrust crosses the center line of the ship* [3] *is above the center of gravity. But if the crossing point is below the vessel's center of gravity a net turning force results that makes the ship capsize* [**F**].

Shipshape
The motion of a passenger ferry is made comfortable even in heavy seas by stabilizers [**B,C**]. *These are winglets that fold out from the side of the hull and twist to provide an upward lifting force that counteracts the rolling motion of the sea when the boat is moving.*

turning speed of a marine propeller. Using gas turbines therefore demands the use power-sapping gearing systems to reduce the rotation speed. The gearing systems detract from the already low efficiency of the gas turbines, making for fuel-hungry propulsion systems. Conventional diesel engines are more efficient, and very large versions are used in the latest cargo ships. They turn so slowly that the engine can be directly coupled to the propeller. An alternative way of using diesel power is to have engines driving generators. The current these produce can then be fed to electric motors that turn the propellers. This arrangement allows the engines to be sited away from the propeller in the best positions for weight distribution.

Sail power is now used almost entirely for leisure. Nevertheless designs are constantly becoming increasingly sophisticated, the process spurred on by yachting races that produce ingenious shapes such as winged keels.

Beneath the Waves

How divers explore the ocean depths

The oceans that cover two-thirds of the earth's surface are a hostile environment for human exploration. The upper hundred feet or so were made accessible by SCUBA equipment, the self-contained underwater breathing apparatus invented by the Frenchman Jacques Cousteau. Traveling to greater depths requires protection in the form of rigid diving suits that are effectively one-person submarines, and the use of special gas mixtures for breathing. The deepest trenches are reachable only in submersibles built around immensely strong metal spheres.

The greatest problem facing any diver or ocean explorer is how to deal with the crushing pressure of the ocean depths. Even in shallow waters, pressure prevents a diver's lungs from expanding properly and taking in adequate air. This problem is overcome by feeding the diver a supply of compressed air. Up to depths of roughly 100ft, divers use a Self-Contained Underwater Breathing Apparatus, or SCUBA gear. Compressed air is stored in tanks carried on the diver's back, and is fed to a mouthpiece through valves that reduce its pressure to match that of the surrounding water.

At greater depths, cold and pressure are more intense. Professional divers wear suits linked to the surface by umbilicals supplying compressed air and hot water, which circulates through the suit keeping the diver warm.

The bends

As a diver breathes air at very high pressures, the nitrogen it contains dissolves into the blood in greater quantities than it would on land. If the diver surfaces too quickly, this excess nitrogen comes out of solution in the same way that bubbles of carbon dioxide form when a soft drink bottle is opened. The result is a painful and sometimes fatal condition called "the bends." The only solution is for the diver to ascend slowly, pausing at intervals to allow the body to adjust and expel excess nitrogen through breathing.

Diver ferry
*A lock-out submersible [**A**] is used to carry divers to and from work sites at depths of up to 800ft. At the front of the vessel, the crew breathe air at normal atmospheric pressure. They steer the craft and can even collect samples with a pair of manipulator arms. An airlock connects the crew compartment to a second chamber for the divers. This is filled with heliox, a mixture of helium and oxygen at high pressure. Divers can leave and enter the craft through a second airlock on the underside. The craft is electrically powered from batteries stored in long side pods.*

Nitrogen and oxygen at extremely high pressures become poisonous. For dives below 150ft, helium is often substituted for nitrogen in the breathing mixture, creating a mix called heliox. This has a strange side-effect on the larynx, giving a diver a squeaky, high-pitched voice.

If divers must remain under water for days at a time, a technique called *saturation diving* is used. Even when they are not working, the divers stay under water in a pressure chamber, breathing a heliox atmosphere. This saves the time and inconvenience of worrying about decompression, but means that, in an emergency, the divers are effectively trapped.

Water workers

For some tasks, diving is too expensive or dangerous. Here, Remotely Operated Vehicles (ROVs) are used. An ROV is used, for example, to clean the underwater parts of oil rigs. Semicircular in shape, it locks onto the pipework exteriors of the rig and removes barnacles and other types of marine fouling. It is linked to an operator on the surface by a cable through which it sends pictures from an onboard TV camera. It can even by put into place by a another ROV that can move under its own power. ROVs are also ideal research tools. One highly maneuverable vessel named *Jason* used its onboard TV cameras to explore the wreck of the ocean liner *Titanic* 13,000ft below the surface of the North Atlantic.

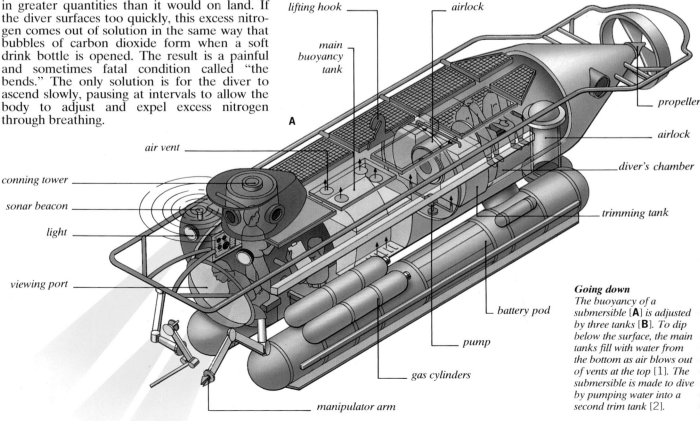

lifting hook

main buoyancy tank

airlock

propeller

airlock

diver's chamber

air vent

conning tower

sonar beacon

light

viewing port

trimming tank

A

battery pod

pump

gas cylinders

manipulator arm

Going down
*The buoyancy of a submersible [**A**] is adjusted by three tanks [**B**]. To dip below the surface, the main tanks fill with water from the bottom as air blows out of vents at the top [1]. The submersible is made to dive by pumping water into a second trim tank [2].*

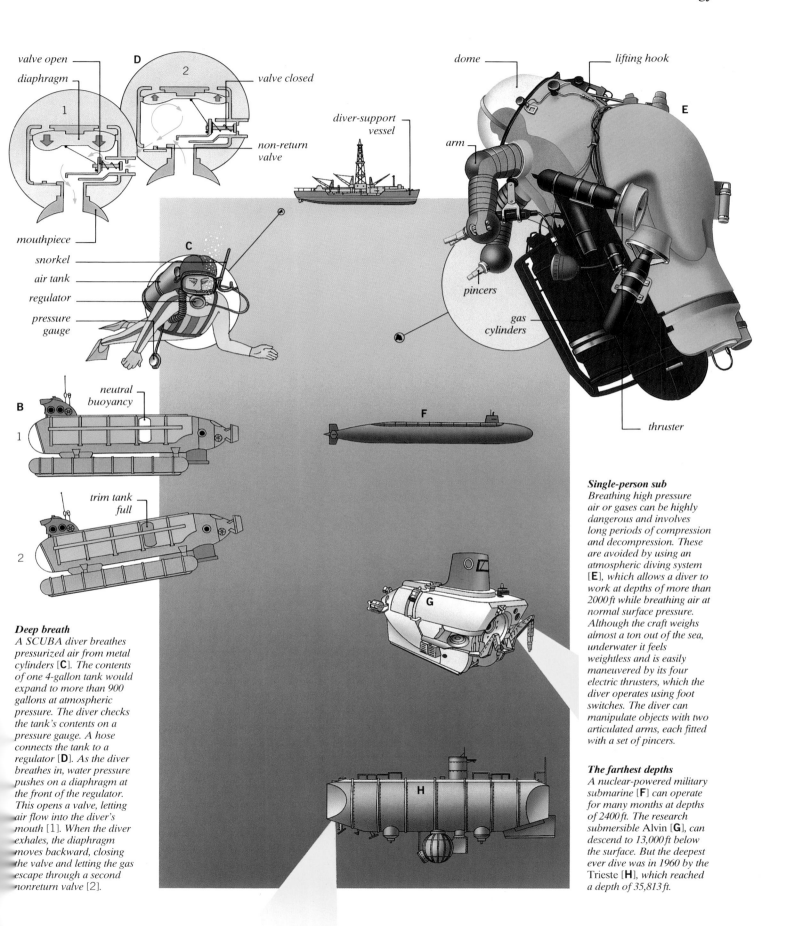

valve open

diaphragm

D

2

valve closed

non-return
valve

mouthpiece

diver-support
vessel

dome

lifting hook

E

arm

snorkel

air tank

regulator

pressure
gauge

C

pincers

gas
cylinders

thruster

B

1

neutral
buoyancy

2

trim tank
full

F

G

H

Deep breath
A SCUBA diver breathes
pressurized air from metal
cylinders [**C**]. The contents
of one 4-gallon tank would
expand to more than 900
gallons at atmospheric
pressure. The diver checks
the tank's contents on a
pressure gauge. A hose
connects the tank to a
regulator [**D**]. As the diver
breathes in, water pressure
pushes on a diaphragm at
the front of the regulator.
This opens a valve, letting
air flow into the diver's
mouth [1]. When the diver
exhales, the diaphragm
moves backward, closing
the valve and letting the gas
escape through a second
nonreturn valve [2].

Single-person sub
Breathing high pressure
air or gases can be highly
dangerous and involves
long periods of compression
and decompression. These
are avoided by using an
atmospheric diving system
[**E**], which allows a diver to
work at depths of more than
2000 ft while breathing air at
normal surface pressure.
Although the craft weighs
almost a ton out of the sea,
underwater it feels
weightless and is easily
maneuvered by its four
electric thrusters, which the
diver operates using foot
switches. The diver can
manipulate objects with two
articulated arms, each fitted
with a set of pincers.

The farthest depths
A nuclear-powered military
submarine [**F**] can operate
for many months at depths
of 2400 ft. The research
submersible Alvin [**G**], can
descend to 13,000 ft below
the surface. But the deepest
ever dive was in 1960 by the
Trieste [**H**], which reached
a depth of 35,813 ft.

The Pathfinders

How aircraft and ships are guided

Taking off and landing in pitch darkness, and flying thousands of miles without sighting the ground beneath, today's airliners achieve what birds can not – blind flight. At sea, on land, and in the air, traditional navigation skills have been supplemented by electronic guidance systems that operate reliably in all weather conditions. Drawing positional information from radio beacons on land or in orbit, these new technologies allow even the novice to pinpoint his position to an accuracy of within thirty feet.

The earliest seafarers employed a crude form of navigation, using landmarks and meteorological signposts to find their way. Navigational techniques soon became increasingly reliable. Sea charts improved in detail, and more accurate marine clocks and sextants allowed sailors to make timed observations of the movements of the sun and stars. Using almanacs – celestial "timetables" – these movements could be translated into positions given by longitude and latitude. Astronomical determination of position was supplemented by the technique of *dead reckoning,* by which a vessel's position could be calculated from its starting point, speed, and magnetic compass bearing since departure. These traditional techniques, perfected by the 19th century, have been improved and, in some cases, superseded by 20th-century technologies.

The magnetic compass, for example, is an inaccurate device because the earth's magnetic poles actually change position over time. Moreover, the steel hulls of modern ships interfere with the magnetic field, further decreasing the reliability of readings. The magnetic compass was first replaced by the *gyroscopic* or *inertial* compass. In this device, a flywheel, rotated at high speeds by an electric motor, is mounted in a framework via a set of swivels. The *inertia* of the spinning flywheel resists any force (such as gravity) that tends to realign its axis of spin. Once the gyroscope is set spinning around a true north–south axis, it maintains this attitude, holding its course irrespective of the motion of the vessel, and can be used to indicate bearing.

Guidance systems

In modern aircraft, the mechanical gyroscope has itself been superseded by the more reliable *laser gyroscope.* This consists of a 0.6-mile-long optical fiber, coiled up on itself. Light from a single laser is split into two beams, each directed down one end of the coiled light "pipe." In a stationary aircraft, the waves of each beam are precisely out of step with one another. Within the light pipe, they interfere, canceling one another out. But if the craft rotates, the path of one beam is effectively lengthened, and the other shortened: the beams do not cancel one another out, and the intensity of the resulting laser light can be translated into navigational information in the cockpit.

Gyroscopes are at the heart of many types of guidance systems, such as the artificial horizon, found in the cockpit of every aircraft. Here the gyroscope spins horizontally, maintaining this alignment as the aircraft moves around it. By reference to the artificial horizon, the pilot can ensure that the aircraft is flying straight and level after a climb, bank, or dive.

Gyroscopes are part of the aircraft's two main computer systems – the *autopilot* and the *Inertial Navigation System (INS).* The autopilot takes care of the routine control adjustments needed for stable flight. Its two gyroscopes detect any deviation from a preset course and feed this information to the computer. This in turn activates sets of motors that control the aircraft's aerodynamic surfaces.

The INS is a sophisticated dead reckoning system that works out an aircraft's position relative to its starting point. It contains a platform held absolutely stable by a set of gyroscopes. Mounted on the platform are motion detectors called *accelerometers* that record the smallest acceleration by the aircraft in any direction. These data are used to calculate the aircraft's position relative to its point of origin.

Dead reckoning systems like INS must be checked periodically against reliable points of reference: otherwise, small errors in position are quickly magnified. Reference points were once visible landmarks and radio beacons, but today's pilots have at their disposal an accurate satellite positioning system, known as GPS.

aircraft in "stack"

localizer beams

glideslope beams

A

ILS distance beacons

omnidirectional VHF beacon

Controlled landing

Airline pilots rely on the Instrument Landing System (ILS) [A] to make safe landings in all weathers.

Two highly directional radio antennae flank the runway: they produce radio "paths" that indicate the centre line of the runway and the correct angle of approach to an incoming aircraft. The fencelike localizer *antenna indicates the centre line. It produces two overlapping horizontal "lobes" of radio energy, each carrying a slightly different frequency (dark and light red). Instruments onboard the aircraft compare the strength of the two signals: when they are*

Connections: Aircraft: Systems 28 Ships 36 Radio 50 Timekeeping 148 Satellites: Orbits 172 Principles 238 242 248 254 256

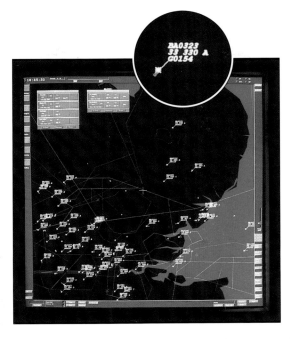

Controlled airspace
Hundreds of miles away
from an airport [**A**], an
approaching aircraft is
given precise information
on distance and bearing by
the airport's omnidirectional
VHF radio beacon. Its
position is tracked by
primary and secondary
radar. If the airport is too
busy for an immediate
landing, the aircraft must
circle in a stack centered on
another radio beacon. A
height of at least 1000 ft
separates the aircraft.

Radar progress (left)
An aircraft's movement
is tracked in the airport
control tower. This is
achieved using two types
of radar. Primary radar is
microwave energy beamed
by a rotating antenna in all
directions. Reflections from
passing aircraft are detected
by the same antenna and
their positions are displayed
on a screen. Secondary
radar is more advanced: it
"interrogates" the aircraft,
which responds, sending
back its four-figure call sign
and its altitude (inset). This
picture shows information
from both types of radar
combined on a display with
the outline of the coast.

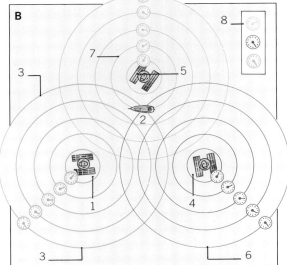

incident radar

reflected
radar

primay radar

secondary radar

control tower

terminal building

runway

glideslope
antenna

localizer
antenna

Ground control
Separate groups of air
traffic controllers direct an
aircraft in the four phases of
its journey – on the ground;
in the airspace within 18
miles of the airport; in the
airspace between 18 and
60 miles away; and then
along the whole length of
its flight "corridor."

Global guidance
The Global Positioning
System (GPS) [**B**] is the
most important navigational
development of the 20th
century. Although it was
developed by the US
Department of Defense, it is
now available to all users.
GPS is made up of 24
Navstar satellites, each just
16.5 ft long, orbiting at a
height of 11,000 miles. From
any point on the earth, at
least four satellites are above
the horizon at any one time.

Each Navstar [1] satellite
has an onboard atomic
clock that measures time
with pinpoint accuracy,
and a transmitter that
broadcasts time signals
down to the earth. As a time
signal intersect the earth
beneath, it spreads outward
describing a circle that
increases in diameter with

time. A GPS receiver on
board a ship [2] picks up
the time signal, and
compares it with its own
inbuilt clock. Because the
speed of radio waves is
known, the ship's GPS unit
can calculate the exact
distance to the satellite. A
computerized almanac in
the unit "knows" the exact
position of the transmitting
satellite, so the navigator
knows that the vessel lies
somewhere on a circle [3]
described on the earth's
surface. Simultaneous
signals from another two
GPS satellites [4,5] provide
two other circles [6,7] that
intersect at a unique point.
Measuring the time signals
received from three satellites
[8] allows GPS to locate a
vessel to within 30 ft.
Aircraft use signals from
three GPS satellites to
pinpoint their position,
adding data from a fourth
satellite to fix altitude.

GPS signals can be
received anywhere in the
world and are largely
unaffected by weather,
season, and location.

exactly equal, the plane is
on the correct course. Any
inequality is translated into
a course correction and
displayed in the cockpit.
The second, glideslope,
antenna works in a similar
way, but its two radio lobes
are vertically oriented (dark
and light blue). By keeping
the strength of the two
signals equal, the pilot

descends at the correct angle
to touch down at the start of
the runway. At intervals
along the ILS path, beacons
transmit signals to aircraft
passing directly overhead,
letting them know the exact
distance to the airport.

Ultrapure gold wires connect the minute components etched onto a silicon chip with pins on the casing of an integrated circuit.

2
Information and Entertainment

Audio Equipment

Telephones

Cellular Phones and Fiber Optics

Radio

Sound Recording

Compact Discs

Television

Video Technology

Television: New Developments

Photographic Cameras

Movie Cameras

Special Effects

Photographic Film

Lasers and Holography

Office Equipment

Pre-Print Technology

Printing

Computers: Structure

Computers: Memory and Storage

Computers: Networks

Computers: New Developments

High Fidelity

How sound is captured and reproduced

As a vocalist sings into a microphone at a rock concert, a chain of energy conversion and reconversion takes place before the crowd hear the amplified music. Most microphones are sensitive enough to pick up whispers, converting the sound into a constantly varying electrical waveform. Amplifiers turn this tiny signal into a current powerful enough to drive huge banks of speakers, as well as boosting certain frequencies to achieve a balanced sound. The speakers come in an array of sizes, from tiny high-frequency tweeters to 16-inch-wide bass drivers.

Microphones and speakers are very similar machines. A microphone uses the varying pressure wave of sound to create an electrical wave: a speaker takes such a wave and converts it back into sound. However, the *amplitude*, or size, of the electrical wave needed by a speaker is much greater than that produced by a microphone. An *amplifier* increases the amplitude of a wave while faithfully preserving its shape.

Almost all microphones contain a *diaphragm* of some description – a thin surface that moves in sympathy with the varying pressure of a sound wave to produce a matching electrical wave. In a *condenser* microphone, the diaphragm forms one part of a capacitor, a device that stores electrical charge. The charge is either permanently trapped in an *electret*, or comes from a *phantom power supply* along the same cables used by the signal, but without affecting its shape. As the diaphragm moves in and out, the voltage across the capacitor varies, in effect becoming an electrical version of the incoming sound wave.

There are other ways of converting sounds to electricity. The diaphragm of a *moving-coil* microphone is attached to a coil of wire suspended in a strong magnetic field. As the diaphragm vibrates, a voltage is generated in the coil, making a current flow. Another design has a diaphragm mounted on a piece of *piezo-electric* material. This produces a varying voltage as it is compressed by the sound waves.

Sound into signals

A microphone converts sound waves into electricity. An electret condenser microphone [A] contains a flexible plastic diaphragm, a fortieth of a millimeter thick, and a fixed metal backplate. The diaphragm carries a permanent positive electric charge that draws negative charge to the backplate. Sound waves make the film vibrate, varying the gap between it and the backplate and so attracting more or less charge. This movement of charge produces an alternating signal in the electrical circuit which corresponds exactly to the shape of the incoming sound wave.

Echo-free (below left)

The walls of an anechoic chamber are lined with wedge-shaped pieces of foam which, coupled with a "floating" floor and ceiling, completely absorb any reflected sounds. This makes it possible to check the fidelity and sensitivity of a microphone.

Mikes of all trades

The varied shapes and sizes of microphones reflect their different sensitivities and uses. A heart-shaped pattern of sensitivity shows that a cardioid microphone [B] picks up sound only from the area immediately in front of it. A hyper-cardioid [C] has an elongated version of the same pattern, making it suitable for picking out one sound source from many. The figure-of-eight pattern of a bi-directional [D] microphone shows that it is sensitive to sound from both sides, and so can be used for recording duets, and also to capture the direct and reflected sounds inside a concert hall.

Pump up the volume

The electrical signals from microphones, compact disk players and other instruments are far too small to feed directly to speakers. An amplifier [G] increases the size or amplitude of the electrical waves without changing the basic shape of their waveform [E]. Any changes to the shape are apparent to the listener as distortions in the sound. One form of distortion is clipping [F], when the amplified signal tries to exceed the maximum and minimum voltages that the amplifier can produce. The peaks are clipped off the waves, producing a gritty sound.

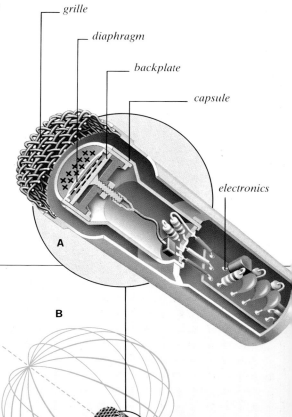

grille

diaphragm

backplate

capsule

electronics

A

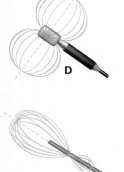

D

C

B

E

F

Wired for sound
*Almost all loudspeakers contain a cone that pushes and pulls air to create the pressure waves that are sound. A hi-fi speaker [**H**] has three units to cover all frequencies. The domed cone of the tweeter produces the highest notes, vibrating up to 20,000 times a second (20kHz). The mid-range unit covers 500Hz up to 4kHz, and the lowest frequencies come from a woofer. Its cone is attached to a coil, which vibrates in a magnetic field as the sound signal flows through it. A crossover splits the main signal into currents for each speaker.*

amplifier

G

H

tweeter

mid-range unit

cross-over

transmission line

woofer

coil

magnet

low-pressure

high-presssure

I

Boosting the signal
Microphones produce only a very weak signal that must be strengthened before it is broadcast, recorded, or reproduced. Amplifiers use the small voltages of the input signal to control the size of a much larger output voltage. Most contain a series of *transistors*, electronic "gates" with three connections. Small changes in the current applied to one connection produce proportional changes in a far higher current passed between the other two.

Early amplifiers used a different device, the *valve*, which was bulky and produced a lot of heat. Valves also did not produce perfect amplification, boosting some frequencies at the expense of others. Sometimes this is desirable – guitar amplifiers often use valves precisely because they do not boost a signal accurately but instead distort it to produce the characteristic "gritty" sound favored by some musicians.

Woofers and tweeters
The loudspeaker works like a microphone in reverse. It has a vibrating diaphragm, the *cone*, which in bass-frequency *woofers* is usually made of paper, but may also be metal or a composite material. One end of the cone is attached to a wire coil, which carries the signal from the amplifier. A magnet surrounds the coil, producing a force in the same way as an electric motor, that pushes and pulls the cone in and out. This generates traveling areas of low and high air pressure – sound waves. Some speakers with small diaphragms, *tweeters*, have dome-shaped cones that project sound more effectively.

An alternative design is the electrostatic speaker. In essence these are like huge *capacitors*, containing a pair of charged plates. In the speaker, one of these plates is free to move in and out as the varying voltage of the sound signal is applied to it. Because of the flat shape of the plates, this movement very accurately re-creates the pressure waves of sound.

In harmony
A speaker cone radiates sound waves from both the front and rear surfaces of its diaphragm. Although each side produces the same sounds, the waves that they produce are exactly out of phase with one another –

*when the front produces an area of high pressure, the rear produces low pressure. If a speaker is open-mounted, the waves from the back can rush around the cone and cancel out those at the front [**I**]. The effect is a particular problem for the long wavelengths of bass sounds. Some woofers sit in cabinets containing a transmission line. This channels the waves from the rear through a tube that has just the right length to ensure that when the waves emerge, high- and low-pressure areas line up with those coming from the front [**J**].*

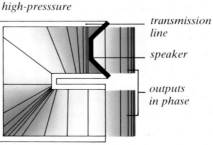

transmission line

speaker

outputs in phase

J

On the Line

How information is carried on the telephone network

The global telephone network is the most complex machine ever built. It handles about 600,000 million calls a year between some 600 million telephones, and is continually growing in size and sophistication. Today's systems rely on digital technology, which allows images and computer data, as well as conversations, to be sent with unprecedented accuracy and reliability. New digital exchanges, routing more than 1.5 million calls per hour through a wide variety of cables, optical fibers, radio and microwave links, provide instant, affordable global communication to all.

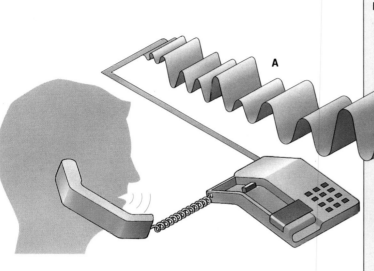

Modern telephone systems, known as Pulse Code Modulation (PCM) systems, are based on converting sound waves into digital information that is then transmitted around the network.

The caller's voice is picked up by a microphone in the telephone mouthpiece, which translates it into a fluctuating electrical current. This *analog* signal is sent down the line to the local exchange. Here it meets an electronic switch, which allows a short pulse or *sample* of the analog signal to pass through every 125 microseconds. The signal is therefore *sampled* 8,000 times a second, enough to convey the subtleties of the human voice.

The amplitude (or intensity) of each of the pulses is then measured against a scale of numbers, usually 1–256. Each number is converted into a binary code composed of eight digits. This *eight-bit* sequence is the shortest code that can cope with numbers up to 256.

These digital codes then pass into a modulator, where the binary signal is imprinted upon an electrical *carrier wave* – high-frequency carriers are used for radio transmission, lower frequencies for cables. The imprinting takes the form of either amplitude or frequency modulation in which the amplitude/frequency of the carrier is varied in response to the code. This whole process is reversed at the other end of the line, where the message is demodulated, decoded and passed to a metal diaphragm in the earpiece via an electromagnet.

Binary conversations
When the handset of a telephone is lifted, an electrical loop is completed between the telephone and the local switching center, or exchange. Electrical power is supplied by the exchange. Access to the network is through the dialing code. As each digit of the code is selected, a unique combination of two single-frequency tones passes down the line to the exchange. The tones activate an automatic switching system that routes every call to its destination. When making a call within a large
town or city, a conversation is typically relayed via two or more exchanges.

Once connected, the caller's voice is converted by the telephone instrument into a continuously varying voltage – or analog *signal* [A]. *This signal travels down copper wires to the local exchange* [B]. *Electronic equipment here samples the signal* [1] *to convert it into digital form.*

The intensity of the analog signal is measured once every 125 microseconds, or 8000 times per second (any fewer and human speech would begin to break up). Each measurement is then translated into a binary code which comprises a set of 8 discrete bits [2] – *electrical pulses that can be either on or off. Each sample takes*

Global linkages

*The telephone network is a hierarchical system [**E**]. At the bottom of the hierarchy are millions of subscribers, directly connected to a local exchange, or end office. In a densely populated area, numerous end offices [1] are interconnected, allowing for the high volume of calls within a town or city. The end offices are also linked to long-distance exchanges [2], which in turn may be linked to other long distance exchanges or to regional exchanges farther up the hierarchy [3].*

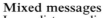

Not all connections are through wire or optical cable: microwave transmission is often employed to link distant exchanges. Finally, international exchanges [4] relay conversations between countries and continents, increasingly via orbiting communications satellites.

Mixed messages

Long-distance lines represent a considerable investment to a telephone company. To maximize the use of lines between exchanges, each channel is made to carry a number of conversations simultaneously. In this process, known as *time-multiplexing*, separate conversations are woven together in such a way that they can be disentangled when they reach their destination.

A conversation is transmitted as a series of digital pulses sent down a line – at the rate of one every 125 microseconds. But each pulse is only about four microseconds long, which means that 121-microsecond "gaps" exist between the information-carrying pulses. An electronic multiplexer fills in these gaps with pulses from other conversations. At present about 25 calls can be multiplexed in this way, but the technology for hundreds of calls on a single line already exists.

To concentrate still further the number of calls transmitted at one time, numerous time-multiplexed groups can be sent simultaneously on carriers of different frequencies, a process known as *frequency-multiplexing*.

The shift from analog to digital telephone systems has allowed information to be transmitted faster and more cheaply than ever before. It has also helped to overcome many of the problems associated with sending electrical signals over long distances – noise, distortion, and the weakening of the signal.

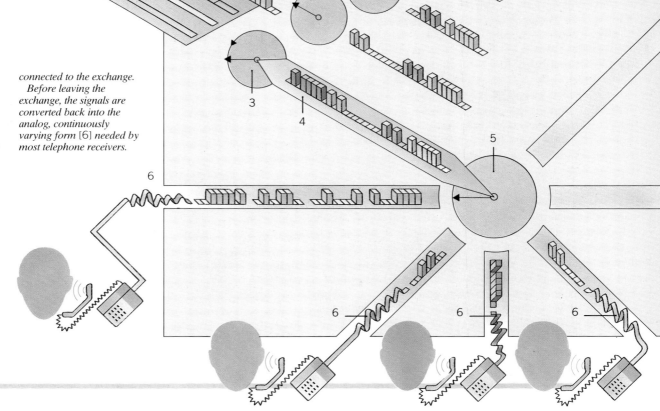

...ust 4 microseconds to record. This means that ...amples from numerous ...eparate conversations can ...e interleaved during one ...25-microsecond period ...nd sent down the same line ...t the same time.

*Here, four conversations ...re shown on the same line ...**C**]. When the stream of ...nterleaved samples reaches ...e second exchange [**D**], an ...lectronic switch [3] "pulls ...ff" one sample from each ...onversation in turn and ...ores it in a short-term ...emory [4]. A second ...witch can access the stored ...amples in any order [5]: ...is allows it to redirect any ...onversation to any of the ...dividual telephones* connected to the exchange.

Before leaving the exchange, the signals are converted back into the analog, continuously varying form [6] needed by most telephone receivers.

The Information Explosion

How the communications network is expanding

A revolution in communications is taking place. Telephones, formerly connected through underground wires, are increasingly becoming hand held devices that transmit via radio waves. Soon people could be reached anywhere on the globe via a network of satellites. Conversely, television, originally broadcast on the same frequencies that cellular phones now use, is set to become a digital medium, providing hundreds of different channels piped into homes along optical fibers – hair-thin strands of glass that carry enormous amounts of information at the speed of light.

A cellular phone conversation is accompanied by unheard digital exchanges (a series of 1s and 0s representing data) as the handset and the computers that control the network process the call. The cells the phones are named after are a mosaic of hexagonal areas, each with a transmitter/receiver or *base station* at its center. More people can use cellular phones than other radio phone systems because the signals used are very weak, so frequencies used in one cell may be reused in another a short distance away.

Every 15 minutes, each base station beams out a message asking all the handsets within its cell to "report in." This enables the central computers to know where to route a call when a handset is phoned.

Digital superhighways

The fast-growing number of fax machines, computer modems, and new telephone users demands transmission lines that can handle far greater numbers of calls than traditional copper wires. This demand is being met by fiber-optic cables, which carry digital messages in the form of rapid bursts of intense laser light. Capable of carrying hundreds of thousands of simultaneous phonecalls down a pair of glass strands, in addition to fax messages, computer data, and television signals, fiber-optic cables are revolutionizing global communication and home

entertainment. Some cable operators already offer a huge choice of channels, interactive games, and even on-demand video films.

Conversations are more intelligible when the two parties can see one another. However, video phones, which make this possible by simultaneously transmitting pictures and speech, are still not widely used. This is because transmitting a complete video signal requires the sending of more than 200 million bits (units of information) a second – 4000 times more than existing cables can handle.

Accepting lower picture quality and using *compression*, a technique by which redundant or repeated bits of data are omitted, the signal can be reduced to 64,000 bits per second. Even this is beyond the capacity of ordinary telephone lines, so current videophones can send only crude, still pictures. One model, sending data at 14,400 bits per second, takes five seconds to send one still picture.

Data highway
An optical trunk cable [**C**] *comprises a bundle of optical fibers around a thicker strengthening wire, contained in layers of protective sheaths. Each fiber has a core, through which light travels, and a cladding, which contains the light in the core. Both are made from silicon glass, with small amounts of boron or germanium added to improve transmission properties. A plastic sheath around the cladding ensures that no stray light passes into other fibers.*

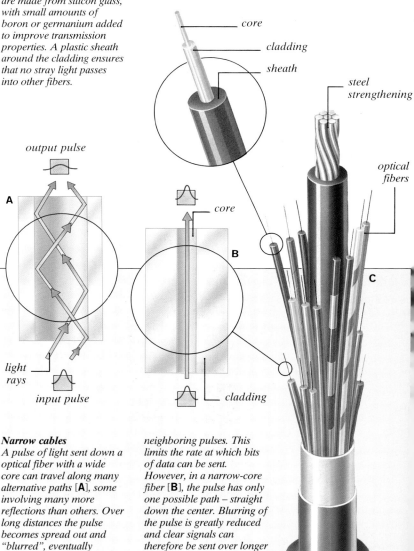

Light Pipes

Optical fibers can transmit digital data in the form of up to 2 billion pulses of laser light a second. This makes them the ideal medium for carrying the rapidly increasing numbers of telephone calls, fax messages, and computer information traveling from place to place. The glass they are made of is so clear that signals can travel for tens of miles before they have to be amplified – ten times farther than traditional copper cables.

A fiber is in fact made up of two concentric layers of ultra-pure bubble-free glass. The cylindrical *core* is surrounded by a *cladding* drawn from glass with a different *refractive index*. Laser light shone into the core is confined in a process called *total internal reflection* – rays hitting the boundary between the two layers at a shallow enough angle are reflected rather than escaping. Because fibers are so thin – narrower even than human hair – they can be bent quite sharply before light "leaks" out.

Narrow cables
A pulse of light sent down a optical fiber with a wide core can travel along many alternative paths [**A**], *some involving many more reflections than others. Over long distances the pulse becomes spread out and "blurred", eventually merging with the edges of*

neighboring pulses. This limits the rate at which bits of data can be sent. However, in a narrow-core fiber [**B**], *the pulse has only one possible path – straight down the center. Blurring of the pulse is greatly reduced and clear signals can therefore be sent over longer distances in such fibers.*

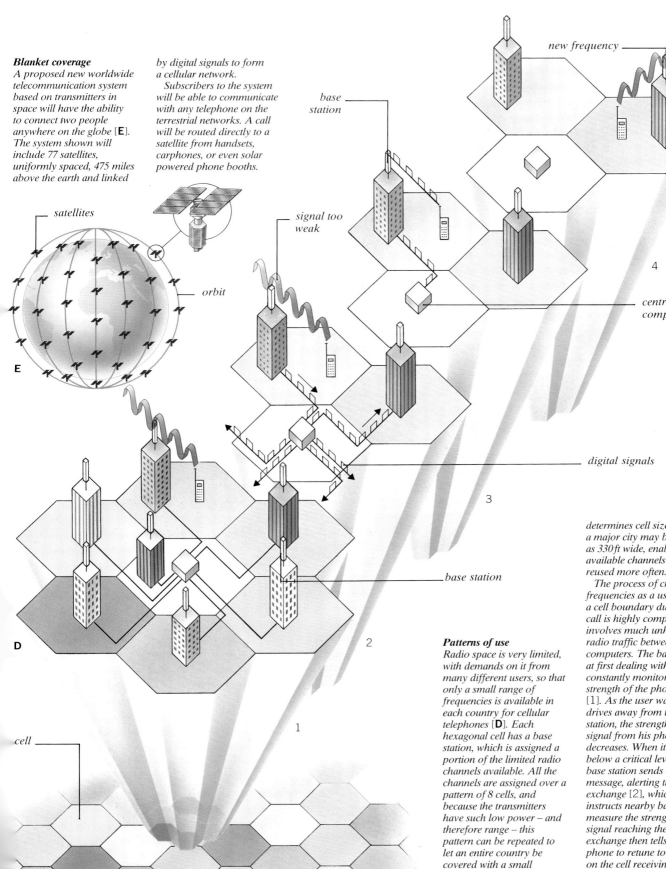

Blanket coverage
A proposed new worldwide telecommunication system based on transmitters in space will have the ability to connect two people anywhere on the globe [**E**]. The system shown will include 77 satellites, uniformly spaced, 475 miles above the earth and linked by digital signals to form a cellular network.

Subscribers to the system will be able to communicate with any telephone on the terrestrial networks. A call will be routed directly to a satellite from handsets, carphones, or even solar powered phone booths.

satellites

orbit

E

new frequency

base
station

signal too
weak

4

central
computer

digital signals

3

base station

cell

2

1

urban
area

D

Patterns of use
Radio space is very limited, with demands on it from many different users, so that only a small range of frequencies is available in each country for cellular telephones [**D**]. Each hexagonal cell has a base station, which is assigned a portion of the limited radio channels available. All the channels are assigned over a pattern of 8 cells, and because the transmitters have such low power – and therefore range – this pattern can be repeated to let an entire country be covered with a small number of channels. The number of local users

determines cell size. Cells in a major city may be as small as 330 ft wide, enabling the available channels to be reused more often.

The process of changing frequencies as a user crosses a cell boundary during a call is highly complex, and involves much unheard radio traffic between computers. The base station at first dealing with the call constantly monitors the strength of the phone signal [1]. As the user walks or drives away from the base station, the strength of the signal from his phone decreases. When it falls below a critical level, the base station sends a digital message, alerting the central exchange [2], which instructs nearby bases to measure the strength of the signal reaching them. The exchange then tells the phone to retune to a channel on the cell receiving the strongest signal [3], and the conversation is resumed [4].

On Air

How radio waves carry signals around the world

Radio is not just a means of transmitting speech and music. Criss crossing the atmosphere at the speed of light, radio waves relay a bewildering range of information – from navigational data that enable aircraft to pinpoint their position to within 30 ft, to high-definition television pictures. Like microwaves, visible light, and X-rays, radio waves are a form of electromagnetic radiation. They are distinguished from these other waves by their longer wavelengths (a million times greater than light) and by the way they are generated.

Radio waves are electromagnetic waves with wavelengths of between 1 cm and 100 km. They are produced by a transmitter, which makes an electric current surge back and forth very quickly along a length of wire (or *antenna*). As the electrons oscillate within the antenna, they give out a simple, regular radio wave of constant frequency and intensity.

The earliest radio messages were sent in Morse code by simply turning this wave on and off. However, to send more complicated messages, such as sound and pictures, this basic (or *carrier*) wave has to be shaped or *modulated* by the information to be transmitted.

The audio or video signal can shape the carrier wave in one of two ways. In AM (*amplitude modulation*), the height (or *amplitude*) of the peaks of the carrier wave follow the signal waveform. FM (*frequency modulation*) instead alters the frequency of the carrier to follow the signal. AM broadcasts need a narrower range of frequencies (or *bandwidth*) compared to FM, but FM signals are of much better quality.

When radio waves reach a receiver, they force electrons in the receiving antenna to oscillate in sympathy with their electromagnetic energy. This produces an alternating current in the antenna, which is then fed into a tuning circuit. The tuner lets through only waves of the required frequency, allowing the listener to select the desired station. Other circuits in the receiver separate the audio signal from the carrier wave (*demodulation*) before the signal is amplified and sent to a loudspeaker.

Specially built transmitters are not the only sources of radio waves. They can be produced by any charged particle moving at the right speed. Lightning strikes produce bursts of radio energy that can be heard as crackling interference on a radio receiver. And *pulsars* – dense, spinning stars – emit pulses of radio waves that are so regular they were once thought to be evidence of extraterrestrial life.

Carving up the air

The radio spectrum is arbitrarily divided up into a number of frequency bands that have different transmission properties and hence different uses. In general, low-frequency waves have longer ranges, but can carry less information.

Very-low-frequency transmissions (3–30 kHz), can cover enormous distances, but only carry simple information. Such frequencies are used in submarine communications because, unlike

The antenna (above)
The most prominent feature of a radio transmitter is its antenna. This works at optimum efficiency if its length is related to the wavelength of the signal it sends (usually one half or one quarter of the wavelength). Antennae of longwave transmitters may therefore be hundreds of feet long, whereas microwave antennae, such as those on mobile phones, are just inches in length. Antennae send radio waves out in all directions: in some cases, a parabolic dish is used to focus the waves, giving a more powerful, but highly directional radio transmission.

shorter wavelengths, they are able to pass through water. High-frequency (3–30 MHz) or shortwave signals can can also travel long distances, but by a different mechanism. These waves are reflected between the ground and a layer of the atmosphere known as the ionosphere: they can "bounce" halfway around the world carrying international communications and amateur broadcasts. The sun's radiation affects the gases in the atmosphere, interfering with the reflection of short waves during daylight hours, so these amateur broadcasts are best heard at night.

Medium frequencies, (300 kHz to 3 MHz) are used for AM radio signals and have a range of a few hundred miles. Very high frequencies (30–300 MHz) have a shorter range still, but can accommodate more complicated signals that need greater bandwidths. This waveband carries FM stereo radio broadcasts as well as television pictures and signals from cellular telephones.

Send and receive
An AM radio broadcast [A] begins with the material to be sent – usually speech or music. In the case of speech, the broadcaster's voice is first translated into an electrical audio signal by a microphone [1]. This signal [2] is an irregular pattern of different frequencies ranging from about 100 to 3000 Hz (the frequency range of the human voice). The signal is amplified [3] before being "imprinted" onto a radio signal, or carrier wave, with a frequency of about one million hertz. The carrier signal is generated by an oscillator [4] – in some transmitters, this is a quartz

Connections: Navigation Aids 40 Audio Equipment 44 Television 56 Magnetic Body Scanners 152

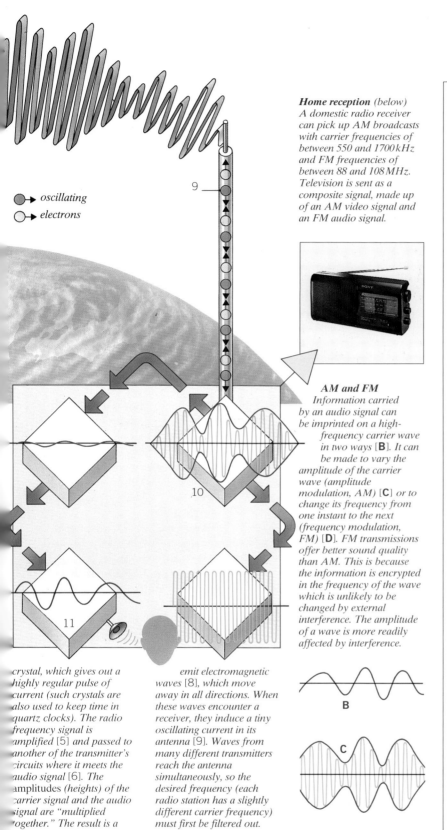

- ● → *oscillating*
- ○ → *electrons*

atellites: Orbits 172 Principles 246 248 250 258 262

Home reception (below)
A domestic radio receiver can pick up AM broadcasts with carrier frequencies of between 550 and 1700 kHz and FM frequencies of between 88 and 108 MHz. Television is sent as a composite signal, made up of an AM video signal and an FM audio signal.

AM and FM
Information carried by an audio signal can be imprinted on a high-frequency carrier wave in two ways [**B**]. *It can be made to vary the amplitude of the carrier wave (amplitude modulation, AM)* [**C**] *or to change its frequency from one instant to the next (frequency modulation, FM)* [**D**]. *FM transmissions offer better sound quality than AM. This is because the information is encrypted in the frequency of the wave which is unlikely to be changed by external interference. The amplitude of a wave is more readily affected by interference.*

B

C

D

crystal, which gives out a highly regular pulse of current (such crystals are also used to keep time in quartz clocks). The radio frequency signal is amplified [5] *and passed to another of the transmitter's circuits where it meets the audio signal* [6]. *The amplitudes (heights) of the carrier signal and the audio signal are "multiplied together." The result is a radio signal of constant frequency whose amplitude is shaped or modulated by the audio signal.*
The oscillating radio signal causes electrons to surge up and down a transmitting antenna [7]. *As the electrons alternate, they*

emit electromagnetic waves [8], *which move away in all directions. When these waves encounter a receiver, they induce a tiny oscillating current in its antenna* [9]. *Waves from many different transmitters reach the antenna simultaneously, so the desired frequency (each radio station has a slightly different carrier frequency) must first be filtered out. This is done by a resonant circuit (see box). The selected signal is then demodulated* [10] – *the audio signal is separated from the radio carrier, and then amplified* [11]. *It is then strong enough to drive the receiver's loudspeaker.*

Tuning in

A radio antenna simultaneously picks up a multitude of different frequencies. The desired station is filtered out by a *tuning circuit*, which is designed to allow current to oscillate at only one frequency. It is made up of a *capacitor* [1] (which stores energy as an electric field) linked to a coil [2] (which stores energy as a magnetic field). First, the capacitor is fully charged [3] – one of its plates is negative, the other positive. Charge naturally flows from one plate to the other: in doing so, it flows through the coil, setting up a magnetic field [4]. When current stops flowing through the coil, the magnetic field collapses. As it does so, it induces an electric current, which flows around the circuit to charge up the capacitor once more [5]. This discharges once again through the coil [6] and so on. An oscillating current of one frequency is therefore set up in the tuning circuit.

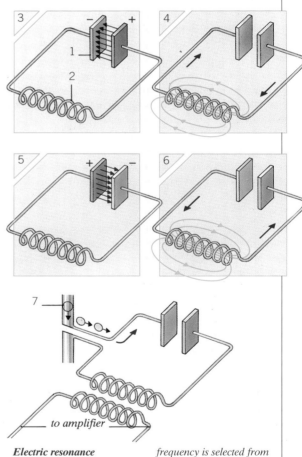

to amplifier

Electric resonance
The alternating current in the receiver's antenna [7] *is fed into the tuning circuit. Only if its rate of oscillation matches the "natural" frequency of the tuning circuit, does current oscillate in the circuit. In this way one*
frequency is selected from the many reaching the antenna. Changing the size of the capacitor in the tuning circuit (by tuning the radio receiver) changes the natural resonant frequency of the circuit and lets a different frequency pass through.

For the Record

How sound is recorded at home and in the studio

The first machine able to record and replay sound was invented by Thomas Edison in 1877. In this instrument, sound waves picked up by a diaphragm were used to press tiny indentations into a sheet of tin foil. Today, sound recording has developed into a high-tech industry. Speech or music produced in acoustically-designed studios is translated into a stream of binary numbers and recorded on a tape as microscopic patterns of magnetization. When replayed, the best recordings can be hard to distinguish from live performances.

Sound is made up of pressure waves traveling through the air. A microphone converts these mechanical waves into electrical waves, the peaks and troughs of which accurately match those of the original sound. These electrical waves are most conveniently recorded on magnetic tape. This is a plastic ribbon, coated with billions of tiny particles, each of which behaves like a tiny magnet with a north and south pole.

The electrical wave is amplified and passed through the coils of a thin electromagnet called a *recording head*. The fluctuating electrical wave produces a varying magnetic field around the head. And as an electric motor draws tape past the head, the particles in the tape realign to match the strength of the field. In this way, the sound is stored as a variable pattern of magnetization along the length of the tape. When the tape is played back, these same magnetic patterns induce an electrical wave in the recording head, which is amplified and passed to loudspeakers. This type of recording is known as *analog*, because the orientation of of the magnetic particles is proportional to (is an analog of) the size of the magnetic field, and therefore of the the sound wave itself.

Analog recording has drawbacks. There is a limit to the size of magnetic field that can be produced and hence to the loudness of sound that can be stored in the tape. This distorting effect, called *saturation*, happens when all the magnetic particles in a stretch of tape have the same alignment. In addition, the quietest passages that can be recorded have to be louder than the *noise* background. This is a hiss produced by the random orientations of the magnetic particles in the unrecorded tape. Even with noise-reduction circuitry, these constraints mean that the *dynamic range* of an analogue recording – the difference between its loudest and quietest parts – is limited. Other problems are *wow and flutter*, small variations in tape speed, which produce unwanted vibrato effects.

For these reasons, professional recorders are *digital* rather than analog, and use Digital Audio Tape (DAT). In digital recording, the electrical waveform from a microphone or musical instrument is *sampled* by electronic circuits. This involves measuring the wave's *amplitude* or height thousands of times each second. These height values are converted into binary numbers made only of 1s and 0s, which are represented by discrete on or off pulses of electricity. These pulses, when fed through

recording heads, produce just two distinct magnetic alignments on a magnetic tape. A digital recording is almost free of distortion because a tape player has only to distinguish between 1s and 0s rather than the infinite range of field strengths in an analogue recording. It can also be re-recorded with no loss of quality.

Cutting a disk
Tape is not the only the medium on which digital sound can be stored. In many professional recording studios, sound is now written directly on to the hard disk of a computer. Unlike tape, which has to be wound backward or forward to find a particular section, the "recording", or read/write, head of a hard disk can be moved to any section of the disk in a matter of milliseconds. This makes editing or otherwise manipulating the sound a much simpler procedure, and one that involves no degradation of clarity.

Mixing down
Most professional recordings consist of several tracks, or paths, of magnetic information laid down on the same tape. A multi-track recorder in a studio can lay down as many as 64 such tracks, each recorded and replayed by its own recording head. This means that a separate band along the tape can be used for each instrument or voice in a song. The instruments do not have to be recorded all at the same time – new tracks can be overdubbed onto the tape without erasing the existing ones, so that a complex arrangement can be built up by just one musician. A simpler machine has just 8 separate tracks [A]. One of these (top) carries a pulse code used to synchronize the tape machine with other recorders and instruments.

After recording, the 8 tracks are mixed down to a master tape carrying just the two tracks (left and right) needed for stereo reproduction. The output from each track first goes through a fader – like the volume control on a radio – which adjusts its level. Next it passes to a pan-pot, which works like a hi-fi balance control, setting the proportions of the signal that are sent to the left and right channels. True mixing desks have many more controls to alter the tonal balance of each sound by accentuating or reducing a particular band of frequencies. Using these controls a sound engineer is able to combine different tracks so that they present the listener with a spatial "picture," in which each instrument has an apparent location in the sound reproduced [B].

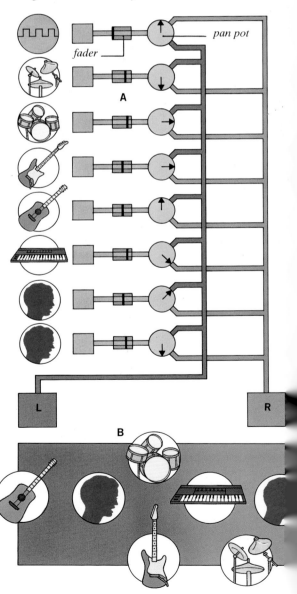

pan pot

fader

A

L   **R**

B

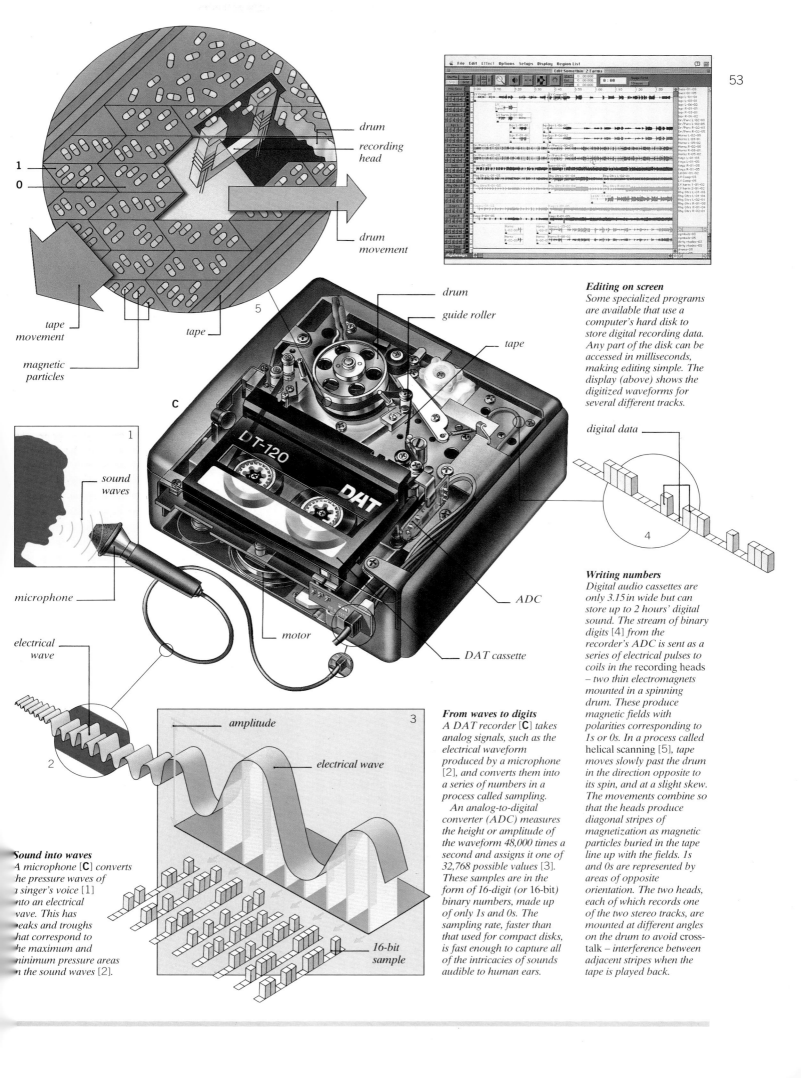

drum

recording head

drum movement

1

0

tape movement

magnetic particles

tape

5

drum

guide roller

tape

C

DT-120

DAT

sound waves

microphone

ADC

motor

DAT cassette

electrical wave

2

Editing on screen

Some specialized programs are available that use a computer's hard disk to store digital recording data. Any part of the disk can be accessed in milliseconds, making editing simple. The display (above) shows the digitized waveforms for several different tracks.

digital data

4

Writing numbers

Digital audio cassettes are only 3.15 in wide but can store up to 2 hours' digital sound. The stream of binary digits [4] from the recorder's ADC is sent as a series of electrical pulses to coils in the recording heads – two thin electromagnets mounted in a spinning drum. These produce magnetic fields with polarities corresponding to 1s or 0s. In a process called helical scanning [5], tape moves slowly past the drum in the direction opposite to its spin, and at a slight skew. The movements combine so that the heads produce diagonal stripes of magnetization as magnetic particles buried in the tape line up with the fields. 1s and 0s are represented by areas of opposite orientation. The two heads, each of which records one of the two stereo tracks, are mounted at different angles on the drum to avoid cross-talk – interference between adjacent stripes when the tape is played back.

From waves to digits

A DAT recorder [**C**] takes analog signals, such as the electrical waveform produced by a microphone [2], and converts them into a series of numbers in a process called sampling.

An analog-to-digital converter (ADC) measures the height or amplitude of the waveform 48,000 times a second and assigns it one of 32,768 possible values [3]. These samples are in the form of 16-digit (or 16-bit) binary numbers, made up of only 1s and 0s. The sampling rate, faster than that used for compact disks, is fast enough to capture all of the intricacies of sounds audible to human ears.

amplitude

electrical wave

3

16-bit sample

Sound into waves

A microphone [**C**] converts the pressure waves of a singer's voice [1] into an electrical wave. This has peaks and troughs that correspond to the maximum and minimum pressure areas in the sound waves [2].

Music by Numbers

How compact discs are read and recorded

The shimmering patterns of light on a compact disc are reflections from billions of pits less than a millionth of a meter wide buried under its surface. It takes the power and precision of a semiconductor laser to extract the digital message of 1s and 0s that these pits contain. Each twelve-centimeter disc contains at least three billion pits in a spiral track more than five kilometers long. Digital storage is versatile – text, videos, and animation as well as sound can be stored on interactive discs, capturing an entire twelve-volume encyclopedia on one small piece of plastic.

Sound from light
*The digital information on a compact disc is recorded as a spiral track of pits buried within its plastic body [**B**]. These are "read" by an infrared laser beam [**A**], which is reflected by a semi-silvered mirror and passes through two lenses before being focused on the track of pits. If the beam strikes a "flat" – the area between two pits – it reflects back through the lens system. When the beam hits a pit, it is dispersed so that almost no light reflects back. After passing through a cylindrical lens, the light falls on four light sensors. These produce an electrical*

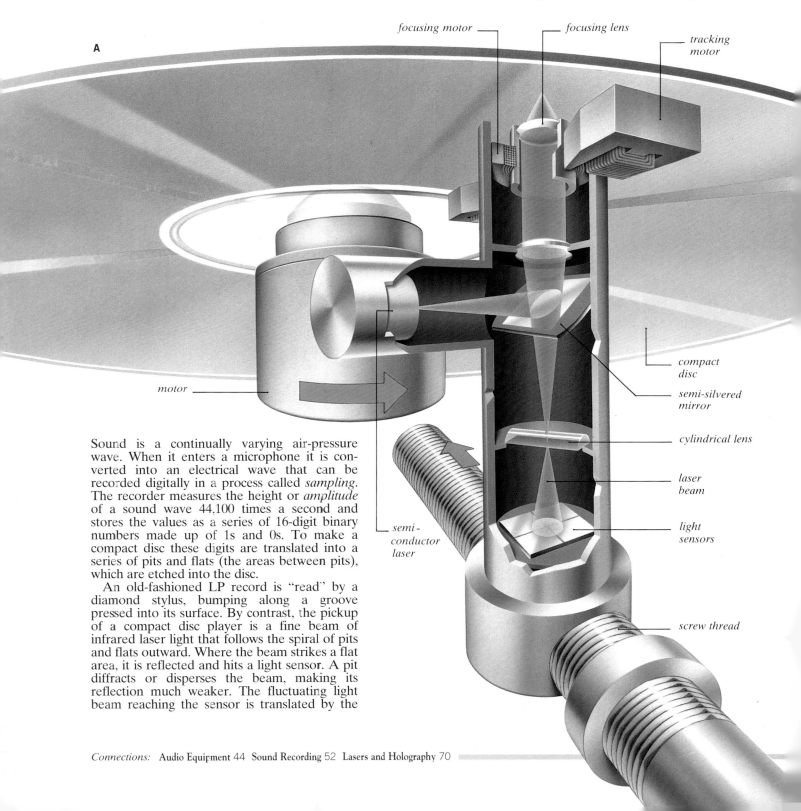

focusing motor · focusing lens · tracking motor

A

motor

compact disc

semi-silvered mirror

cylindrical lens

laser beam

light sensors

semi-conductor laser

screw thread

Sound is a continually varying air-pressure wave. When it enters a microphone it is converted into an electrical wave that can be recorded digitally in a process called *sampling*. The recorder measures the height or *amplitude* of a sound wave 44,100 times a second and stores the values as a series of 16-digit binary numbers made up of 1s and 0s. To make a compact disc these digits are translated into a series of pits and flats (the areas between pits), which are etched into the disc.

An old-fashioned LP record is "read" by a diamond stylus, bumping along a groove pressed into its surface. By contrast, the pickup of a compact disc player is a fine beam of infrared laser light that follows the spiral of pits and flats outward. Where the beam strikes a flat area, it is reflected and hits a light sensor. A pit diffracts or disperses the beam, making its reflection much weaker. The fluctuating light beam reaching the sensor is translated by the

Connections: Audio Equipment 44 Sound Recording 52 Lasers and Holography 70

pulse when they are illuminated, so the presence or absence of the beam gives an on/off digital signal, which is converted by the CD player into the waveforms of the original recording. As the disc spins, a screw thread moves the entire optical system out from the center.

C

Molding music
When a sound recording is made, it is stored on magnetic tape [1]. A digital version of the recording, which includes complete error correction codes, is also stored on tape and used to record the compact disc [**C**]. The tape controls a laser beam that burns a pattern into a special photoresist *coating on a blank glass plate* [2]. Where the laser strikes the photoresist, the properties

of the material are altered so that it is vulnerable to acid. The affected material is etched away, together with the glass beneath, to form the pits of the track [3]. This glass master disk is very slowly plated with nickel metal until a suitable thickness has built up [4]. Then an etchant removes the remaining glass leaving a nickel mother – a "negative" of the original. The mother is used to stamp out multiple copies of the clear plastic disks that become CDs [5]. The top of the disk is coated [6] with a thin layer of aluminum to reflect the light from the player's laser, followed by a protective layer of lacquer onto which the the ink of the label is printed.

track of pits

B

D

lens

laser beam

sensors

Keeping on course
An elegant subsystem ensures that the CD player's laser is always correctly focused and aligned on the track of pits and flats. After striking the CD, the laser beam is reflected down on to a set of four light sensors [**D**]. If focus and tracking are correct, the reflected light forms a circular spot, spread equally over the four sensors. If focus or tracking are "out" the spot formed is either skewed [1] or uneven in intensity [2]. These irregularities in the distribution of light over the four sensors are translated into movements of two tiny motors that move and refocus the beam.

player into a sequence of "on" and "off" electrical pulses, or *bits*, which are translated back into sound and fed to the loudspeakers.

A remedy for errors
Each pit is only three-fifths of a micrometer across – about one-hundredth of the breadth of a human hair. With data so finely spaced on the disc, a small particle of dust could block large amounts of data and cause major problems. *Error correction* – ways of eliminating or hiding errors – enables a player to give a smooth and accurate output. In the recording, extra information is added to the samples written on a disc. These *parity bits* can be used by the microprocessor inside the player to check if any data is missing and fill in small errors. Large errors are avoided by splitting up and *interleaving* the 16 digit words, so that a scratch only obliterates small pieces of several samples rather than one in its entirety.

Although the sound quality that CDs give is very good, it is still not perfect. Sampling – the process of converting a smooth sound wave into a series of 1s and 0s – itself can introduce specific distortions. One is called *quantizing error*. During recording, the amplitude, or size, of a sound wave is measured and assigned one of more than 32,000 values, 44,100 times each second. Despite this huge range, most often the true value falls somewhere between two possibilities. The resulting random rounding-up or rounding-down is heard as *noise* – a disturbing high-frequency hiss behind the music. The only way to mask this is to add another, less disturbing type of noise, although this is still noticeable to very careful listeners.

More than music
Any form of data – written words, numbers, sounds, or pictures – can be represented in the form of numbers and stored on a CD. One CD can store over 100 million words of text, the equivalent of a thousand novels.

Similarly, instead of representing letters of the alphabet, the numbers can represent the brightness and color of pixels – the picture cells that make up a TV image. A CD only has enough capacity to store about a minute's worth of a complete video signal, and even its playback rate of 1.3 million bits per second is not fast enough to display the pictures. So various forms of *data compression* are used to store pictures in much less space without losing quality.

This creates the possibility of a *multimedia* format, combining text, images, and sound. Using a keypad, users can choose their own routes through the material on the disc. A multimedia work contains text that can be rapidly and easily searched for all items containing a particular word or phrase. Its "pages" also contain embedded drawings, sounds, animations, and video clips as well as text, although the access time of three seconds can be a drawback with some kinds of subject matter.

Looking at TV

How a television receiver works

Color television is taken for granted: watching "the box" is the single most significant leisure activity in developed countries. This commonplace appliance – the product of over fifty years of research – is truly remarkable in its ability to handle and display information. Each image on its screen is made up of over 100,000 picture elements, arranged in several hundred lines: and the image displayed changes every few hundredths of a second. To show a 15-minute newscast, therefore, the television set must accurately process more than one billion units of information.

Television technology has been made possible by a quirk of human vision – that images are retained on the retina of the viewer's eye for a fraction of a second after they strike it. So by displaying images piece-by-piece at sufficient speed, the illusion of a complete picture can be created: and by changing the image on the screen 25–30 times per second, movement can be realistically represented.

Building an image
The inside of a television screen is coated with millions of tiny dots of a *fluorescent* compound – one that emits light when struck by high-speed electrons. A narrow beam of electrons produced in the television tube is fired at the screen in a preset pattern, causing the fluorescent dots to light up in sequence. For each new picture, the beam is first directed to the top left of the screen: it then scans along horizontally until the top right of the screen is reached. The beam is then momentarily switched off before reappearing just below its original starting position, when another horizontal line is "painted" on to the screen. Each complete picture is made up of 525 (or 625 in Europe) of these horizontal scans or "lines," and each set of 525 lines is painted onto the screen in around one-thirtieth of a second.

Color, light, and timing
Each line scan is completed within one ten-thousandth of a second. In this tiny space of time the intensity of the electron beam is made to fluctuate, causing some fluorescent dots to shine brightly and others to emit no light at all. In this way, light and dark areas of a picture can be reproduced on the television screen.

Color television sets have not one but three electron "guns," which fire parallel streams of electrons at the screen. Each gun is responsible for illuminating red, blue, or green components of the image. By mixing the output of the electron guns most colors can be reproduced on the screen. A combination of red and green gives yellow; blue and green gives a bluish color called cyan; red and blue together give magenta; and all three colors combined result in white. As a line is scanned, the output of each gun is controlled to produce a given color at each point along its length. The output of each gun, and therefore the brightness and color at each point of each scanned line, is controlled by the television signal, which is picked

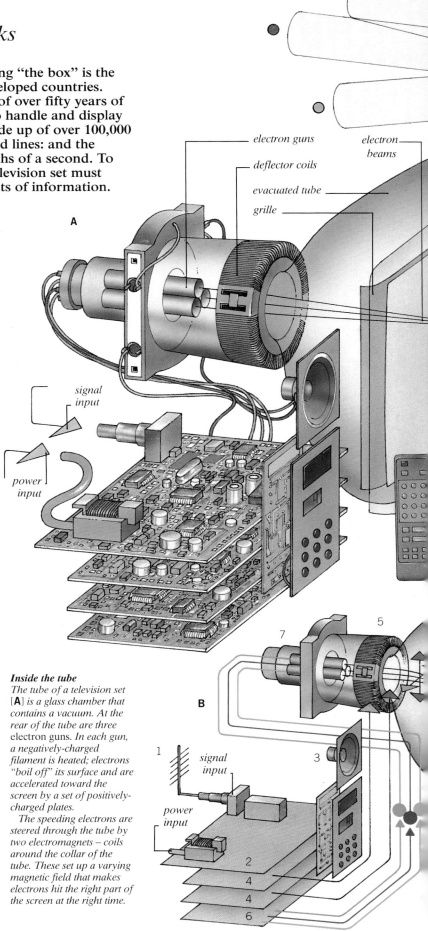

electron guns

electron beams

deflector coils

evacuated tube

grille

A

signal input

power input

B

signal input

power input

Inside the tube
The tube of a television set [**A**] *is a glass chamber that contains a vacuum. At the rear of the tube are three electron guns. In each gun, a negatively-charged filament is heated; electrons "boil off" its surface and are accelerated toward the screen by a set of positively-charged plates.*

The speeding electrons are steered through the tube by two electromagnets – coils around the collar of the tube. These set up a varying magnetic field that makes electrons hit the right part of the screen at the right time.

Connections: Audio Equipment 44 Radio 50 Video Technology 58 Television: New Developments 60 Principles 244 246 248 250 254 258 262 266

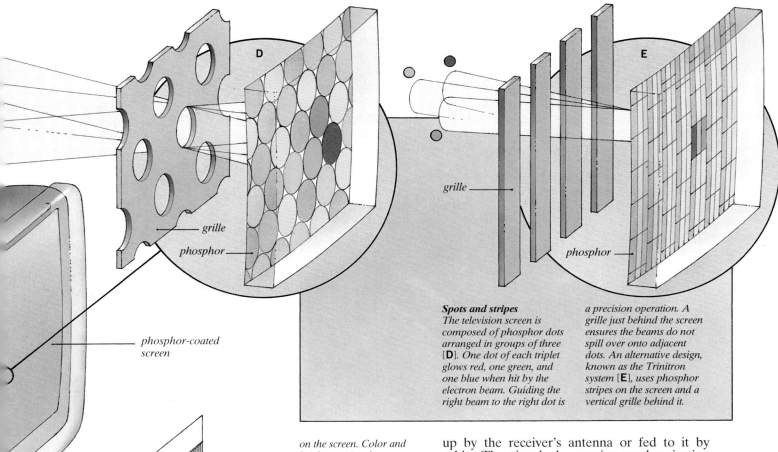

grille

phosphor

phosphor-coated
screen

C

From signal to screen
*Television is carried by a
composite signal [**B**]. The
picture is defined by signals
carrying information about
color and brightness; sound
is carried by a separate
audio signal; and timing
signals ensure that the
picture elements are put*

*together in the correct order.
This complex transmission
is picked up by an antenna
[1] and passed to a detector
circuit [2], where the desired
station is tuned into. The
signal is then split: sound
signals pass to a speaker
[3]. Timing signals pass to
synchronization circuits [4],
where they are split into two
components before being
fed to the electromagnets
around the collar of the tube
[5]. One component
(yellow) defines the
horizontal position of the
beam; the other (orange)
defines its vertical position*

*on the screen. Color and
brightness signals pass to a
decoder circuit [6] before
reaching the three electron
guns in the base of the tube
[7]. Each gun produces a
separate beam which
corresponds to red, green,
or blue parts of the image.
Brightness and color are
controlled by moderating
the power of the beam
moment by moment.*

*As the electron beams
emerge from their respective
guns, they are guided by the
magnets around the collar
of the tube, which are in
turn controlled by the
timing pulses in the
television signal. The beam
is deflected from left to right
and from top to bottom,
painting rows of horizontal
lines on the screen.*

*The number of pictures,
or frames, shown by the
television is limited to 30 per
second. This rate is too slow
to fool the eye into seeing a
continuous image, and the
picture appears to flicker.
To reduce flicker each
frame is in fact shown twice
[**C**]. As the electron beam
passes down the screen, it
paints every other line. One
thirtieth of a second later, it
paints the same image on
the lines missed out in the
first pass. Each frame of the
image is the result of two
passes over the screen by the
three electron guns.*

up by the receiver's antenna or fed to it by
cable. The signal also carries synchronization
pulses to ensure that the electron guns fire at
the right part of the screen at the right time.
Without this information, the television picture
would be no more than an irregular jumble of
colored spots.

Quality and quantity

The huge volume of coded information needed
to transmit a television picture imposes limits
on its resolution – the amount of detail that can
be held by an image. Each television channel
requires a *bandwidth* of 6 Megahertz – 600
times that occupied by a radio station. Increasing resolution requires the channel to carry a
still greater quantity of information, which in
turn increases its bandwidth. The frequencies
available in any one area for television transmission are limited: increasing image quality
therefore limits the number of stations available
to viewers. Today's television service is therefore a compromise between image quality and
channel choice.

The advent of cable television, and advances
in fiber-optic technology, will allow many more
channels to operate simultaneously without
interfering with one another, thereby lifting the
present restrictions on resolution. High-definition television (HDTV) – with image quality
comparable to that of movie film – may soon
become a standard fixture in our living rooms.

The development of new television systems
will depend not only on the availability of
advanced technology, but on the economics of
connecting millions of households to a network
based on fiber-optic cable. Any future HDTV
system will also need to be able to transmit programs made in today's format, to satisfy
viewers' demand for repeats of old shows.

Electric Pictures

How camcorders store moving images

Early television cameras had such poor contrast that actors had to wear black lipstick for their mouths to be seen. A modern camcorder can record detailed pictures in candlelight and contains all the components of a miniaturized TV studio. Its zoom lens concentrates light onto a detector chip that breaks down a picture into 300,000 points and produces an electric signal. This is passed to whirring tapeheads which cram information about the scene's brightness, color, and sounds into a strip of tape as thin as a human hair.

A

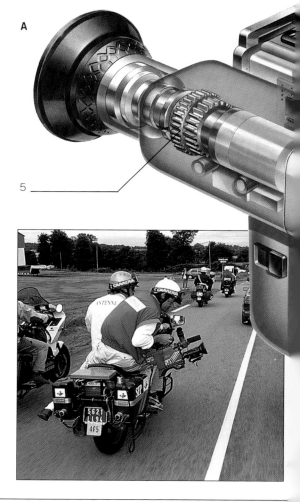

5

Camcorders, like film cameras, freeze a moving scene into a series of still pictures. In the case of the camcorder the stills are made up of a grid of *pixels* – each corresponding to a dot of light on a television screen. A zoom lens focuses light from the scene being recorded on to a CCD (*charge-coupled device*). This is a silicon chip, its surface engraved with a grid of hundreds of thousands of light sensors. When these are read off row by row they form an electric waveform, in which the peaks and troughs echo the bright areas and shadows of the scene.

Coloring in

The camera uses the special properties of red, green, and blue light to record color. These are called *primary* colors because in different combinations they can reproduce all the other colors in the rainbow. A tiny filter in front of the CCD, made up of stripes only a fiftieth of a millimeter wide, splits the light into these colors, which are individually sensed by detectors in groups of three. The signal these give contains information to light up corresponding red, green, and blue dots on a TV screen. More sensitive cameras use a system of prisms to split light up into its primary components, which are then sensed by three separate CCDs.

If too much light gets through to the CCD it *saturates*, giving a completely white picture. The processor can prevent this either by closing up an *iris* in the lens (just like the one in a human eye) or by making the CCD collect light for a shorter time each frame.

Getting it on tape

All tape recorders use small electromagnets – *tapeheads* – to store a signal as patterns of magnetization within a length of plastic-coated tape. For audio recordings, the tape moves past the heads at 1.9in/s. But audio signals contain much less information than video, which needs a *writing speed* of many feet per second. If the signal was recorded in a straight line along the tape, many miles of tape would be needed to store even a short programme. To avoid this, video recorders use a trick called *helical scanning*. Very narrow recording heads are mounted on drums, which spin more than 2000 times per minute. The tape is slowly drawn diagonally over the drum surface, so that the signal is recorded as a series of thin diagonal lines across the tape. This pattern can accommodate the density of data needed for video.

Moving pictures
Light entering a camcorder [A] is focused by the many glass facets of a zoom lens [1]. The beam then passes through a dichroic prism [2] *– glass blocks coated with color filters. This splits the light into its primary colors of red, blue, and green. Each is directed to an individual CCD light-sensing chip [3], which converts the image into an electronic signal, sent onto the video head [9] and recorded on magnetic tape with sound from the microphone [4]. The viewfinder [5] – a miniature television – displays pictures as they are recorded.*

The same principles are at work inside a professional camcorder, the light weight of which makes it possible to record even from the back of a motorcycle (right).

Charge-coupled Devices

A CCD is a light-sensitive sandwich of silicon above a base of metal electrodes. In between is a layer of insulating silicon dioxide. The device works by repeating two steps 25 times a second. In the first, *photons* (particles of light) strike silicon atoms in the body of the CCD, releasing negative electrons. An electrode, mounted above each pixel and separated from the silicon by an insulating layer, traps the electrons with a positive voltage. As more light falls onto the CCD a "picture" builds up, the bright areas represented by pixels containing large numbers of electrons. The second stage is to "read" this picture off the chip, a pixel at a time, to give a video signal.

CCDs are used in devices such as photocopiers and fax machines, as well as in video cameras. However, the biggest and most sophisticated are used in astronomical telescopes and contain more than 4 million image-sensing elements (pixels) in a 2.2in² area. These are more than 25 times larger than a camcorder CCD, are sensitive to much fainter light than photographic plates, and have a wider range of wavelengths.

Charging up
As incoming photons (packets of light) hit the silicon atoms within a CCD chip, they knock out electrons in what is known as the photoelectric effect. The electrons, which are negatively charged, are attracted toward a positive voltage on the middle of the three electrodes beneath each pixel [1]. Thus the light intensity of the scene being recorded is captured in terms of the number of electrons in each pixel.

Periodically, the amount of charge is measured. The positive voltage is made to ripple along the electrodes of each column of pixels, "dragging" the electrons from one pixel to another [2]. Those in the lowest sensor of each column are transferred onto a single

Connections: Sound Recording 52 Television 56 Television: New Developments 60 Telescopes 138 Principles 244 246 248 254 264 266

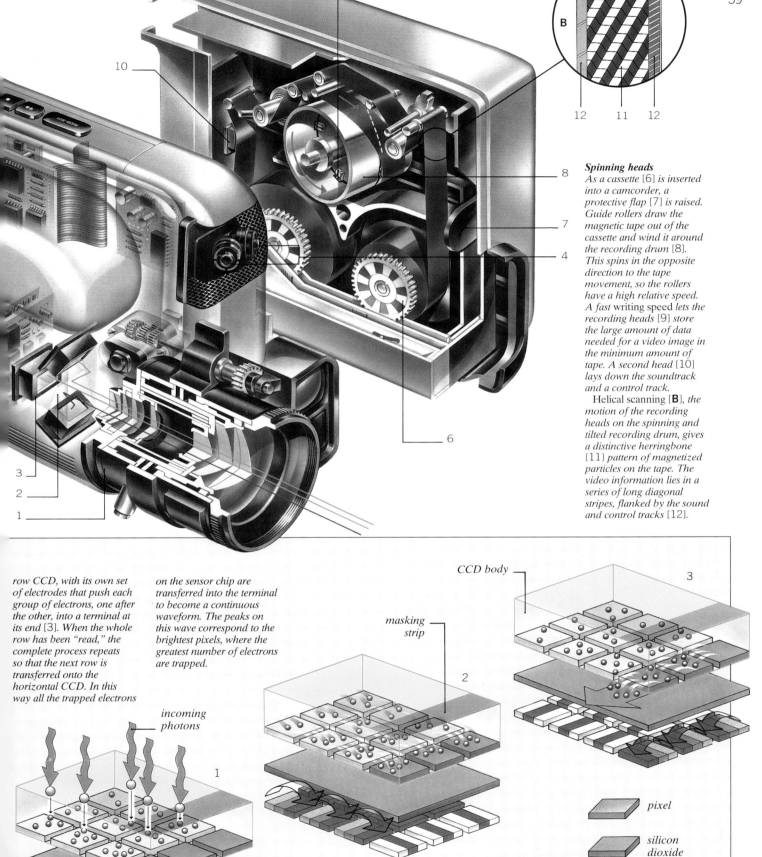

12   11   12

Spinning heads

As a cassette [6] is inserted into a camcorder, a protective flap [7] is raised. Guide rollers draw the magnetic tape out of the cassette and wind it around the recording drum [8]. This spins in the opposite direction to the tape movement, so the rollers have a high relative speed. A *fast* writing speed *lets the recording heads* [9] *store the large amount of data needed for a video image in the minimum amount of tape. A second head* [10] *lays down the soundtrack and a control track.*

Helical scanning [**B**], *the motion of the recording heads on the spinning and tilted recording drum, gives a distinctive herringbone* [11] *pattern of magnetized particles on the tape. The video information lies in a series of long diagonal stripes, flanked by the sound and control tracks* [12].

9

10

8

7

4

6

B

3

2

1

row CCD, with its own set of electrodes that push each group of electrons, one after the other, into a terminal at its end [3]. When the whole row has been "read," the complete process repeats so that the next row is transferred onto the horizontal CCD. In this way all the trapped electrons

on the sensor chip are transferred into the terminal to become a continuous waveform. The peaks on this wave correspond to the brightest pixels, where the greatest number of electrons are trapped.

CCD body

masking strip

incoming photons

pixel

silicon dioxide

silicon atom

electrode

electron

1

2

3

Future Visions

How television will be transmitted and viewed in the future

More than 22,000 miles above the earth, a television satellite beams down a score of stations to an entire continent. Under the streets of our towns and cities run fiber-optic cables, piping yet more choices to the viewer. Television itself is changing – high-definition digital services will provide cinema-style wide-screen pictures and surround sound. The race is on to perfect the technology for completely flat screens that can be hung on the wall. Via this new generation of screens people will have access to new media on disks, from games to holiday photographs.

A domestic satellite dish can receive signals from a constellation of transmitters circling the earth in geostationary orbit. Each station is beamed down to a satellite's footprint – the area it is aimed at – with only 100 watts of power, the same amount used by a lightbulb. By the time it reaches the surface the signal is weaker than the current in a watch motor, yet a single dish can gather waves from scores of different stations.

Even this explosion in the choice of viewing available is tiny compared with the capabilities of a digital network. Rather than a conventional television wave, which varies continuously to represent the brightness of a particular point, a digital signal uses numbers to represent the brightness and color of each point on a screen. Fiber-optics can carry hundreds of these signals into the home, and can be used interactively – responding down the same path that delivered a program, a viewer can vote, play games, or order action replays.

Wider, sharper, flatter

These digital signals are essential to the next generation of TV displays – high-definition TV, or HDTV. At present, TV pictures in the US and Japan have 525 lines (625 in Europe). In an HDTV picture there will be 1250 lines (1125 in Japan), almost doubling the number of pixels per inch to give a far sharper picture. At the same time, screens will become wider to match the shape of those in cinemas.

Most existing large TVs are built around cathode ray tubes (CRTs). These contain bulky and power-hungry electron guns, making them unsuitable for truly portable models. Large color LCD (liquid-crystal display) panels, expanded versions of a typical laptop computer screen, are the holy grail of current research. One problem is that these screens do not respond quickly enough to cope with a TV signal containing 50 frames a second. Another pitfall is the huge expense of building large screens: even one the size of a small CRT set costs more than US$10,000. However, prices are likely to fall once mass production begins.

Bigger, although fainter, pictures are possible with projection systems based on either CRTs or LCDs. An entire wall in a room can be covered with an image using these systems, although a picture any larger than 40in across shows all too clearly the poor resolution of current TV standards.

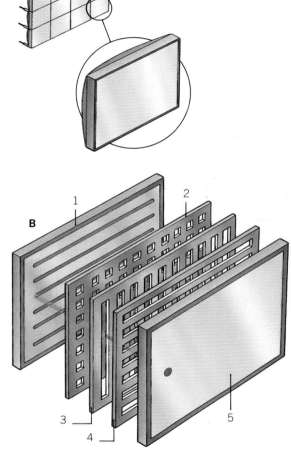

Slimline TV
Some companies, rather than concentrating on developing liquid-crystal displays for flat-screen television, have reinvented the cathode ray tube. A conventional CRT produces an electron beam that is deflected by a magnetic coil to scan across the screen. The flat-screen version [A] divides the screen into a matrix of about 10,000 unit cells. Each cell [B] is a sandwich of six layers. At the back, strip cathodes [1] generate electrons that are then strengthened and focused into beams by the next plate [2]. The beams pass through horizontal [3] and vertical [4] deflection plates that move the beams from side-to-side and up-and-down, to scan the screen [5]. As each electron beam is responsible for a tiny area, rather than the whole screen, the result is a slimmer television: 4in thick compared with 13in.

Liquid-Crystal Displays

Liquid crystals (LCs) are compounds that flow like liquids, but whose molecules have crystalline structure. The usefulness of LCs in displays arises from the effect they have on light. Light is made up of transverse waves that vibrate at right angles to the direction they travel. Ordinarily, light waves vibrate in all possible directions, but a *polarizer* blocks out all waves except those vibrating in one particular direction. Liquid-crystal molecules naturally align themselves in a spiral, and this pattern twists the polarization of light.

In a liquid-crystal display, light that has passed through a polarizer has its polarization turned through 90° by the LC. This means that the light can pass through a second polarizer at right angles to the first and be seen by the viewer. A voltage across the LC makes the molecules untwist so the light is blocked.

LCD
One of the simplest applications of liquid crystals is a seven-segment numeric display. The LC is sandwiched between two glass plates each of which is embossed with seven clear electrodes, which, when activated, create electric fields through the LC.
A polarizer allows only light waves that vibrate in one direction to pass through the liquid [1]. The spiraling molecules of the liquid crystal twist the polarization of the light through 90° [2], and the light can pass through a second polarizer, on to the display. When a voltage is applied across a segment,

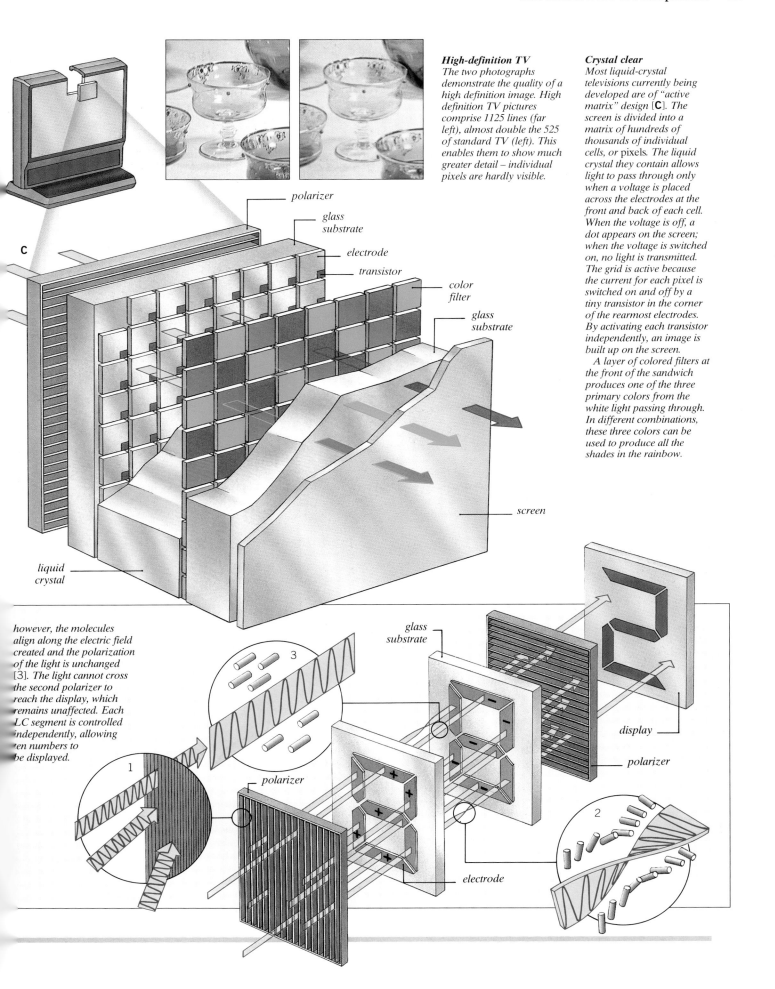

High-definition TV
The two photographs demonstrate the quality of a high definition image. High definition TV pictures comprise 1125 lines (far left), almost double the 525 of standard TV (left). This enables them to show much greater detail – individual pixels are hardly visible.

Crystal clear
Most liquid-crystal televisions currently being developed are of "active matrix" design [**C**]. The screen is divided into a matrix of hundreds of thousands of individual cells, or pixels. The liquid crystal they contain allows light to pass through only when a voltage is placed across the electrodes at the front and back of each cell. When the voltage is off, a dot appears on the screen; when the voltage is switched on, no light is transmitted. The grid is active because the current for each pixel is switched on and off by a tiny transistor in the corner of the rearmost electrodes. By activating each transistor independently, an image is built up on the screen.

A layer of colored filters at the front of the sandwich produces one of the three primary colors from the white light passing through. In different combinations, these three colors can be used to produce all the shades in the rainbow.

polarizer

glass substrate

electrode

transistor

color filter

glass substrate

screen

liquid crystal

however, the molecules align along the electric field created and the polarization of the light is unchanged [3]. The light cannot cross the second polarizer to reach the display, which remains unaffected. Each LC segment is controlled independently, allowing ten numbers to be displayed.

glass substrate

display

polarizer

polarizer

electrode

Getting the Picture

How a camera makes a photograph

Launched in 1888 with the slogan "You push the button, we do the rest," the Kodak camera opened up the art of photography to the masses. Photography is now the world's most popular pastime, and taking a picture has never been simpler. Modern cameras rely on microchip technology to calculate exposure and focus. Fitted with refined optics, they put professional results within reach of the amateur. Yet all cameras, however sophisticated, share the same basic design, which has changed little since the earliest days of photography.

A camera is essentially a light-proof box with a lens at one end and film at the other. Most cameras also allow some control over focus and *exposure* – the amount of light falling on the film – in order to maximize the sharpness and fidelity of the image formed.

Cameras are produced in a variety of sizes, or *formats,* (determined by the size of film used) to suit different needs. In general, the larger the film, the better the resolution: for this reason landscape and advertising photographers tend to use bulky cameras loaded with single sheets of film measuring up to 10x8in. Amateurs favor more portable cameras that do not need to be reloaded after each exposure. The most popular cameras of this type use rolls of 35mm-wide film that can record 36 separate exposures.

There are two categories of 35mm cameras – *compact* and *single lens reflex* (SLR). Compact cameras are simpler: they have fewer moving parts because they have two lenses. The main lens forms the image, while a smaller viewfinder lens, offset to one side, shows what the camera will record. The drawback of this design is that there is a discrepancy between what is seen and what is recorded, called the *parallax* error. An SLR is mechanically more complex, but more versatile: because it has just one lens through which the scene is both viewed and recorded, parallax error is avoided. SLR lenses can also be interchanged and the viewfinder always shows what will appear in the photograph.

Light work
*Exposure is controlled by changing aperture or shutter speed. But altering either affects the image recorded in subtle ways. Setting a small aperture [**C**] increases depth of field – the depth of the scene that appears to be in focus. This happens because when a point in the scene is out of focus on the film, it is recorded as a disk of small diameter, and so still seems relatively sharp. With a wider aperture [**D**], this point becomes a wide disk that overlaps adjacent disks giving a fuzzy image.*

The shutter of a 35mm SLR comprises two blinds just in front of the film. When the shutter is released,

*the blinds move from left to right. The film is exposed though the gap between the two blinds. A narrow gap corresponds to a fast shutter speed, which is capable of "freezing" motion [**E**]. A wide gap corresponds to a slow speed, which blurs motion [**F**].*

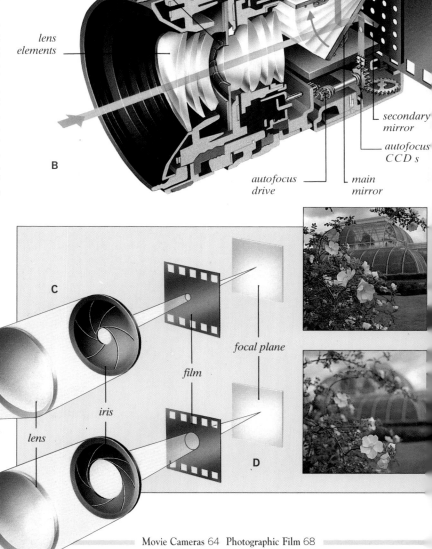

light metering cells
film advance lever
shutter speed control
flash hotshoe
viewfinder
shutter release
pentaprism
aperture control ring
iris
lens elements
secondary mirror
autofocus CCDs
autofocus drive
main mirror
B
C
film
iris
lens
focal plane
D

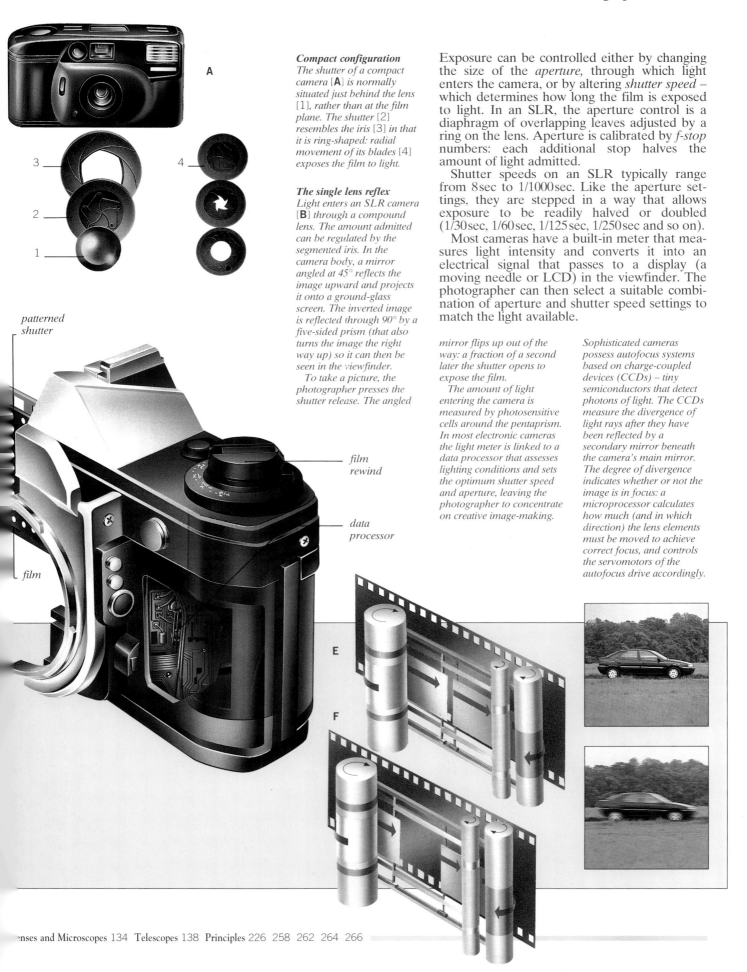

A

3

2

1

4

Compact configuration
*The shutter of a compact camera [**A**] is normally situated just behind the lens [1], rather than at the film plane. The shutter [2] resembles the iris [3] in that it is ring-shaped: radial movement of its blades [4] exposes the film to light.*

The single lens reflex
*Light enters an SLR camera [**B**] through a compound lens. The amount admitted can be regulated by the segmented iris. In the camera body, a mirror angled at 45° reflects the image upward and projects it onto a ground-glass screen. The inverted image is reflected through 90° by a five-sided prism (that also turns the image the right way up) so it can then be seen in the viewfinder.*
  To take a picture, the photographer presses the shutter release. The angled

patterned
shutter

film
rewind

data
processor

film

Exposure can be controlled either by changing the size of the *aperture*, through which light enters the camera, or by altering *shutter speed* – which determines how long the film is exposed to light. In an SLR, the aperture control is a diaphragm of overlapping leaves adjusted by a ring on the lens. Aperture is calibrated by *f-stop* numbers: each additional stop halves the amount of light admitted.

Shutter speeds on an SLR typically range from 8 sec to 1/1000 sec. Like the aperture settings, they are stepped in a way that allows exposure to be readily halved or doubled (1/30 sec, 1/60 sec, 1/125 sec, 1/250 sec and so on).

Most cameras have a built-in meter that measures light intensity and converts it into an electrical signal that passes to a display (a moving needle or LCD) in the viewfinder. The photographer can then select a suitable combination of aperture and shutter speed settings to match the light available.

mirror flips up out of the way: a fraction of a second later the shutter opens to expose the film.
  The amount of light entering the camera is measured by photosensitive cells around the pentaprism. In most electronic cameras the light meter is linked to a data processor that assesses lighting conditions and sets the optimum shutter speed and aperture, leaving the photographer to concentrate on creative image-making.

Sophisticated cameras possess autofocus systems based on charge-coupled devices (CCDs) – tiny semiconductors that detect photons of light. The CCDs measure the divergence of light rays after they have been reflected by a secondary mirror beneath the camera's main mirror. The degree of divergence indicates whether or not the image is in focus: a microprocessor calculates how much (and in which direction) the lens elements must be moved to achieve correct focus, and controls the servomotors of the autofocus drive accordingly.

E

F

Moviemakers

How a movie camera captures a moving image

The movie camera is the backbone to arguably the most important, and certainly the most lucrative, art form of the twentieth century. Within one month of its release in 1993, Steven Spielberg's blockbuster movie *Jurassic Park* had made 236 million dollars at the box office in the United States alone. The movie camera has revolutionized education and news gathering, and changed the way we think about our world. It has allowed us all to witness the most important, moving, and terrifying moments in recent history.

The human eye is unable to distinguish separate, successive images presented to it at a rate greater than 16 images per second: instead it sees a continuously evolving scene. This *persistence of vision* is the secret of the movie camera – an instrument designed to capture movement as a sequence of still photographs or *frames* along a strip of film. When projected onto a screen at the same rate at which they were recorded (typically 24 frames per second), these images create the impression of smooth motion.

In common with a conventional stills camera, the movie camera possesses a lens to focus light onto the film, a lens diaphragm to vary the aperture (and thus the amount of light admitted), and a shutter to expose the film at the required instant. It differs from a still camera in that the movement of film and shutter must be precisely coordinated to produce a series of evenly spaced and correctly exposed frames.

The format of a movie camera is named after the width of the film it takes. Of the three most common, Super 8 – an amateur format that uses 8mm-wide film – has been almost completely superseded by video; 35mm is the standard format for feature films; and 16mm is where amateur and professional meet, often being used by makers of documentaries.

Stop-start photography

For use in a professional 35mm camera, unexposed film is cut to a 300-meter length, wound onto a metal spool, and loaded into a light-tight *magazine*. The magazine is attached to the camera body, and the film fed into the camera *gate*. The gate is an assembly of guide rails and pressure plates that position the film precisely behind a rectangular aperture in line with the lens. During shooting, the shutter opens, exposing the frame in the gate to light for around 1/50th second. At this instant, the film is held absolutely still: the slightest movement during exposure produces a blurred image. The shutter then closes, blocking out light. In less than 1/20th second the film is wound on by a fixed distance, ready for the next exposure.

The movement of the film in the camera must be accurate as well as fast: the smallest discrepancy between the position of one frame and the next causes a "jittery" image when the movie is projected. The film's complex stop-start motion is choreographed by a system of claws and pins that engage with the regularly-spaced perforations along one edge of the film.

Movement and light
*The heart of a 35mm movie camera [**A**] is a mechanical assembly that directs and coordinates the movement of film and shutter.*

The camera's electric motor turns a main driveshaft. A gear links this to the shutter – a semicircular mirror – that is made to rotate 24 times per second. As the mirror spins, it intermittently allows light from the camera lens through a rectangular aperture on to the film.

Film is drawn past the shutter by a pulldown claw, which engages with the perforations on the edge of the film. The claw moves downward, pulling the film along by a fixed distance, before moving back to its original position.

The claw is driven via a geared connection to the main driveshaft. The gearing system ensures that the claw moves the film down only when the shutter is closed. When the shutter opens, a registration pin hooks into a perforation on the edge of the film, holding it absolutely still during exposure.

When closed, the shutter reflects light from the camera lens, projecting an image onto a ground-glass screen. A series of lenses and prisms carries this image to the viewfinder so that the camera operator can see what is being filmed.

Coordinated camera
*When the shutter is closed, the pulldown claw engages and drags the film down by a distance corresponding to the height of one frame [**B**]. The claw then disengages and begins to move back to its starting position; while this is happening, the shutter opens, exposing the*

*stationary film to light [**C**]. When the exposure is almost complete [**D**], the claw reengages, ready to pull the film down to the next frame as soon as the shutter closes fully. The registration pin (not shown) engages the film, holding it absolutely still and in the correct position during exposure.*

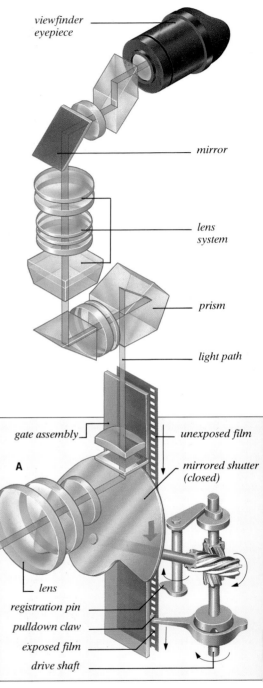

viewfinder eyepiece

mirror

lens system

prism

light path

gate assembly

unexposed film

mirrored shutter (closed)

A

lens

registration pin

pulldown claw

exposed film

drive shaft

B

Connections: Audio Equipment 44 Photographic Cameras 62 Special Effects 66 Photographic Film 68

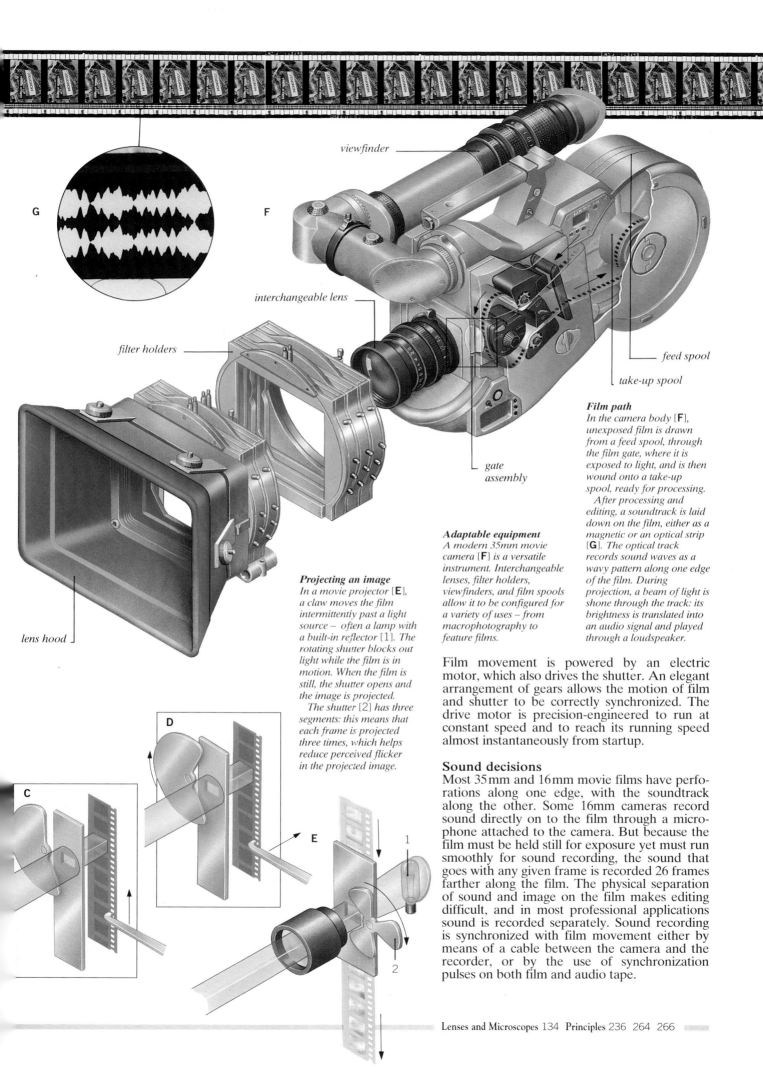

G

F

viewfinder

interchangeable lens

filter holders

feed spool

take-up spool

gate assembly

lens hood

Film path
In the camera body [**F**], unexposed film is drawn from a feed spool, through the film gate, where it is exposed to light, and is then wound onto a take-up spool, ready for processing.

After processing and editing, a soundtrack is laid down on the film, either as a magnetic or an optical strip [**G**]. The optical track records sound waves as a wavy pattern along one edge of the film. During projection, a beam of light is shone through the track: its brightness is translated into an audio signal and played through a loudspeaker.

Adaptable equipment
A modern 35mm movie camera [**F**] is a versatile instrument. Interchangeable lenses, filter holders, viewfinders, and film spools allow it to be configured for a variety of uses – from macrophotography to feature films.

Projecting an image
In a movie projector [**E**], a claw moves the film intermittently past a light source – often a lamp with a built-in reflector [1]. The rotating shutter blocks out light while the film is in motion. When the film is still, the shutter opens and the image is projected.

The shutter [2] has three segments: this means that each frame is projected three times, which helps reduce perceived flicker in the projected image.

C

D

E

1

2

Film movement is powered by an electric motor, which also drives the shutter. An elegant arrangement of gears allows the motion of film and shutter to be correctly synchronized. The drive motor is precision-engineered to run at constant speed and to reach its running speed almost instantaneously from startup.

Sound decisions

Most 35mm and 16mm movie films have perforations along one edge, with the soundtrack along the other. Some 16mm cameras record sound directly on to the film through a microphone attached to the camera. But because the film must be held still for exposure yet must run smoothly for sound recording, the sound that goes with any given frame is recorded 26 frames farther along the film. The physical separation of sound and image on the film makes editing difficult, and in most professional applications sound is recorded separately. Sound recording is synchronized with film movement either by means of a cable between the camera and the recorder, or by the use of synchronization pulses on both film and audio tape.

Lenses and Microscopes 134 Principles 236 264 266

Film and Fantasy

How movie special effects are created

The ten-foot-long model of the liner *Titanic* – the star of the 1980 movie *Raise the Titanic* – reportedly cost more to build than the original ship. Yet the main use of special effects is to save money on expensive sets and locations. Most good special effects therefore go unnoticed – for example, the lavish plantations and houses in *Gone With the Wind* were actually painted on glass – and it is usually only in sci-fi spectaculars that it becomes obvious how much time and money has been spent to create a believable fantasy.

Using today's advanced computer technology it is possible to digitize a filmed sequence and manipulate the picture in almost any way imaginable, or even to create entirely new digital characters. This was how actor Tom Hanks was able to meet dead US presidents in *Forrest Gump*, and how the liquid metal man was brought to life in *Terminator 2: Judgment Day*. Yet these high-profile digital effects are generally used sparingly, being very expensive and time consuming: some frames in *Terminator 2* took up to two hours to generate (there are 24 frames in each second of action). Most films, for reasons of economy, time, or verisimilitude, combine computer manipulation with a variety of other lower-tech effects.

Perhaps the most useful tool in the filmmaker's workshop is the *optical printer*. This is simply a movie camera that records one piece of film onto another. The optical printer is widely used for subtle effects such as *dissolves*

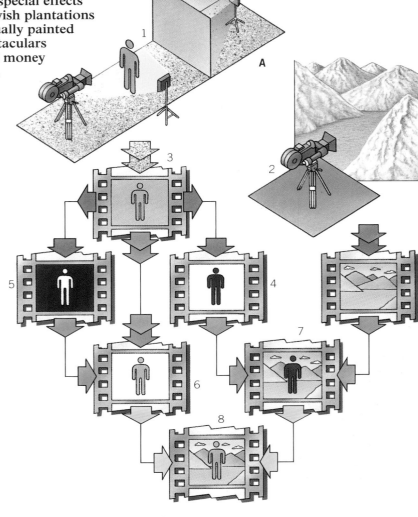

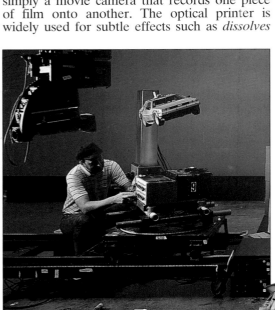

Composite image
*Using a traveling matte (or mask) a moving object or actor can be superimposed on any background [**A**]. The technique is widely used to save money on location shoots, and for effects in which actors or objects apparently "fly" through the air (right). The*

foreground object (above) is first filmed against a blue background [1] (blue is used because this color is absent from skin tones). This film goes through several steps in the optical printer to produce masks that can then be combined with a separately filmed background [2].

The positive film of the foreground action [3] is photographed in the optical printer to produce two black and white strips of film: the "male" matte [4], which is clear except for a black silhouette of the actor, and the "female" matte [5], which is black with a clear silhouette of the actor. Sandwiching together the female matte and the film of the actor creates a print with the actor on a clear background [6]. The male matte is sandwiched with the film of the background, providing a landscape with an "actor-sized" hole [7]. These two films are then sandwiched together [8] to make the composite image.

Connections: Photographic Cameras 62   Movie Cameras 64   Photographic Film 68   Computers 78 80 84   Principles 264 266

Digital manipulation
The computer effect called morphing is used to blend real-life action into computer animation. In the film Lawnmower Man *it was used to change the face of an actor into a computerized model, which was stretched and distorted before transforming back into the real actor.*

To achieve this effect, the actor's face is covered with contoured grid lines and photographed (bottom). This image is scanned into a computer to create a "wire frame" – a grid that can be manipulated readily in digital form (below). The wire frame is then covered by a computer-generated

and *fades,* which are vital for determining the pace of a film. A dissolve suggests a brief passage of time, a fade a longer period: although these are merely cinematic conventions, they have come to be accepted by film audiences the world over. A fade involves photographing the master film onto a duplicate film while slowly opening or closing the shutter of the optical printer. In a dissolve the new scene is optically faded in, while the old scene is faded out on the same film strip.

The optical printer can also be used to superimpose one image on another to ghostly effect. It was in this way that director Alfred Hitchcock momentarily placed a skull over Norman Bates's head in the last scene of *Psycho.* Another more conspicuous effect that relies on the optical printer is the *traveling matte,* which combines two or more different scenes, possibly shot in different parts of the world or at different scales (see main illustration).

"skin." An operator can move the grid lines to change the shape of the face. As the grid lines move, the skin moves with it and the face becomes distorted (below). Reflections and shadows are added to the image to give it context and add credibility.

At this point, each computer-generated frame is a static image: "blur" must be added by computer to make the projected image appear to move naturally. Where the digital image appears alongside live action, the computer-generated components must be "dirtied up" to match the grain of the film on which the live action is shot.

Traditional tricks

Other established movie effects rely on the skill of makeup artists, set and model makers. Models are brought to life using *stop-motion* photography, in which the object is exposed on a single frame, then moved very slightly, before being photographed on the next frame. Playing the film at full speed gives a believable impression of movement. A modern development of this technique is *go-motion,* in which the skeleton of the model is moved by motors and wires controlled by a computer. Scale models often appear in the same scene as an actor. Here, the credibility of the image depends on the false perspective imparted by the one-eyed camera. In the movie *Superman,* a scene in which Christopher Reeve pushes a huge boulder up a hill was created by getting Superman to mime a pushing action while standing some distance beyond a 10 ft, hydraulically lifted model rock, thus making the rock appear gigantic.

Computer realities

Digital image manipulation relies on computing power originally developed for flight simulators and engineering design. Its techniques allow filmmakers infinite flexibility. Characters and scenes can be generated entirely by computer animation and then made to coexist on screen with real-life actors. One of the most spectacular effects, used in films such as *Terminator 2, The Mask,* and *Lawnmower Man* is *morphing.* This allows, for example, a real actor to be seamlessly transformed into a computer-generated animation, which can then be made to perform impossible feats, before morphing back to the conventionally filmed image.

But many uses of computer graphics are more mundane. Optical mattes are being replaced by computerized systems, and computer manipulation is used to "remove" safety wires necessary for stunts, and to clean up damaged or scratched film.

Small worlds *(far right) Models are cheaper and easier to build than full-scale sets. The 1993 film* Attack of the 50 Foot Woman *used miniatures along with forced perspective, in which a camera's one-eyed view convincingly combines objects of different scales.*

Silver Sight
How photographic film captures an image

Inexpensive, convenient, and immediate, photography allows anyone to create "instant history" at the push of a button. Central to the process is photographic film – a light-sensitive sandwich less than one-tenth of a millimeter thick. Today's films are capable of "memorizing" an image in a fraction of a second and reproducing it with a definition millions of times superior to a television screen. Yet these films rely on the same fundamental chemistry as that used by the French inventor Nicéphore Niepce to take the first true photograph in 1826.

Film is produced commercially in a bewildering variety of sizes and types. Large sheets of black-and-white film record medical X-rays, giant spools of color negative film are used by cinematographers, while miniature film cassettes are favored by amateur photographers. But regardless of their format and the uses to which they are put, most films work in basically the same way.

Photographic film is a composite material made up of several different layers sandwiched together. The "filling" of the sandwich is the *emulsion* – one or more coats of light-sensitive chemicals. Black-and-white films have just one light-sensitive layer, which records the presence or absence of light, whereas color films have three layers, each sensitive to a different color, or wavelength, of light.

Viewed through a microscope, the emulsion of a black-and-white film appears as a mass of irregular grains embedded in a matrix of gelatin

Film structure
A typical color negative film [**A**] *has eight layers. Outermost are protective coatings that shield the film from abrasion and ultraviolet (UV) light. Beneath these coatings are three layers of emulsion, sensitive to blue, green, and red light respectively. A yellow filter beneath the blue-sensitive layer removes any remaining blue light that could affect the green- and red-sensitive layers. Beneath the emulsion is a coating that absorbs stray light and prevents it from reflecting back up to the emulsion. A thick base gives the film strength and flexibility.*

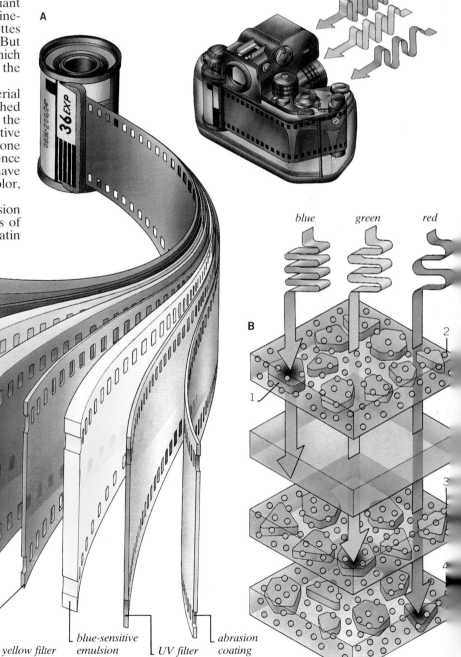

5 microns

film base
non-reflective coating

Film speed *(above)*
"Fast" films, used for action photography, are more sensitive to light than general-purpose films. They contain broader crystals of silver bromide, which present a larger target to light. Their drawback is that the bigger crystals produce a grainy image with poor resolution. "Slower" films give better definition.

red-sensitive emulsion

green-sensitive emulsion

yellow filter

blue-sensitive emulsion

UV filter

abrasion coating

blue   green   red

Connections: Photographic Cameras 62 Movie Cameras 64 Special Effects 66 Lasers and Holography 70

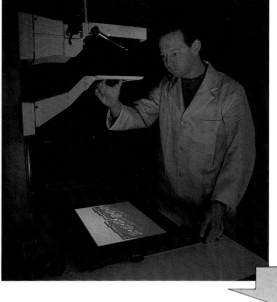

Color printing (left)
An enlarger, comprising a light source and lens system, is used to project the negative image onto light-sensitive paper. Filters in the enlarger head are used to adjust the color of light given out, enabling a skilled darkroom technician to make a print in which colors appear natural and "in balance."

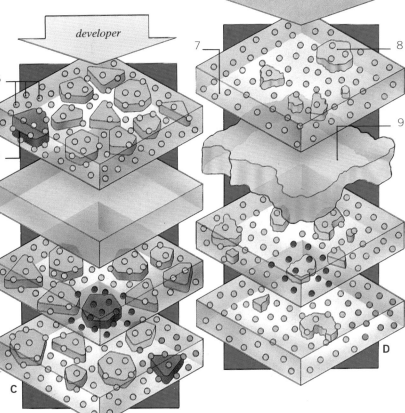

(the substance used to "stiffen" jelly). Each grain is in fact an individual crystal of the light-sensitive compound silver bromide. Billions of silver bromide crystals distributed through the emulsion, each around one-thousandth of a millimeter in diameter, are capable of accurately recording the patterns of light and shade falling on the film.

Crystal images

A crystal of silver bromide consists of bromine ions (negatively charged, because they carry a surplus of electrons) interspaced with silver ions (positively charged, due to a deficit of electrons). When the crystal is struck by light, electrons are knocked away from the bromine ions and grabbed by silver ions, forming atoms of metallic silver, which are visible as microscopic black specks on individual crystals. The number of silver atoms produced depends on the intensity and duration of illumination.

Develop and fix

Although the film has now recorded an image, this image is hidden, or *latent*. To make it visible to the naked eye, the number of silver atoms present is greatly multiplied by immersing the film in a liquid developer. Chemicals in the developer "recognize" crystals that contain traces of metallic silver and "pump" electrons into them, with the result that all the silver ions in the crystal are converted to visible silver atoms. Crystals containing no silver atoms are unaffected by the developer. After development, the remaining crystals of silver bromide are washed away by treatment with a chemical fixer. This makes the developed film insensitive to light, and therefore stabilizes the image.

The film now holds a negative image – one in which dark areas correspond to bright parts of the subject. A positive print is made by projecting the negative image onto a sheet of paper coated with a photographic emulsion. The paper is processed in the same way as the film to complete the photographic image.

Colorful chemistry

Color film has a three-layered emulsion. Each layer contains silver bromide crystals chemically sensitized to a different wavelength of light: the first records blue, the second green, the third red. After exposure, the film is developed as above, but goes through an additional stage in which colored dyes are substituted for the exposed grains of silver. In each layer of emulsion, the dye color produced is complementary to the color of the incoming light. So, the blue-sensitive layer produces yellow dye; the green-sensitive layer produces magenta dye; and the red-sensitive layer makes cyan dye. In this way, the film records every color as its "negative." Just as with a black-and-white film, the negative is printed onto a sheet of paper sensitized with the three color emulsions, and chemically processed to form a positive image.

Process chemistry
When the film is exposed [**B**], each emulsion layer records a different wavelength of light. Silver bromide crystals [1] in the blue-sensitive layer [2] respond only to blue light, forming a latent image – tiny specks of metallic silver on the illuminated crystals. Latent images are formed by green and red light on the green- [3] and red-sensitive [4] layers beneath. The latent image made visible by immersing the film in liquid developer [**C**]. This is done in total darkness. The developer multiplies the silver content of those crystals exposed to light, converting them entirely to dark metallic silver [5]. As the developer works on a crystal, it causes dye molecules immediately around it [6] to take on color – yellow in the top layer, and magenta and cyan in the emulsion layers below. The film is then immersed in another bath – the bleach/fix [**D**]. This dissolves all the silver [7] and silver bromide crystals [8] in the film, leaving behind just areas of colored dye, which make up the color negative image. It also dissolves the yellow filter [9] in the film.

An Added Dimension

How laser light is used to make holograms

Holograms are not tricks or psychological effects, but three-dimensional images of objects as they really are. Although perhaps most familiar as the security devices on credit cards, holograms have many other industrial and medical applications. Dentists have begun to use them as accurate records of their patients' teeth: measurements can be taken directly from the holograms, doing away with the need for bulky plaster casts. And computer-generated holograms can transform architects' plans into realistic three-dimensional models.

To the human eye, an object appears to have a particular shape because of the unique way that its surface scatters light. A conventional photograph is a straightforward record of the light scattered by the object. Like a photograph, a hologram is recorded on film. But in the case of a hologram, the film is made to record the *way* in which the object's surface scatters light. This is achieved by using the properties of laser light.

A laser produces *coherent* light. This means that light waves emitted by the laser have the same wavelength and are in step with one another – the peaks and troughs of each wave coincide exactly.

Images through interference

The most common way of making a hologram is to split the light from a single laser into two beams. One beam, known as the *reference beam*, is shone straight onto a photographic film. The other – the *object beam* – is directed at an object, and bounces off its surface before hitting the film. Because one beam has been scattered by the object, the light waves of the two beams are no longer in step when they reach the film. This causes waves from each beam to interfere, resulting in a complex pattern of light and dark bands, or *fringes*, which is recorded on the film.

This type of hologram, in which reference and object beams strike the same side of the photographic film, is called a *transmission* hologram. It is viewed by shining laser light (of the same wavelength as that used to record the hologram) through the film. The laser light is *diffracted* by the fringes on the film. This effectively re-creates the light-scattering properties of

the object and projects an image of it in three dimensions. Holograms made by directing the reference and object beams to opposite sides of the film can be viewed in natural light. They are called *reflection* (or white-light reflection) holograms, and work by filtering out all light except that which is suitable for viewing the hologram. Reflection holograms are embossed on credit cards for added security and are used decoratively on badges and magazine covers.

Holograms are put to more serious work in the testing of engineering components, such as aircraft rivets. A hologram of the component is superimposed on a second hologram of the same component under stress. The resulting image reveals lines of stress corresponding to microscopic faults that cannot be detected by conventional examination.

Making a hologram
Laser light (in this case the red light from a ruby laser) is first split into two beams [A]. This is done by passing it through a semi-silvered mirror, which allows half of the light through and reflects the remainder. Lens systems direct one beam (the reference beam) straight onto a photographic film; the second, object beam is directed at the object to be recorded. Laser light reflected from the object's surface interferes with the light of the reference beam, forming a series of fringes, which are recorded on the film as sets of tangled microscopic lines.

lens systems

coherent laser light

semi-silvered mirror

mirror

A

semi-silvered mirror

ruby rod

xenon flash tube

mirror

power supply

Laser light
Inside a laser, atoms of a specially selected medium are stimulated by external energy, often in the form of light or electricity, and made to emit coherent light. The wavelength (and color) of this light depends on the medium used. The ruby laser gives out red light.

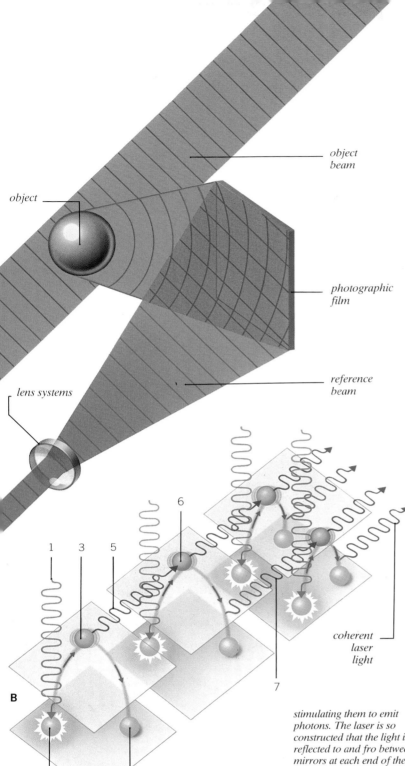

object

object
beam

photographic
film

lens systems

reference
beam

6

1   3   5

coherent
laser
light

7

B

2   4

Keeping still
*During exposure, the object,
laser, mirrors, and film must
be held absolutely still. A
movement of less than one
thousandth of a millimeter
would change the way the
object and reference beams
interfere on the film, and the
pattern recorded, thereby
spoiling the image.*

Holographic art (above)
*The technique of
holography was developed
in 1948 by the British
scientist Dennis Gabor. The
advent of inexpensive and
convenient lasers has made
the medium accessible to
artists, enabling them to
produce arresting three-
dimensional images.*

Light fantastic

Lasers are valuable because they produce a
special type of light, which differs from that
emitted by a normal lightbulb in three respects.
Firstly, it is highly directional and concentrated
or *collimated*: laser beams have parallel edges
that diverge very little as the beam travels.
Secondly, lasers produce "pure" light of a single
color, or wavelength. Thirdly, laser light is
coherent: all its constituent light waves are synchronized, like ranks of soldiers marching in
step. Because of these properties, laser light is
intense, and can carry large amounts of energy,
or encoded information, over long distances
with minimal scattering or distortion.

The intensity of laser light is defined by the
power output of the laser. For example, a 40W
laser produces as much power as is consumed
by a 40W lightbulb. But whereas the bulb's
energy spreads out in all directions, a laser
focuses all the energy into a tiny beam of light,
resulting in high power density. Industrial lasers
used for precision cutting of steel can produce
power densities of more than 1 million watts
per square centimeter. Powerful lasers are also
used in nuclear research to generate the high
temperatures needed to start nuclear fusion.
Low-power lasers are used to "read" and transmit information with great accuracy, and are
put to work in CD players, laser printers, and
fiber-optic networks.

To produce laser light, atoms of a suitable
medium, (which may be a solid, such as ruby
crystal, a liquid, or a gas), are first raised to a
high energy level. This is achieved by pumping
the medium with energy, in the form of light or
electricity. As each atom falls back to its original energy level, it gets rid of its excess energy
by emitting a photon, or "packet," of light. If
this photon goes on to strike a second excited
atom, it can stimulate that atom to emit a
second photon. This process, repeated billions
of times, produces a beam of laser light.

Stimulated emission
*In a ruby laser [**B**] a flash
of light [1] from a Xenon
discharge tube excites atoms
in a ruby crystal [2]. These
atoms are first elevated to a
high energy level [3], and
then spontaneously fall back
to the lower energy level [4],
releasing their extra energy
in the form of a photon of*

*light [5]. This photon is
"tuned" to a specific
frequency, so that when it
hits another excited atom, it
stimulates that atom [6] to
emit a photon of its own
light [7]. What is remarkable is
that the second photon
travels in the same direction
as the one that triggered it,
and that the two photons are
exactly in step, or in phase.
The photons continue their
passage down the ruby
crystal, colliding with other
excited atoms, and*

*stimulating them to emit
photons. The laser is so
constructed that the light is
reflected to and fro between
mirrors at each end of the
ruby crystal [**A**]. Because
the light effectively travels a
long distance through the
ruby rod, it interacts with
many atoms, creating a
growing cascade of photons.
When the intensity builds up
sufficiently, laser light
emerges through one of the
mirrors, which has a semi-
silvered surface.*

*The process by which
light is generated inside a
laser explains the
instrument's name:* **laser** *is
short for* **l**ight **a**mplification
by the **s**timulated **e**mission
of **r**adiation.

Connections: Special Effects 66 Photographic Film 68 Computers: New Developments 84 Principles 246 258 262 264 266

Paper Images

How photocopiers and fax machines reproduce images

The fax machine and photocopier are as essential to the modern office as pen and paper. Thanks to the fax, documents and photographs can be sent across the world in a matter of seconds; while the photocopier gives faithful reproductions of any image at the touch of a button. Although communication by computer may be the shape of the future, the "paperless office" has yet to become a reality. As long as doubts remain about the integrity of electronic information systems, there will be a need for the duplication and transmission of paper documents.

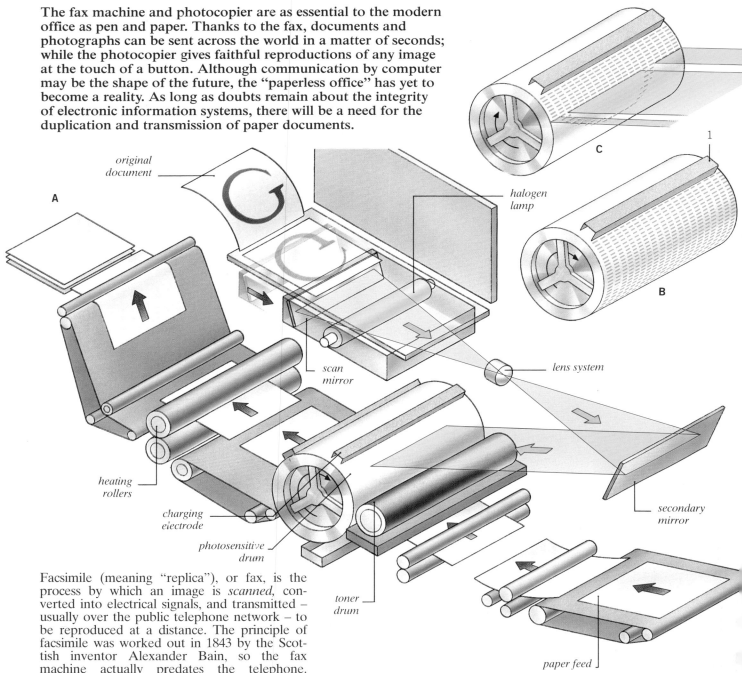

original document

A

halogen lamp

C

1

scan mirror

lens system

heating rollers

charging electrode

photosensitive drum

toner drum

B

secondary mirror

paper feed

Facsimile (meaning "replica"), or fax, is the process by which an image is *scanned*, converted into electrical signals, and transmitted – usually over the public telephone network – to be reproduced at a distance. The principle of facsimile was worked out in 1843 by the Scottish inventor Alexander Bain, so the fax machine actually predates the telephone. However, the fax did not appear in its modern incarnation until the 1960s. The most important event in its development as a business tool was standardization, when all machines were built to accepted international specifications. This allowed every fax machine to "talk" to any other fax machine, thereby opening up a global facsimile network.

A document to be faxed is first converted, strip by strip, into a series of electrical pulses. An electric motor advances the document through the fax machine in 0.13 mm increments. At every increment, the paper halts and a

The copy machine
A document in a black-and-white photocopier [A] is illuminated by a halogen lamp and scanned, strip by strip, by a moving mirror. Blank parts of the document reflect light, while printed parts reflect little. A second mirror directs the reflected light, via a lens system, onto

a revolving drum coated with a photosensitive polymer. This converts the pattern of light and shade on the drum into a pattern of electric charge, which is used to attract particles of black toner. Paper fed through the machine by conveyor belts picks up the toner and the image is

transferred to the sheet. A color copier works in essentially the same way as above, except that the original document is scanned three times and resolved into its yellow, magenta and cyan components. The three images are superimposed to give a color reproduction.

Connections: Telephones 46 Pre-Print Technology 74 Computers: Networks 82 Lenses and Microscopes 134 Principles 244 248 250 252 254 264 266

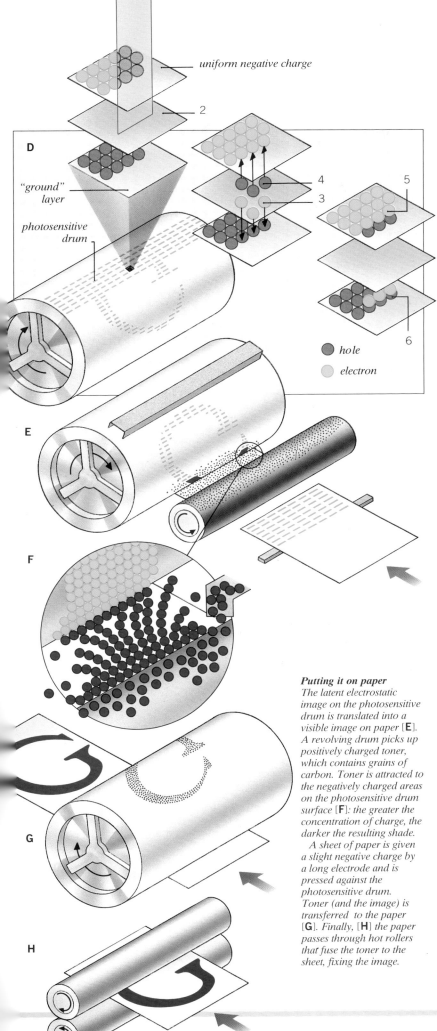

uniform negative charge

D

"ground"
layer

*photosensitive
drum*

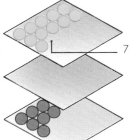

● *hole*

○ *electron*

E

F

G

H

Electric image
*The heart of a photocopier
is an aluminum drum with a
photoconductive surface –
one that conducts electricity
only when exposed to light.
Before scanning [**B**], an
electrode [1] applies a
uniform negative charge to
the surface of the drum. The
image is then projected onto
the drum. Wherever light
strikes, charge "leaks"
away, creating a latent
image [**C**] made of residual
areas of negative charge.
At the molecular level
[**D**], light energy [2] excites
the molecules of photo-
conductive polymer,
splitting them into electrons
[3] and corresponding areas
of positive charge, or holes
[4]. The holes migrate
upward to the surface layer
of negative charge [5], while
electrons migrate to the
positively charged ground
layer [6]. Holes and
electrons cancel each other
out, giving patches of no
charge [7] corresponding
to white parts of the image.*

narrow strip is illuminated by a row of light-
emitting diodes (LEDs). Light reflected by the
document is then focused onto 1728 tiny optical
sensors arranged across the width of the sheet.
The sensors are made of semiconductor material and respond to light by producing a
low-voltage electrical pulse when they "see"
bright light, and a higher-voltage pulse when
they "see" dim light. Each sensor looks at a
very small spot, or *pel,* on the document, determines if it is black or white, and generates a
correspondingly strong or weak electrical
current. The whole row of sensors "records" the
pattern of black and white in a 0.13mm-deep
strip of the document.

An electronic clock, which "ticks" about a
million times per second, reads the current from
each sensor in sequence. The paper then moves
up by 0.13mm, and the next strip is scanned and
read, and so on. The stream of electrical pulses
produced is then converted to digital form, and
can be sent down a telephone line. Speed of
transmission varies with the type of fax: most
modern machines transmit at 9600 *bits* – units
of information – per second, and can send an
A4 document in less than one minute.

Message received
The receiving fax decodes the transmission and
feeds the signals to a printer, which re-creates
the document in precisely the same order as it
was scanned. Many machines use thermal printing, in which the print head comprises a row of
some two thousand pinpoint heating wires
arranged across the width of the paper. The
slick thermal paper used has a chemical coating
that turns black when a hot wire is applied to it.
The application, or not, of each wire to the
paper corresponds to the incoming series of on-
off pulses. Although this system is cheap,
convenient, and reliable, the image deteriorates
if the paper is exposed to light or heat. The
most expensive models have laser printers.

Putting it on paper
*The latent electrostatic
image on the photosensitive
drum is translated into a
visible image on paper [**E**].
A revolving drum picks up
positively charged toner,
which contains grains of
carbon. Toner is attracted to
the negatively charged areas
on the photosensitive drum
surface [**F**]: the greater the
concentration of charge, the
darker the resulting shade.
A sheet of paper is given
a slight negative charge by
a long electrode and is
pressed against the
photosensitive drum.
Toner (and the image) is
transferred to the paper
[**G**]. Finally, [**H**] the paper
passes through hot rollers
that fuse the toner to the
sheet, fixing the image.*

Spotted Images

How pictures are prepared for the printing press

For centuries after its development in second-century China, printing remained an extremely laborious process. Each printed page was made by engraving characters and images by hand into wood or metal, and pressing the inked engraving against a sheet of paper. This contrasts starkly with today's methods of preparing material for the printing press. A color picture can be "captured" electronically in a matter of seconds, broken down into its component colors, and converted into a pattern of minute dots on a printing plate.

The process of converting a photograph or illustration into a form that can be printed onto a page is called *origination*. For a black-and-white photograph, the aim is to translate subtleties of tone in the original picture into a pattern of dots on a printing plate. This is necessary because tones in the original are continuous – the picture is made up of an infinite number of shades of gray. A printing press, however, can lay down only one shade of (black) ink. Converting the image into dots of equal density but unequal size fools the eye into seeing continuous tones, provided the dots are sufficiently small in size.

This conversion is often carried out using a *process camera*, which makes a photographic image of original artwork on a sheet of film. Pressed against the film is a *halftone* or *contact* screen – a transparent sheet carrying a grid of equally sized, equally spaced dots. When photographed under white light, the original is broken down into a series of dots recorded on the film. A photochemical process then converts the (negative) pattern of dots on the film into a positive pattern on a flexible metal plate. Ink applied to the plate "sticks" only to the dots, which are then printed on a sheet of paper. In shadow areas the dots are large and densely grouped; in highlight areas, they are smaller and less frequent. This gives the impression of variable tone when the image is printed.

The resolution of the printed image is determined by the number of dots on the halftone screen. By convention, this is measured in terms of the number of rows (or lines) of dots per inch. A black-and-white newspaper picture has around 100 lines of dots per inch, while a photograph in a glossy magazine has at least 200.

Living color

Color origination is a more involved but fundamentally similar process. Every color reflected from the surface of a painting or photograph can be described in terms of the proportions of red, green, and blue – the primary colors – that it contains. In color origination, a process camera fitted in turn with blue, green, and red filters, records the blue, green, and red components of the image on three separate sheets of film. In each case the image is screened, and the end result is a film on which the dot pattern is a record of the intensity of one of the three primary colors.

Scanner technology
A modern desktop color scanner [F] contains a laser [1] that produces an intense beam of light. This is focused [2] and directed at a spinning multifaceted, mirror [3]. As the mirror turns, the laser beam is deflected rapidly from left to right. Mirrors [4] direct the scanning beam on to the original [5]. The reflected light follows a near-identical path back to the spinning mirror, which reflects it onto a photosensor [6]. Here the intensity and color of the beam is translated into electronic signals, which are then passed to a computer for processing.

In camera
A process camera [A] is used to produce halftone film from a black and white original. The original is laid flat on the copy board and illuminated with white light. The lens is raised or lowered to enlarge or reduce the image recorded on the film in the film carrier.

Screened image
Immediately beneath the film [1] in the process camera is the halftone screen [2]. This is a photographically produced grid of dots. Each dot is effectively a cone of dense silver [3]. Bright light, reflected from a highlight on the original [4], can pass through the "shoulders" of the cones, forming a large circle on the film above. Dim light from a shadow area of the original [5] can only penetrate where the shoulders of the cones are low, so forming a small dot on the film. The screened, printed image [B] contrasts with the continuous tone of the original [C].

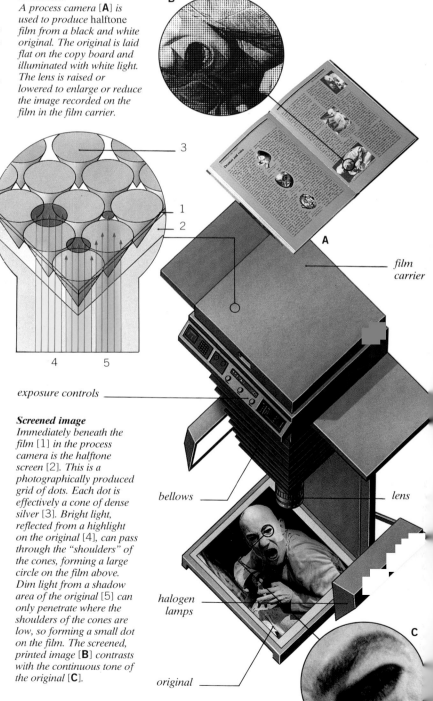

film carrier

exposure controls

bellows

lens

halogen lamps

original

Connections: Photographic Cameras 62 Photographic Film 68 Lasers and Holography 70

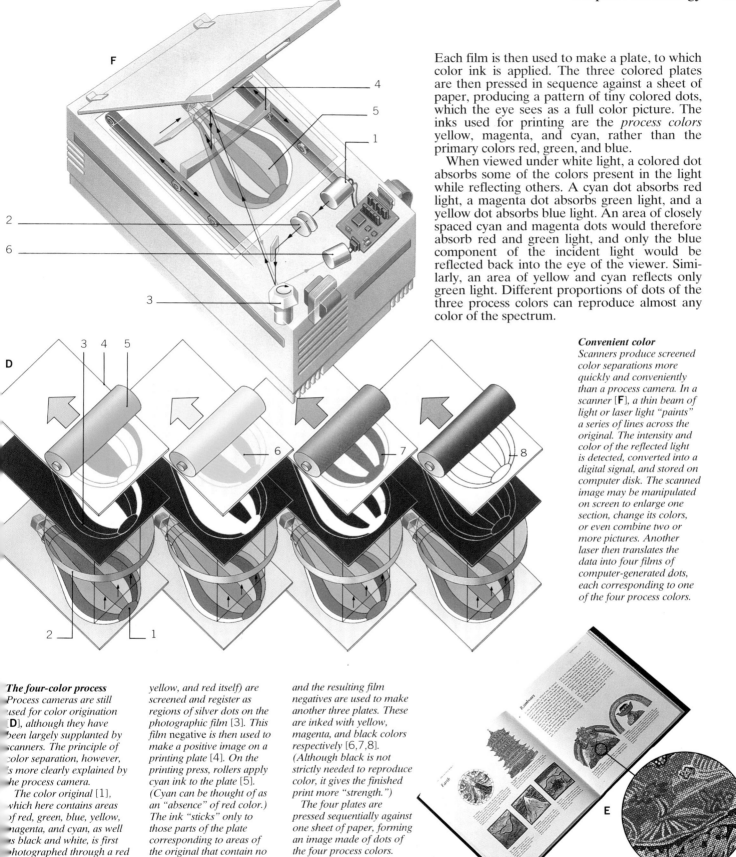

Each film is then used to make a plate, to which color ink is applied. The three colored plates are then pressed in sequence against a sheet of paper, producing a pattern of tiny colored dots, which the eye sees as a full color picture. The inks used for printing are the *process colors* yellow, magenta, and cyan, rather than the primary colors red, green, and blue.

When viewed under white light, a colored dot absorbs some of the colors present in the light while reflecting others. A cyan dot absorbs red light, a magenta dot absorbs green light, and a yellow dot absorbs blue light. An area of closely spaced cyan and magenta dots would therefore absorb red and green light, and only the blue component of the incident light would be reflected back into the eye of the viewer. Similarly, an area of yellow and cyan reflects only green light. Different proportions of dots of the three process colors can reproduce almost any color of the spectrum.

Convenient color
Scanners produce screened color separations more quickly and conveniently than a process camera. In a scanner [F], a thin beam of light or laser light "paints" a series of lines across the original. The intensity and color of the reflected light is detected, converted into a digital signal, and stored on computer disk. The scanned image may be manipulated on screen to enlarge one section, change its colors, or even combine two or more pictures. Another laser then translates the data into four films of computer-generated dots, each corresponding to one of the four process colors.

The four-color process
Process cameras are still used for color origination [D], although they have been largely supplanted by scanners. The principle of color separation, however, is more clearly explained by the process camera.

The color original [1], which here contains areas of red, green, blue, yellow, magenta, and cyan, as well as black and white, is first photographed through a red filter [2]. Those areas of color that contain red namely white, magenta, *yellow, and red itself) are screened and register as regions of silver dots on the photographic film [3]. This film negative is then used to make a positive image on a printing plate [4]. On the printing press, rollers apply cyan ink to the plate [5]. (Cyan can be thought of as an "absence" of red color.) The ink "sticks" only to those parts of the plate corresponding to areas of the original that contain no red color. The original is then photographed through green, blue, and gray filters,* *and the resulting film negatives are used to make another three plates. These are inked with yellow, magenta, and black colors respectively [6,7,8]. (Although black is not strictly needed to reproduce color, it gives the finished print more "strength.")*

The four plates are pressed sequentially against one sheet of paper, forming an image made of dots of the four process colors. This is made more visible on this enlargement of a printed page [E].

Pictures on a Page

How a color printing press works

The ubiquity of magazines and books, as well as paper money, packaging, and stamps, attests to the continuing importance of printing in our society. However, Johann Gutenberg, who introduced movable type to Europe in 1439, would have difficulty recognizing modern printing technology. Today's presses are capable of printing high-resolution color images at a rate of more than 25,000 copies per hour. Computer control has transformed the printing process: some presses are fully automated, able even to clean themselves between jobs.

The starting point of most commercial printing processes is a sheet of photographic film, which carries an image of the type and pictures to be printed. The film image is transferred to a carrier – usually a printing plate or cylinder – on which the image areas are distinguished from blank (non-image) areas by virtue of their mechanical or chemical properties. Ink is applied to the carrier, where it "sticks" only to the image areas: the carrier is then pressed against paper, card, or other printing medium. There are three basic printing processes – *letterpress*, *lithography*, and *gravure* – which differ mainly in the way that image and non-image areas are defined on the carrier. Lithography is the most versatile of these techniques, accounting for over 40 percent of all printed material.

Images cast in stone
The technique of lithography emerged during the 19th century. Originally a slab of limestone (*lithos* is the Greek word for "stone") was used as the image carrier, but today's presses use a thin plate of zinc or aluminum.

Before printing, the plate is specially prepared to make it receptive to an image. It is first treated with a chemical that renders its surface porous, and then coated with a layer of photographic emulsion. The plate is brought into contact with the image-bearing film, exposed to light, and developed. Development washes away the emulsion from the non-image areas, revealing the porous plate beneath, and hardens the emulsion on the image areas.

The plate is then clamped onto the curved surface of a metal cylinder and fixed into the press. Sets of rollers coat the plate first with water, then with a grease-based ink. Water is drawn to the porous, non-image areas of the plate and repelled from the image areas: conversely, image areas pick up the ink, while the film of water covering the rest of the plate prevents ink from spreading into non-image areas.

The image is printed (*offset*) onto a rubber blanket that is wrapped around a second cylinder. Only then is it transferred onto a sheet of paper, which is pressed between the blanket cylinder and a third *impression* cylinder. The use of a rubber blanket, which has a certain amount of "give," allows images to be printed on both card and low-quality paper.

Paper may be fed into the press as individual cut sheets, which are later trimmed to size and bound in a separate process.

Printing processes
The earliest printing technique, letterpress [A], relies on ink-carrying surfaces that stand proud of non-image areas. Thick paste inks are applied to the printing surface by rollers before the plate is pressed against a sheet of paper.

In gravure [B], the image is etched or engraved into the surface of a metal plate. Ink is spread over the whole plate, which is then scraped with a blade, known as the doctor blade, leaving behind ink-filled depressions that carry the image to the paper.

In lithography [C], a thin layer of water prevents grease-based inks from spreading into non-image areas on the printing plate.

The machinery of mass media
For long print runs, in which tens or even hundreds of thousands of copies are needed, paper is supplied as a continuous roll, or *web*, rather than as individual cut sheets. Instead of having separate blanket and impression rollers, web presses have two blanket rollers, each one acting as the impression roller for the other. In this way both sides of the web can be printed simultaneously, saving valuable time.

To produce a consistent printed image by lithography, the balance between ink and water applied to the plate must be precisely maintained. For very long print runs (more than 500,000 copies) the gravure process is preferred to lithography because the ink is held in engraved depressions in a metal cylinder, and the image is therefore more stable. The high cost of etching and preparing the cylinder is justified by the longer print runs possible.

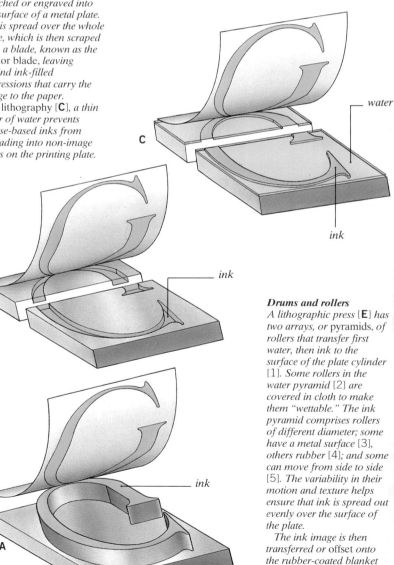

water

ink

C

ink

B

ink

A

Drums and rollers
A lithographic press [E] has two arrays, or pyramids, of rollers that transfer first water, then ink to the surface of the plate cylinder [1]. Some rollers in the water pyramid [2] are covered in cloth to make them "wettable." The ink pyramid comprises rollers of different diameter; some have a metal surface [3], others rubber [4]; and some can move from side to side [5]. The variability in their motion and texture helps ensure that ink is spread out evenly over the surface of the plate.

The ink image is then transferred or offset *onto the rubber-coated blanket*

Connections: Photographic Film 68 Pre-Print Technology 74 Principles 232 266

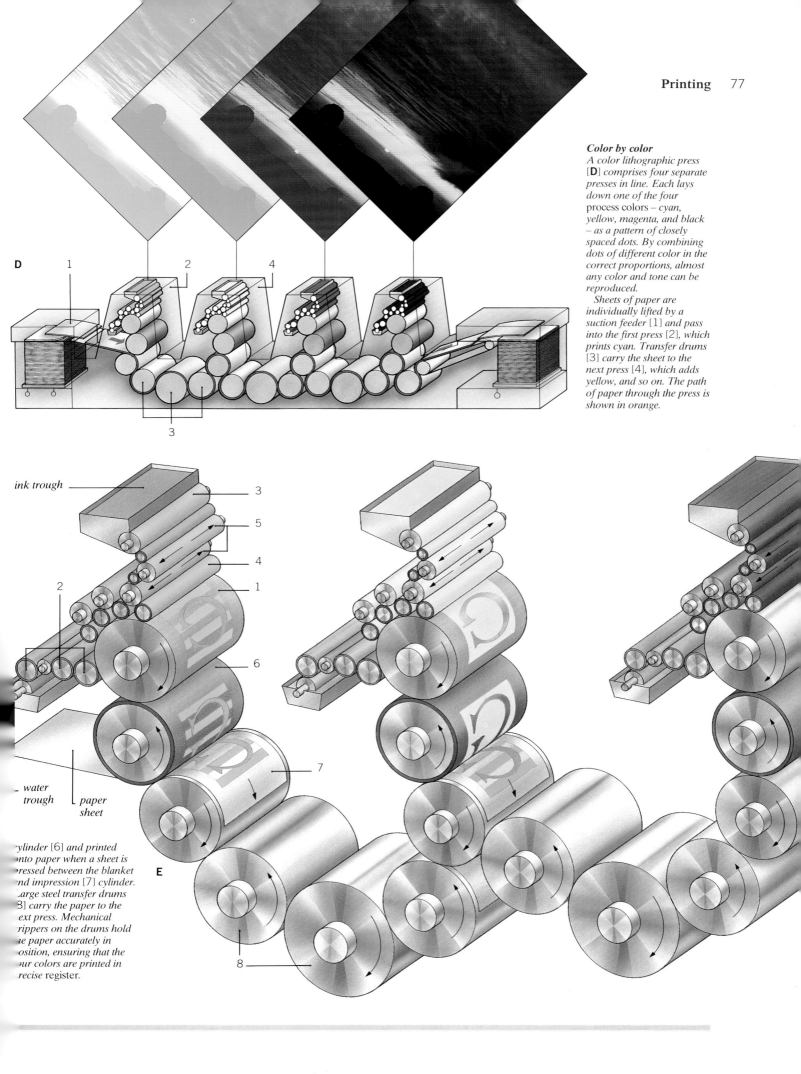

Color by color
A color lithographic press [D] comprises four separate presses in line. Each lays down one of the four process colors – cyan, yellow, magenta, and black – as a pattern of closely spaced dots. By combining dots of different color in the correct proportions, almost any color and tone can be reproduced.

Sheets of paper are individually lifted by a suction feeder [1] and pass into the first press [2], which prints cyan. Transfer drums [3] carry the sheet to the next press [4], which adds yellow, and so on. The path of paper through the press is shown in orange.

D

1 2 4

3

ink trough

3
5
4
1
2
6

water
trough

paper
sheet

7

E

8

...ylinder [6] and printed ...nto paper when a sheet is ...ressed between the blanket ...nd impression [7] cylinder. ...arge steel transfer drums ...3] carry the paper to the ...ext press. Mechanical ...rippers on the drums hold ...e paper accurately in ...osition, ensuring that the ...our colors are printed in ...recise register.

Inside Information

How the components of a computer work together

A glance inside the casing of a personal computer reveals little about the way it functions. But inside the myriad plastic cases are microprocessors and memory circuits – chips made from flakes of silicon that can house as many as three million tiny transistors. Each acts as an electronic switch, turning pulses of electricity on and off. These pulses comprise the digital data that the computer manipulates. Each second, more than 100 million separate logical operations are performed, their results displayed on a screen in "human-friendly" colors and pictures.

A personal computer is a digital device. It works by adding together and performing logical operations on digital numbers – series of 1s and 0s. The numbers may represent many things, among them letters, locations in the computer's memory, and positions on the screen. The mathematical manipulation of these numbers is carried out in the *central processing unit* (CPU), the nerve-center of the computer. It is fixed in a *motherboard*, which contains the computer's built-in electronics – memory chips, logic circuits to link everything together, and expansion slots. The latter are used for adding *cards* – extra circuits that control specific *peripherals*, such as a high-resolution monitor. Linking all the different chips, expansion slots, and outputs on the motherboard is the *system bus*, a network of wires that carries up to three million pieces of data every second.

Some expansion cards and storage devices need data fed to them from the CPU at an even more rapid rate than this. A second, faster bus known as the *local bus* is capable of moving data 50 times faster than the system bus.

Safe storage

To store its data and programs, a computer needs two types of memory – *read-only* (ROM), and *random access* (RAM). The basic unit of data is the byte – a sequence of eight 1s or 0s corresponding to a number between 0 and 255. The high memory capacity of modern computers is usually measured in kilobytes (actually 1024 bytes) or megabytes (1,048,576 bytes). ROM is preprogrammed with the routines that the computer needs to start up and keep running. RAM, in contrast, can be written into and erased to store data and programs. A typical PC has 8 megabytes of RAM, each byte referred to by a specific address.

More permanent storage is offered by the computer's disk drives. The hard disk drive can store hundreds of millions of data bytes on its magnetic surfaces, but access time is very slow compared to electronic memory. The floppy drive uses similar magnetic methods to store files and programs on a small disk that is removable. A third form of storage is the *CD-ROM*, just like an audio compact disk, but carrying billions of · bytes of computer code instead of digitized sound. The name points out a major limitation – unlike other forms of storage, CD-ROMs are read-only devices, onto which a computer cannot write data.

Computer architecture
The rear of a computer [A] is a tangle of different wires connecting the main unit to the peripherals – the keyboard, screen, printer, and mouse [B]. The mouse is a pointing device that moves a cursor around the screen. Movements from side to side and up and down make a rubber ball inside the mouse roll. This in turn spins rollers connected to encoder wheels, which convert the movements into electrical pulses. A chip inside the mouse collects the information and sends it to the computer.

Inside the main unit itself [C] is a maze of chips and other devices, all of which are connected to the motherboard. A power supply converts a.c. electricity to the steady voltages required by the computer chips. The hard disk drive is fitted inside an airtight box to prevent the entry of dust particles. A floppy disk drive is used to read and write onto removable magnetic disks. Both disk drives connect into the system bus, *wiring that connects the various components together. Signals pulse along the bus at a speed set by the* clock rate *of the computer. The faster the clock speed, the faster the computer can process data – and the more powerful the computer is.*

The relationship between the types of chips inside the computer can be most easily traced by following the steps of a simple process – registering a single keystroke. The key itself works like a simple switch – pressing it closes contacts and allows a current to flow [1]. A low-power processor inside the keyboard detects this current and generates a

scan code *corresponding to the key, as well as a second code when the key is released. This passes along the keyboard cable to the BIOS (Basic Input/ Output Set) chips [2], which translate the two scan codes into a single ASCII code (American Standard Code for Information Interchange), the language used to represent letters and symbols in most computer*

communications. The code passes to the CPU [3], which processes the data to display the letter on the screen. Routines in ROM [4] also tell the CPU to store a record of the keystroke in the RAM, held in SIMMS (Standard In-line Memory Modules) [5].

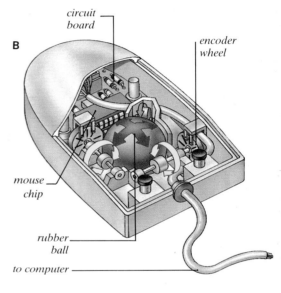

monitor

mouse

keyboard

computer

circuit board

encoder wheel

mouse chip

rubber ball

to computer

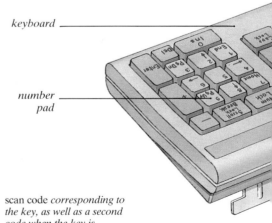

keyboard

number pad

Processing power

As a program runs, separate units inside the CPU police the processing and transmission of data. A *prefetch unit* determines which bits of code will be needed next and asks the *bus interface unit* to fetch them from RAM. Other units translate between the addresses used by the CPU and RAM chips. The heart of the CPU is the *arithmetic logic unit*, which performs simple calculations and comparisons between binary numbers, sending the results back to the bus interface unit for storage in RAM. The power of a personal computer stems from the fact that some 100 million of these steps are performed every second.

The latest processors use *Reduced Instruction Set Computing* (RISC). Because they work with a smaller set of commands than older processors, their chips contain fewer circuits and are able to run programs much more quickly.

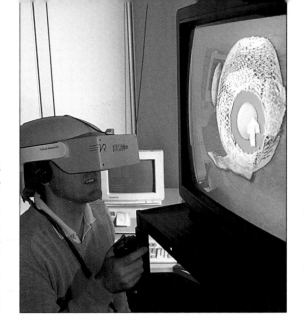

Another world
The keyboard, mouse, and monitor are not the only ways of interacting with a computer. A virtual reality headset and joystick let a scientist explore the complicated structure of a protein molecule via a computer-generated 3-dimensional model (left).

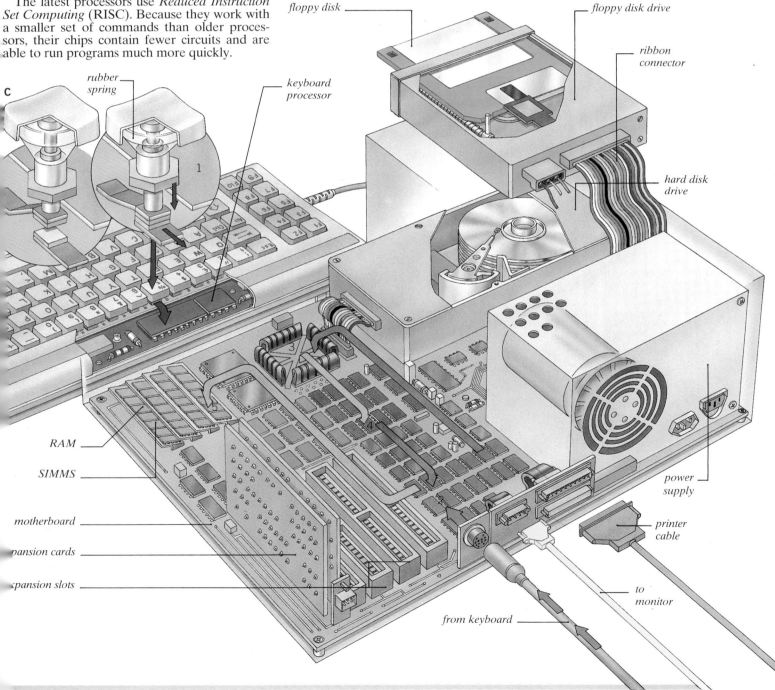

rubber spring

keyboard processor

floppy disk

floppy disk drive

ribbon connector

hard disk drive

RAM

SIMMS

motherboard

expansion cards

expansion slots

power supply

printer cable

to monitor

from keyboard

Total Recall

How computers store data and programs

The hard disk inside a personal computer is able to store billions of bytes of data on its magnetically coated surfaces. The data are recorded by read/write heads, tiny electromagnets that are held one two-thousandth of a millimeter from the disk surface and are able to move to any location in less than fifteen thousandths of a second. But even this retrieval time is too slow to feed the computer's processor. So while a program is running, it is copied from the disk into fast memory chips, where its instructions can be accessed much more quickly.

The performance of a personal computer is often defined in terms of the speed at which its CPU, or *central processing unit*, can perform mathematical calculations. Just as important is the computer's memory – the electronic area in which the CPU stores the data used for its calculations and their results. In addition, the memory must hold the program currently running – the list of instructions that tells the CPU in which order to perform the calculations.

Memory comes in two main forms. ROM, *read-only memory*, cannot be written into, but contains the programs that a computer needs to run when it starts up, in addition to routines that govern the way it interprets signals from the keyboard and other peripherals. *Random access memory*, RAM, can be written into as well as read from. It stores applications programs, such as word processors, when they are being run, along with the files that they are manipulating.

Magnetic memory
The hard disk unit of a computer [A] works in a similar way to an old record player. Instead of a single disk it contains a stack of platters spinning 100 times a second. Their faces have a coating that contains tiny magnetic particles. These are aligned in patterns representing the 1s and 0s of digital data by read/write heads – tiny electromagnets positioned by actuator arms. The arms, the spin of the disk, and the data sent to the heads are all governed by the disk controller, a set of electronics mounted in one of the computer's expansion card slots.

To record data, the head first moves to a track, a particular area on a disk's surface. A current pulses through a wire coil around the head, creating a magnetic field. Particles in the disk surface are at first randomly orientated [1], but align with this field as they pass underneath the head to record a 1 [2]. After the disk has turned through a tiny distance, another pulse of current in the opposite direction reverses the magnetic field [3], aligning the particles in the opposite direction to represent a 0. The next pulse is in the same direction and so another 0 is laid down [4].

To read data back, the head is moved to the relevant part of the disk [5]. The changing magnetic fields in the coating below induce corresponding currents in the coils around the head. These flow back to the disk controller where they are reconverted into digital data.

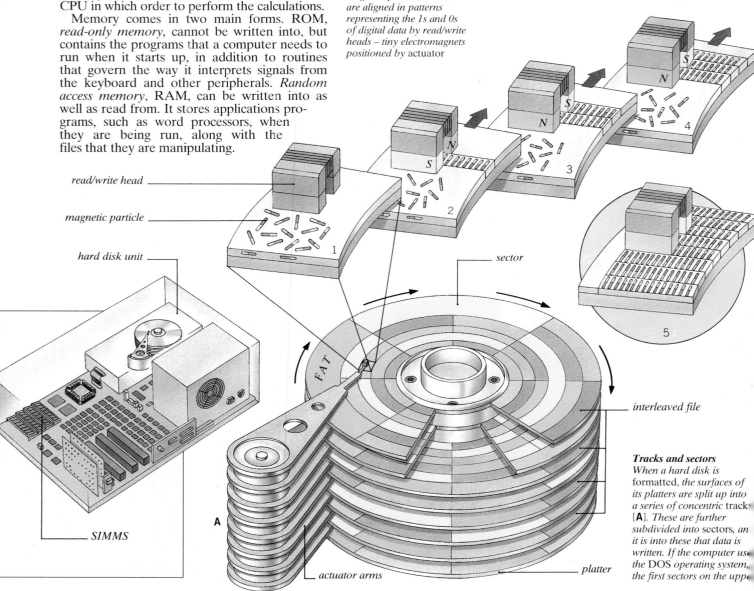

read/write head

magnetic particle

hard disk unit

SIMMS

sector

FAT

interleaved file

A

actuator arms

platter

Tracks and sectors
When a hard disk is formatted, *the surfaces of its platters are split up into a series of concentric tracks [A]. These are further subdivided into sectors, an it is into these that data is written. If the computer us the DOS operating system, the first sectors on the upp*

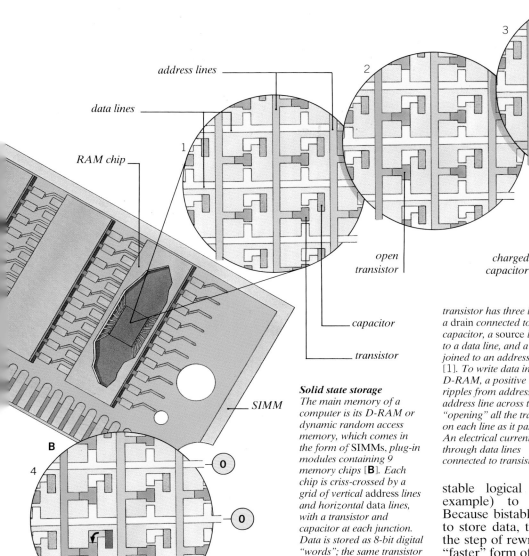

address lines

data lines

RAM chip

SIMM

B

4

0

0

1

current

1

2

3

0

0

1

open
transistor

capacitor

transistor

charged
capacitor

current

*where a 1 will be stored, and
charges up the capacitors
attached to them [3]. In this
way data is written into the
whole of the chip. When the
computer needs to read the
stored data, voltages again
pulse across the address
lines on the chip. As they
open the transistors on each
row, the charge stored in
them is allowed to flow
down the data lines [4].
Each current pulse detected
is translated into a 1.*

*Charge in the capacitors
gradually leaks out, so data
has to be periodically
rewritten, making D-RAM
a "slow" form of memory.*

*transistor has three links:
a* drain *connected to the
capacitor, a* source *linked
to a data line, and a* gate
*joined to an address line
[1]. To write data into
D-RAM, a positive voltage
ripples from address line to
address line across the chip,
"opening" all the transistors
on each line as it passes [2].
An electrical current is fed
through data lines
connected to transistors*

Solid state storage
*The main memory of a
computer is its D-RAM or
dynamic random access
memory, which comes in
the form of SIMMs, plug-in
modules containing 9
memory chips [B]. Each
chip is criss-crossed by a
grid of vertical address lines
and horizontal data lines,
with a transistor and
capacitor at each junction.
Data is stored as 8-bit digital
"words"; the same transistor
location on each chip is
used to store one bit from
each word, with the ninth
chip storing a parity bit
used to detect errors. Each*

*disk surface form the file
allocation table (FAT).
This is a record of which
sectors are being used to
store each file.
The heads are held a tiny
distance from the platter
surfaces, and are so close
that any stray dust particle
could be trapped and cause
serious damage. For this
reason the disk is sealed
inside an airtight box.
Additional protection
against data loss is given by
interleaving. Longer files
may occupy more than one
sector. Rather than storing
them side by side on the
same track, they are split up
and distributed over all the
disk, so that an error will
not entirely destroy the file.*

Static and dynamic
RAM itself has two forms. Most of a personal computer's memory is of a type called *Dynamic RAM* or D-RAM. The 1s and 0s of binary numbers are stored in D-RAM as charge held in hundreds of thousands of tiny capacitors etched into special silicon chips. Each capacitor is connected to a transistor, an electronic switch that can be open or shut to allow charge in or out of the capacitors. When D-RAM is read, the charges flow out of the capacitors and are sensed as currents. Unfortunately, during this process the memory is also erased, so after every readout from D-RAM the data has to be rewritten into the circuits.

Static RAM (or S-RAM) is an alternative to D-RAM. Instead of capacitors, it contains electronic circuits called *bistables* (more popularly *flip-flops*). The popular name precisely describes their action. Pulses of electricity fed to their inputs cause these devices to flip from one

stable logical state (representing a 0, for example) to another (representing a 1). Because bistables do not depend on capacitors to store data, they do not have to go through the step of rewriting bits. This makes S-RAM a "faster" form of memory than D-RAM.

Unfortunately, S-RAM is also bulkier and more expensive to manufacture, and so tends to be used only as *cache memory*. This special memory area is used to store data and program segments that the CPU is likely to want to use quickly. Software monitors the flow of data in and out of the main memory, letting it anticipate which sections the computer will need to use next. These sections are loaded into the S-RAM cache, from where the CPU can read them much more quickly than from standard memory, making a program run faster.

Both types of RAM have a major drawback – when the power is turned off, the data they store disappears. A computer must therefore have a place to store permanently its software and files. A *hard disk drive* is made up of several metal platters with a coating that contains microscopic magnetic particles. Data is stored on the platters in the same way that a cassette deck records music, as patterns of alignment in these tiny magnets. A disk occupying only a small area inside a computer has a capacity far in excess of that offered by a similar volume of RAM – sometimes many gigabytes (billions of eight-digit binary numbers).

Electronic Webs

How networks link computers

Every time a bank customer uses an automatic telling machine, or ATM, he or she is taking advantage of a computer network. The ATM – in effect a small computer that counts bank notes – is connected by electronic lines to a central computer that logs all the transactions, which in turn connects to thousands of similar machines around the country. A customer from a foreign bank can also use the same machine, using connections between different bank networks. Electronic links between computers now span the world and link millions of people via the internet.

Computer networks come in a variety of different forms, from small *Local Area Networks* (LANs) to *Wide Area Networks* (WANs), which can span continents or the entire planet. LANs can range in size from only a few users clubbing together to share a printer, to large organizations with many hundreds of devices, or *nodes*, including computers with extra-large hard disks called servers, and a huge variety of different printing and plotting devices.

The interconnections in a LAN can have different physical arrangements or *topologies*. One such arrangement is the *bus* (also known as an *ethernet*). Here the nodes branch off a long cable (the bus). Every node is fitted with a card that translates digital data into a form suitable for transmission, and reconverts received messages into computer code. The bus itself can be a co-axial wire, similar to a television aerial connection, or a twisted-pair cable, made of two intertwined copper conductors. Because this type of wire is already installed in many offices to carry telephone connections, buses are the most popular form of LAN.

An alternative topology is the *token-ring* network. The token is a binary code carrying data that passes from neighbor to neighbor in one direction around a complete circuit. Another configuration is the *star* network, in which the nodes are connected to one powerful server machine that operates like a signal box, directing traffic from node to node.

On the bus

Most local area networks, such as the one in this bank transaction room (below left), are connected together by an ethernet bus [**A**]. *This is a cable with branches to cards installed in every* node, *whether a computer or printer. The CPU (central processing unit) of a computer sends data to be transmitted to the card, which adds labels giving the destination, and the name of the sender* [1]. *The message travels in both directions along the cable and is inspected by each node that it passes. A node ignores a message that does not bear its address. But if it is the destination, the node extracts the data and then sends an acknowledgment to the sender* [2].

Sometimes two nodes can send messages at the same time, resulting in a collision [3]. *This is sensed by a nearby node, which sends a signal to all branches to stop broadcasting* [4]. *After waiting a random period, the senders try again.*

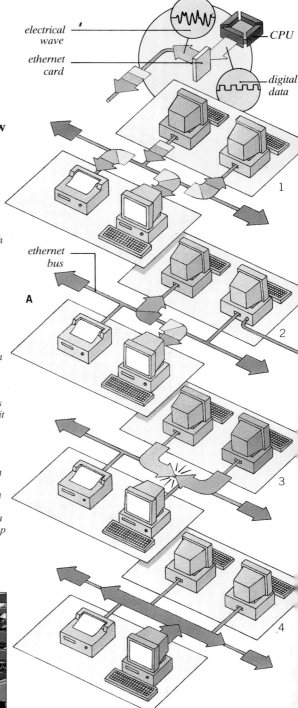

electrical wave
ethernet card
CPU
digital data
1
ethernet bus
A
2
3
4

Modem link

Telephone lines are designed to carry human voices in the form of electrical waves. For them to carry computer data it first must be converted into similar waves by a modem [**B**]. *This cuts up digital words, made up of 1s and 0s, into two-digit blocks.*

There are four possible permutations: each is converted into a particular frequency of electrical wave by the modem. The digital message, in its translated form, is sent down the telephone cable. At its destination, a second modem converts the waves back into digital data.

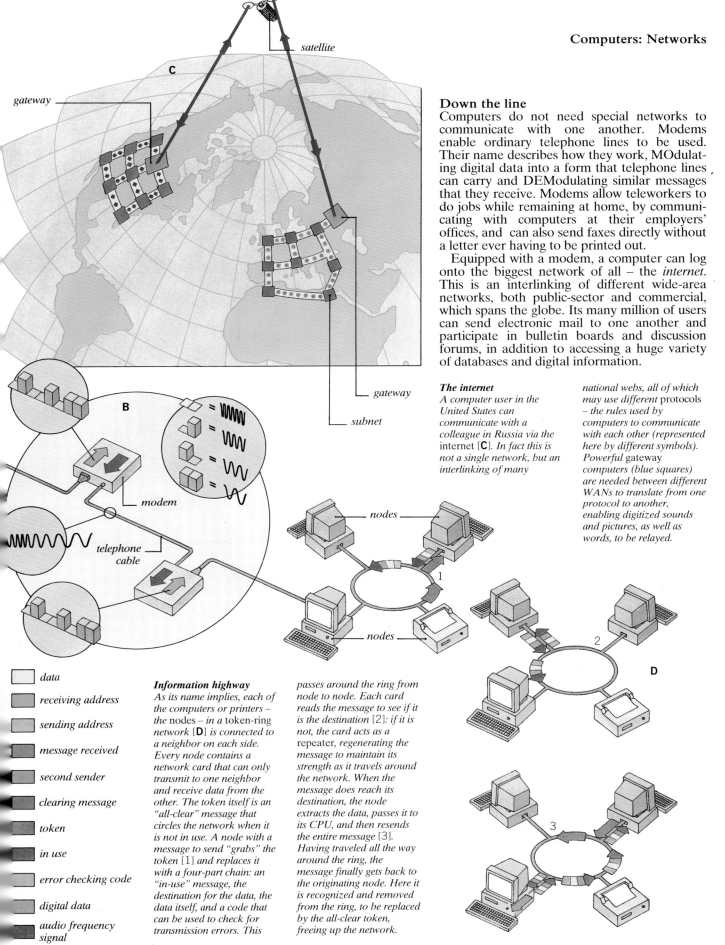

satellite

C

gateway

gateway

subnet

B

modem

telephone cable

Down the line

Computers do not need special networks to communicate with one another. Modems enable ordinary telephone lines to be used. Their name describes how they work, MOdulating digital data into a form that telephone lines can carry and DEModulating similar messages that they receive. Modems allow teleworkers to do jobs while remaining at home, by communicating with computers at their employers' offices, and can also send faxes directly without a letter ever having to be printed out.

Equipped with a modem, a computer can log onto the biggest network of all – the *internet*. This is an interlinking of different wide-area networks, both public-sector and commercial, which spans the globe. Its many million of users can send electronic mail to one another and participate in bulletin boards and discussion forums, in addition to accessing a huge variety of databases and digital information.

The internet
*A computer user in the United States can communicate with a colleague in Russia via the internet [**C**]. In fact this is not a single network, but an interlinking of many*

national webs, all of which may use different protocols – the rules used by computers to communicate with each other (represented here by different symbols). Powerful gateway computers (blue squares) are needed between different WANs to translate from one protocol to another, enabling digitized sounds and pictures, as well as words, to be relayed.

nodes

nodes

1

2

D

3

data

receiving address

sending address

message received

second sender

clearing message

token

in use

error checking code

digital data

audio frequency signal

Information highway
As its name implies, each of the computers or printers – the nodes *– in a token-ring network [**D**] is connected to a neighbor on each side. Every node contains a network card that can only transmit to one neighbor and receive data from the other. The token itself is an "all-clear" message that circles the network when it is not in use. A node with a message to send "grabs" the token [1] and replaces it with a four-part chain: an "in-use" message, the destination for the data, the data itself, and a code that can be used to check for transmission errors. This*

passes around the ring from node to node. Each card reads the message to see if it is the destination [2]: if it is not, the card acts as a repeater, *regenerating the message to maintain its strength as it travels around the network. When the message does reach its destination, the node extracts the data, passes it to its CPU, and then resends the entire message [3]. Having traveled all the way around the ring, the message finally gets back to the originating node. Here it is recognized and removed from the ring, to be replaced by the all-clear token, freeing up the network.*

Thinking Ahead

How computers are likely to evolve

The first computers could be used only by a select band of people versed in the complex mathematical language of programming. Today, almost anyone can communicate with a computer via "user-friendly" interfaces based on icons or simple commands. But the more intuitive an interface is to use, the more complex a computer needs to be to interpret the instructions it receives. Future computers, which will be able to recognize the human voice and handwriting, will demand huge increases in speed, processing power and storage capacity.

At the heart of a modern computer is a microprocessor – a circuit made up of millions of components etched onto the surface of a silicon chip. In the next generation of computers silicon is likely to be replaced by gallium arsenide (GaAs). Like silicon, this is a semiconductor, but smaller components can be etched onto its surface to produce denser and faster circuits. GaAs also has a faster *switching speed* than silicon. Electrons within its crystal structure can move more freely than those in silicon, allowing transistors – electronic switches – on a GaAs chip to open and close more times each second and perform a greater number of logical operations in a given time.

Gallium arsenide allows tiny devices called quantum wells to be built, in which the 1s and 0s of digital data are represented by the presence or absence of single electrons, rather than by the bulk flow of electrons. Chips based on quantum wells are more compact and run faster and at lower temperatures than silicon chips.

At present, messages are sent around a computer as electrical pulses. Although electricity travels quickly (one third of the speed of light), obvious gains in speed could be made by using light itself as the signaling medium. Light beams can also travel through one another's paths without distorting the information they carry, unlike electrical signals, each of which needs a

Deep disk

*Optical disk drives already exist, but only store data on their surfaces. Using holograms computers will be able to store pieces of data stacked one on top of the other throughout the whole depth of a thicker disk [**B**]. To read data, laser light is focused on to the disk by an acoustooptic deflector. Different holograms are selected by varying the angle at which the beam strikes the disk. Light is absorbed or reflected by the hologram; the pattern this creates is detected by a light sensor chip and converted into pulses of electricity.*

Box of tricks

*Light has an advantage over electric currents as a means of transmitting data. Light rays can pass through each other without effect, whereas each electrical signal needs its own insulated path. A future computer may contain components that work and communicate using light signals [**A**]. Its "brain" may be several discrete processing units (DPUs), swapping data in bursts of laser light. The data may be stored by lasers on a holographic disk or in a cube of 3-dimensional optical RAM material. Signals to and from displays and input devices will travel via infrared transceivers.*

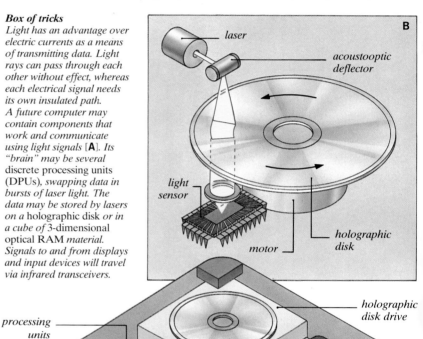

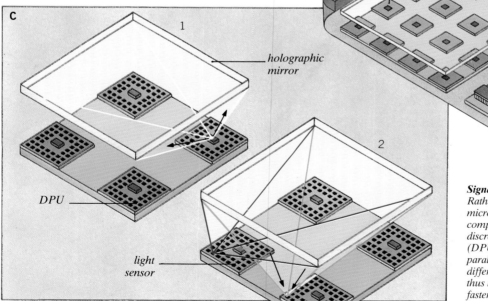

Signaling with light

*Rather than using a single microprocessor, a future computer may have several discrete processing units (DPUs) [**C**]. These work in parallel, each unit tackling a different part of a problem, thus making the computer faster. A laser in the center of each DPU sends signals to other DPUs in the form of rapid bursts of light, which are reflected by a holographic mirror [1]. This reflects different wavelengths in different directions [2], separating out data streams destined for different DPUs.*

Connections: Lasers and Holography 70 Computers 78 80 82 Silicon Chip Manufacture 126 Principles 252 254 256 258

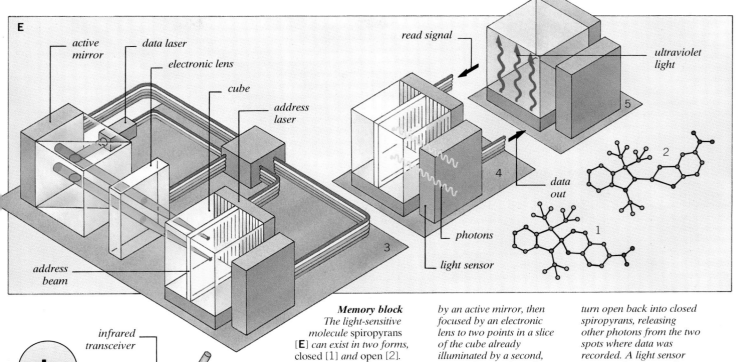

active mirror
data laser
electronic lens
cube
address laser
address beam
read signal
ultraviolet light
5
2
4
data out
photons
1
light sensor
3

infrared transceiver
abc
2
stylus
LCD
D
abc
1
tablet

Memory block
The light-sensitive molecule spiropyrans [E] can exist in two forms, closed [1] and open [2]. When a closed molecule is hit by photons – light particles – of two particular colors, one of its chemical bonds breaks, producing the open form. But if this is hit by another photon the bond reforms and the molecule emits a photon. This process is the basis of a 3-D optical random access memory (ORAM), which contains a cube "doped" with spiropyrans. To store data [3], the light from a data laser is split into two beams by an active mirror, then focused by an electronic lens to two points in a slice of the cube already illuminated by a second, address laser. Where the beams cross, the open form of the molecule is produced. To read data back [4], the address beam is again switched on. Its photons turn open back into closed spiropyrans, releasing other photons from the two spots where data was recorded. A light sensor converts the photons into electrical pulses, which are passed back to the computer. The data stored in the cube can be wiped by ultraviolet light [5].

Writing pad
A keyboard is an awkward way of entering information into a computer. Future, computers [D] may take their input directly from handwriting – a difficult task, because no two people's writing is the same.

The interface consists of a stylus that is used to write on a touch-sensitive liquid crystal display. The computer interprets the lines of the handwriting [1] and converts them into typewritten words [2]. Such an interface could form part of a cordless "tablet style" computer. Equipped with infrared transmitters and sensors this could communicate with other computers as the user moves from place to place.

separate path. This could greatly reduce the extent of circuitry needed by a computer.

In computer applications such as virtual reality simulation there is a need to record and recall huge amounts of information very quickly. Present-day hard disks are able to store the required gigabytes (billions of bytes) of data, but the speed at which this data can be "read" is too slow to provide smoothly changing, and thus credible, graphics.

Faster recall
A new, experimental form of memory called 3-dimensional optical random access memory (3-D ORAM) has a similarly high capacity but can be read thousands of times faster. The data are recorded by laser beams that intersect within a cube of light-sensitive material. The device works in a *highly parallel* way – instead of recording and reading data one bit at a time like a conventional chip, ORAM stores and reads thousands of bits simultaneously. This makes it an ideal medium for the storage of moving images. A similar material can be made into a disk and used to store information in the form of a hologram. It is estimated that one such disk – about the size of a conventional hard disk drive – will be able to store more than twelve hours of high-definition video images.

The processor at the heart of a future computer may not be a single chip but a set of discrete processing units, communicating with each other via pulses of light. These *subchips* will probably have a RISC (Reduced Instruction Set Computing) architecture. Processors in today's PCs use CISC (Complex Instruction Set Computing) architecture – they work by processing a large number of specialized commands. A RISC processor "understands" a far smaller number of simple commands, but carries these out very quickly. More complex (but less frequently used) tasks are emulated by a few simpler ones, so RISC chips process data far more quickly than traditional chips.

Existing computers work by performing calculations that distinguish between the logical states "true" and "false". But in everyday language, distinctions are not so stark and it is often easier to work in terms of degrees of truth (quite hot, very windy, etc.). This type of reasoning is known as *fuzzy logic*. Combined with *neural networks* – systems that learn to solve problems in a way similar to the human brain, fuzzy logic could be used to make a computer that emulates human thought and creativity.

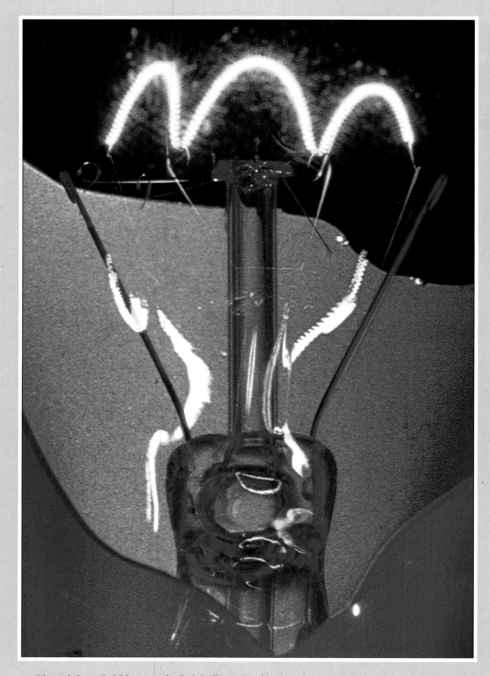

The tightly-coiled filament of a lightbulb, made of high melting point tungsten metal, becomes incandescent when heated to around 2500K by an electric current.

3
Domestic Technology

Climate Control

Food Preparation

Cleaning

Domestic Power

The Intelligent Home

Home Comforts

How machines give us control over our environment

The numerous devices that control our climate are often taken for granted. We usually only notice a heating system or air conditioner when it breaks down. Yet these systems, which make life easier in temperate regions, have also allowed humans to colonize almost every part of the globe, from the fringes of the polar ice caps to the hottest deserts. Human comfort or discomfort depends not only on temperature, but on the combination of heat, humidity, and air quality, all of which can be closely regulated in a modern home.

Although some homes are still warmed by individual gas or electric "fires," it is now more usual to employ a heating system that circulates heat from a single source. This heat may be generated in a *furnace* and carried around the house by the movement of warm air. Alternatively, the heat generator may be a *boiler* that heats water, which is pumped in a continuous circuit through a series of pipes and radiators. Radiators are actually misnamed because they radiate only around 20 percent of their heat, the remainder being carried away by convection currents. Most radiators are made from pressed steel, which would normally corrode when in contact with water and oxygen. The reason radiators do not rust is that the water flowing through them is part of a closed circuit. Any oxygen dissolved in the water is "used up" in a small initial amount of corrosion, after which it cannot be replenished.

Safety and control

Some boilers use an electric element to heat water, but more often gas or oil are burned to provide heat. In these cases, the boiler must be supplied with air for combustion, and the waste gases must be safely vented out of the room. This is accomplished by a *balanced* or *room sealed flue*. The boiler is sited against an outside wall. Air for combustion is drawn from outside, and the exhaust gases (which include carbon dioxide and toxic carbon monoxide) are removed via a parallel route. The result is that the air and gases inside the boiler are always separated from the air in the room.

A domestic heating system is controlled by an interior *thermostat*. This contains a strip (usually wound into a spiral) of two different metals sandwiched together. The metals expand and contract at different rates when the temperature changes, making the *bimetallic* strip bend one way or the other. As it does so, it makes or breaks an electrical circuit that pumps fuel into the boiler and causes it to ignite. A single thermostat, however, is not very efficient at maintaining a constant interior temperature. It switches on the boiler only when temperature drops below a certain threshold value. By the time the boiler fires up and begins to warm the house, interior temperature may have fallen by several more degrees. This "lag" is removed by using a second exterior thermostat that detects sudden drops in outside temperature and so "anticipates" drops in interior temperature.

Wet system
Today's heating systems are designed to be fuel efficient and controllable, and to provide both space heating and hot water. The wet system [A], which uses water to distribute heat, is the most common type of system in European countries. Water is heated by a gas-fired boiler [1]. A pump [10] drives the water around a pipework loop [8] via a series of radiators. These have corrugated faces to increase the surface area available for heat transfer. The hot water pipe also coils through a storage tank [5] where it transfers heat to a separate body of water. This heated water supplies domestic needs through taps, showers etc. The storage tank is topped up by water from a second large tank in the loft [7], which is in turn fed from the water main [6]. As water is heated in the boiler, it expands slightly: a small tank [9] takes up its increase in volume, protecting the pipework from rupture.

In hot countries, solar energy [3] is increasingly used to heat domestic water. A pump [4] drives water through pipes that run across a solar panel [2]. This is painted black and covered with glass to maximize heat absorption. The heated water passes through a heat exchanger in the storage tank [5].

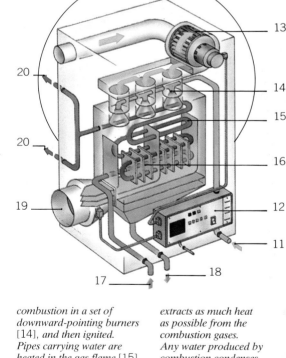

Gas boiler
Modern gas boilers [B] are of an efficient condenser *design. The amount of gas drawn in [11] is controlled by an electronic flow regulator [12]. Air is drawn in from the exterior by a centrifugal fan [13]. Air and gas are mixed together in optimum proportions for* combustion in a set of downward-pointing burners [14], and then ignited. Pipes carrying water are heated in the gas flame [15], and the hot water leaves the boiler [20]. Before being heated in the gas flame, cold water entering the boiler [17] is warmed in a finned heat exchanger [16], which extracts as much heat as possible from the combustion gases. Any water produced by combustion condenses on the plates of the heat exchanger and must be constantly drained out of the boiler [18]. Waste combustion gases [19] leave the boiler through a flue.

Connections: Food Preparation 90 Intelligent Home 96 Oil: Extraction 104 Principles 224 234 244 246

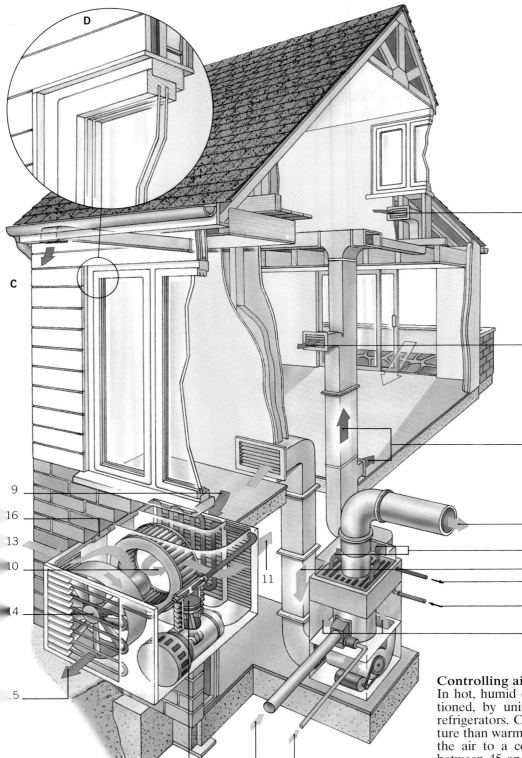

Furnaces

In most houses in the United States [C], domestic heat and hot water are provided by a furnace. Air [1] and fuel [2] for combustion are drawn into the furnace from outside. The fuel can be oil or gas: an oil furnace is shown here. The oil is pumped into the furnace through a narrow jet and ignited electrically. Waste gases pass to the exterior through a flue [3]. A fan at the base of the furnace sucks in cool air from within the house via a broad metal duct [4]. This passes around the furnace walls, picking up heat [5]. The warmed air is carried around the house by a system of ducts [6]. Cold water [7] is pumped through a "jacket" surrounding the furnace, and hot water emerges [8].

Heat loss from the house is reduced by insulating the loft space with fiberglass and by incorporating a fiberglass layer into wall cavities. Double-glazed windows [D], with a gap between the two panes and a coating of tin oxide, which reduces the transmission of infrared radiation (heat), can cut heat losses by up to one half. Heating costs are reduced most dramatically simply by turning down the thermostat. A reduction in the setting of just one degree reduces costs by an estimated 7 percent.

Controlling air quality

In hot, humid climates, air is cooled, or conditioned, by units that are similar to domestic refrigerators. Cool air is able to hold less moisture than warm air, so chilling also dehumidifies the air to a comfortable relative humidity of between 45 and 55 percent. Conversely, in dry climates, humidity is added to the air to improve the local environment. This is done by a humidifier, which forces air over a moving belt of wetted fabric, or which boils water and adds the steam to the moving airstream.

The 45 lb or so of air that we breathe each day is full of dust, pollen, and other minute particles. These may be removed from the airstream by an electrostatic filter. A fan draws air through a positively charged grid, which ionizes the particles, stripping away some of their electrons. A second, negatively charged grid attracts and collects the ionized particles. The cleaned air then passes into the room.

Air conditioner

A sealed room with a window-mounted air conditioning unit is, in effect, a giant refrigerator [C]. Cold liquid refrigerant at low pressure flows through coils on the room side. A centrifugal fan draws warm air from the room [9] over the coils. The cooled air [10] is returned to the room [11]. The warmed refrigerant evaporates, and then passes into a compressor [12], where it is pressurized. The hot, pressurized gas enters a second set of coils on the exterior side. A second fan [14] draws cool external air [13] over the hot coils to dissipate their heat [15]. In this process, the refrigerant is cooled to below its boiling point and condenses into a liquid. The refrigerant then passes through an expansion valve [16], where its pressure is suddenly reduced. As this happens, its temperature drops, and the cooling cycle begins again.

A Matter of Taste

How science is applied to the food we eat

Food processing is perhaps the oldest of all technologies. Cooking, drying, refrigeration, and fermentation date from prehistoric times, but the greatest advances in food technology were brought about during the Industrial Revolution to supply the needs of growing urban populations isolated from productive farmland. Today, virtually all the food that fills supermarket shelves has been treated in some way to preserve it from deterioration, to enhance its quality, taste, and texture, and to improve its nutritive value.

Food spoilage is caused by two main biological processes. As soon as a crop plant is harvested, or an animal slaughtered, enzymes within its cells begin a process of self-destruction, or *autolysis*, breaking down the cell's structures from within. In some cases (such as the tenderization of game) a limited degree of softening is desirable, but in most cases the quality of the food is reduced. Spoilage is also caused by microorganisms – bacteria and fungi – that split the cells' complex organic molecules into smaller compounds that can be absorbed and used to fuel growth and multiplication. Some of these organisms, particularly bacteria belonging to the groups *Clostridium, Campylobacter, Salmonella, Listeria,* and *Staphylococcus,* produce toxic by-products, which are responsible for most cases of food poisoning. The aim of food preservation is to inhibit the activity of microorganisms and autolytic enzymes through the agency of heat, cold, radiation, drying, or chemical additives.

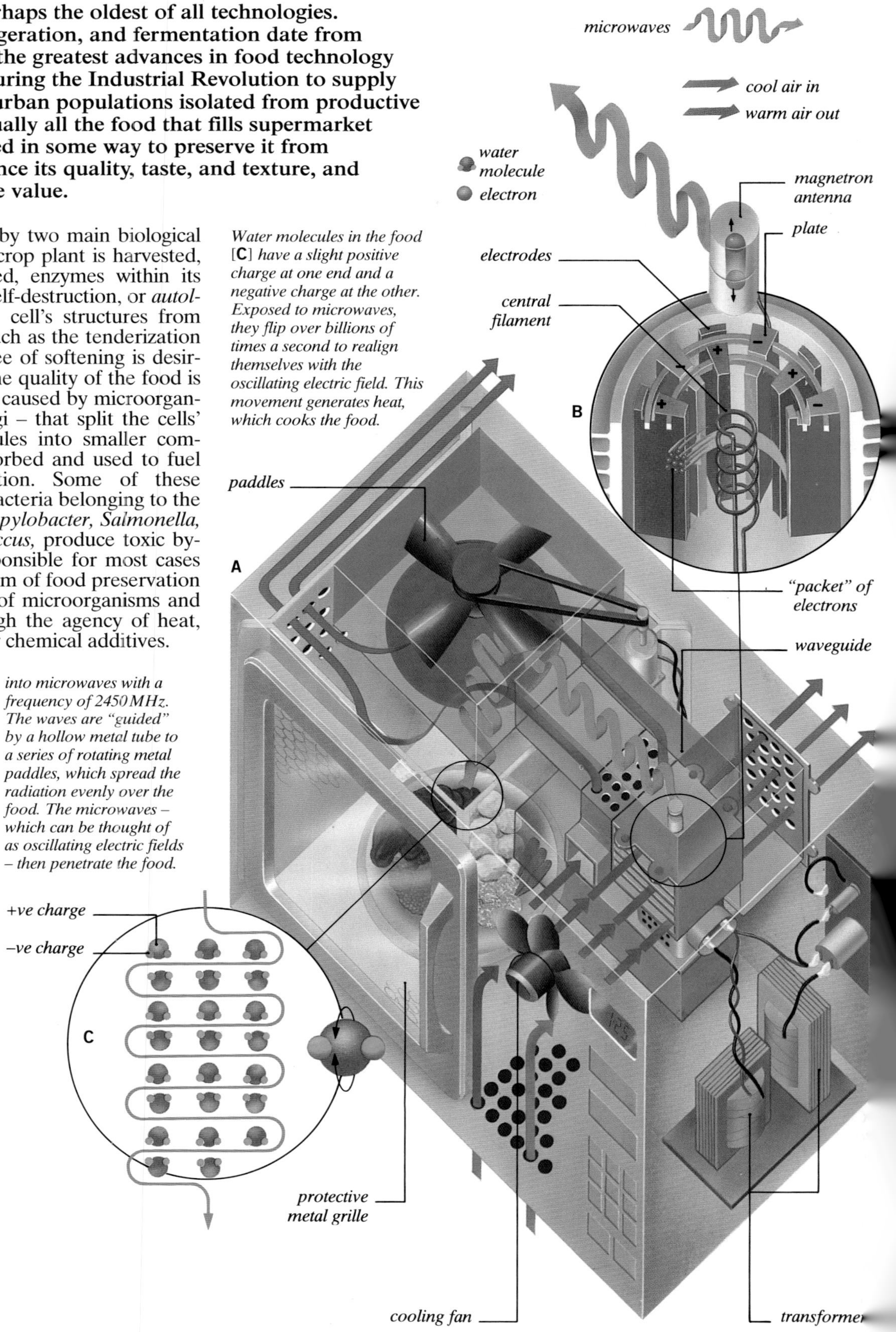

Cooked from within

A microwave oven [A] cooks rapidly by using high-frequency electromagnetic waves (microwaves) to agitate water molecules within food. Inside the oven, transformers produce a voltage high enough to supply a magnetron [B] – a type of cathode ray tube that can generate microwaves. The coiled central filament of the magnetron (the cathode) emits electrons. Magnetic and electric fields within the magnetron make the freed electrons bunch up into a "packet" and move quickly around a circular path, passing by a series of metal plates. As the electron packet nears a plate, it induces in it an opposite (positive) charge, and a negative charge is produced in its neighbors. Because the electron packet moves rapidly, the charge on each plate oscillates between positive and negative billions of times per second. A short antenna connected to one of the plates converts this oscillation of electrons

into microwaves with a frequency of 2450 MHz. The waves are "guided" by a hollow metal tube to a series of rotating metal paddles, which spread the radiation evenly over the food. The microwaves – which can be thought of as oscillating electric fields – then penetrate the food.

Water molecules in the food [C] have a slight positive charge at one end and a negative charge at the other. Exposed to microwaves, they flip over billions of times a second to realign themselves with the oscillating electric field. This movement generates heat, which cooks the food.

microwaves

cool air in
warm air out

water molecule
electron

magnetron antenna
plate

electrodes

central filament

B

"packet" of electrons

waveguide

paddles

A

+ve charge
−ve charge

C

protective metal grille

cooling fan

transformer

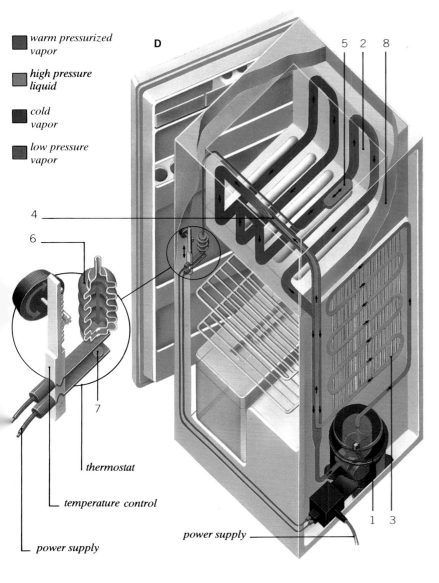

warm pressurized vapor

high pressure liquid

cold vapor

low pressure vapor

D

5 2 8

4

6

7

thermostat

temperature control

power supply

power supply

1 3

Turning up the heat

At high temperatures, the autolytic enzymes present in foods are deactivated and most microorganisms are killed. Cooking can therefore be considered to be a form of short-term preservation. For longer-term storage, heat-treated food is sealed in an airtight steel can (coated with tin against corrosion) to prevent reinfection by microorganisms. Most commercial canneries operate as continuous production lines. First, the open cans are automatically filled with a weighed amount of prepared food. The filled cans are sterilized by pressurized steam at a temperature of around 250 °F: heating also expands the contents and drives air out of the can, preventing spoilage by oxidation. The can is sealed and allowed to cool. As the contents contract, a partial vacuum is created, which firmly fixes the lid.

For certain foodstuffs, lower temperatures are sufficient to reduce the number of microorganisms to safe levels and extend shelf life. Milk, for example, is usually *pasteurized* by raising its temperature to at least 161 °F for a minimum of 15 seconds, before rapid cooling to below 50 °F.

Chilling and drying

Refrigeration and freezing are perhaps the most widely used preservation methods because they retain the taste, quality, and nutritive value of the fresh food. Refrigerating food to between 32 and 40 °F retards the activity of autolytic enzymes and slows the growth of microorganisms. These effects are greater if the food is frozen (at between 0 and –30 °F), partly because water is converted into ice and becomes unavailable to microorganisms. Freezing also kills certain parasites outright.

Most foods are *quick-frozen* – cooled from 32 to 25 °F in less than 30 minutes. This allows only tiny ice crystals to form within the food: these are too small to affect its texture or appearance.

Foods with a moisture content of less than 15 percent are too dry to support the growth of molds and bacteria, so dehydrating a food to below this level is an effective preservation measure. Sun-drying, salting, and smoking are dehydration methods that have been practiced for centuries. A more recent technique, applied to fruit, vegetables, and coffee, is *freeze-drying*. Here, the food is frozen and placed in a vacuum cabinet. When it is then heated under this reduced pressure, the ice *sublimes* – changes directly into vapor, leaving behind a dry, porous food, little changed in shape. The food can be rapidly rehydrated in cold water.

Another high-tech preservation method is *irradiation*, in which food is exposed to gamma rays or high-speed electrons. The radiation causes extensive ionization that kills most organisms, and is a cheap, effective sterilizing agent. But because the radiation has limited penetrating power, this method is best suited to foods such as grains and spices, which can be spread out into a thin layer.

Cool cabinet
A domestic refrigerator [D] is a machine designed to move heat energy from one place to another. The heat is carried by a refrigerant – a liquid with a low boiling point (around 112 °F at atmospheric pressure). Until recently, refrigerants were chlorofluorocarbons (CFCs). These compounds, however, were found to catalyze the depletion of ozone in the earth's upper atmosphere and are being replaced by less destructive HCFCs, which break down before reaching the altitude at which ozone is present.
The refrigerant is cycled by a compressor [1] through a pipe that runs in a loop around the icebox [2] and down the back of the refrigerator. The cycle starts with the refrigerant as a low-pressure vapor. This enters the compressor, emerging as a warm, pressurized vapor before being pushed through cooling coils at the back of the machine [3]. Here, the vapor gives up heat to the atmosphere and condenses into a liquid. The (high-pressure) fluid then passes through an expansion valve [4]. Under reduced pressure, it is vaporized and its temperature drops. The cold vapor absorbs heat from the interior of the freezer, chilling it to –4 °F. Warmer vapor flows back to the compressor and the whole cycle begins over again. Convection currents carry cold air from the freezer down to the refrigerator, which is cooled to around 37 °F. Internal temperature is controlled by a thermostat: this consists of a sealed, air-filled tube terminating in the freezer [5]. As air within the tube (and freezer) warms up, it expands, pushing out a set of bellows [6]. The expanding bellows close an electrical switch [7], which switches on the compressor. The refrigerator cabinet is made of polyurethane foam [8]: this functions as an insulator as well as giving mechanical strength.

Keep it Clean

How simple machines make life easier and safer

Not all technologies are glamorous or exciting. There are machines that have played as important a part in transforming modern lives as the automobile, yet which are often undervalued or ignored completely. The washing machine is essentially a very simple mechanism, but it is one that has freed millions of people from hours of drudgery each week. Although both the washing machine and the vacuum cleaner are well established, newer designs are constantly being created that make better and more efficient use of water and power.

On its own water, does not easily remove dirt and grease from clothes. The scientific way to describe this is to say that water – surprisingly – is not a good wetting agent. It has high surface tension, causing it to form droplets when in contact with grease. A cleaning agent lowers this surface tension and allows water to penetrate a fabric.

Until the middle of the 20th century, the principal cleaning agent in use was soap. This is made from natural oils and an alkali – originally the ash of certain trees and plants, but nowadays caustic soda, made industrially from common salt. Soaps do not work in acid water, and so have to contain an amount of alkali. More seriously, they do not work well in hard water, forming an insoluble scum that leaves a ring around a bath and a white film over glassware. In addition, the natural oils and fats needed for the manufacture of soap are not always readily and cheaply available.

Modern detergents are made from long-chain organic molecules derived from crude oil. At one end of each detergent molecule is an ionic "head" that is attracted to water. Conversely, the main (organic) part of the molecule is attracted to substances like oil and grease that are insoluble in water. Detergent molecules work by surrounding an oil or grease particle with a coating that is attracted to water. The dirt can then be lifted free, passing into suspension in the water.

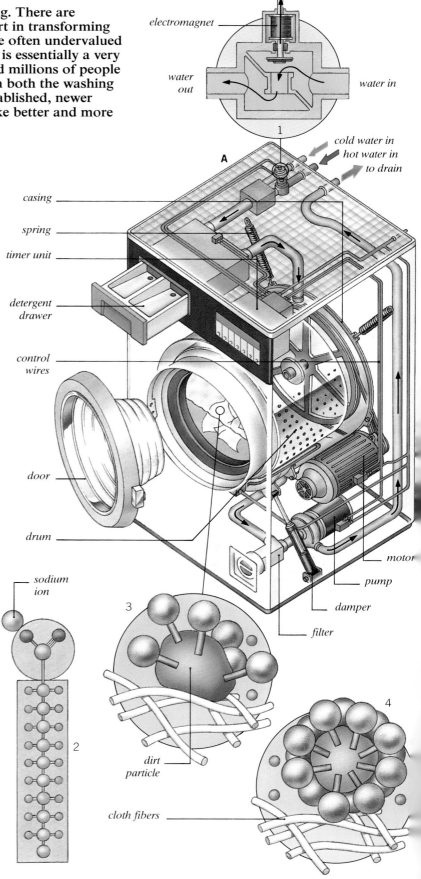

electromagnet

water out

water in

1

cold water in
hot water in
to drain

A

casing

spring

timer unit

detergent drawer

control wires

door

drum

sodium ion

dirt particle

cloth fibers

motor

pump

damper

filter

3

4

2

Cleaning power

*The brain of a washing machine [**A**] is its timer unit. This opens electromagnetic solenoid valves [1] that let hot and cold water flow through a drawer where they pick up detergent. From here the water sprays onto dirty clothes in a stainless steel drum, itself surrounded by a watertight casing. Other leads from the timer go to the motor, which at first turns the drum slowly. Paddles agitate the clothes as they are rotated ensuring that the water soaks into every fold.*

Detergent molecules have two distinct ends [2]. One gives up a sodium ion and bonds with water. The other

end prefers to attach itself to substances that normally do not dissolve in water, such as dirt. Dirt particles in the the fibers of a piece of clothing [3] are gradually surrounded by detergent molecules until they are lifted off [4]. In the final cycle the timer opens another valve and pumps the dirty water out via a filter. The washing cycle can then be repeated or clean water be let in to rinse the clothes. When all the detergent has been removed, the motor spins the drum at high speed to impel water out of the clothes. Springs and dampers (shock absorbers) limit the violent shaking that this creates.

Connections: Domestic Power 94 Power Stations 106 Waste Treatment 130 Principles 218 222 240

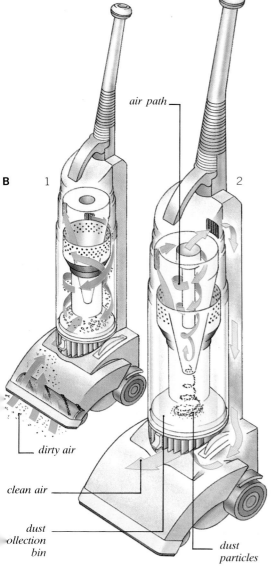

air path

B 1   2

dirty air

clean air

dust
collection
bin

dust
particles

All in a spin
*A new design of vacuum cleaner [**B**] uses centrifugal force rather than a paper filter to remove dust and pollen from the air. At the front, revolving brushes beat the carpet so that dust particles can be sucked up by the cleaner's powerful fan. The shape of the inner chambers twists the dirty airstream into a tapering vortex, called a cyclone, as it flows. In the outer cyclone air spirals downward at 185mph, fast enough to fling out large particles [1]. The airstream then passes inside to a second much narrower cyclone, in which the air rushes around at around 600mph, so fast that the tiniest dust particles are ejected and settle in the dust collection bin [2].*

The same principle is also used on some diesel-engined vehicles, which can emit fine soot particles dangerous to health. A cyclone device fitted in the exhaust system spins the gases as they leave the vehicle so that the particles drop out.

○○ water

○ carbonate ion

○ sodium ion

● calcium ion

○ chloride ion

⇨ soft water

⇨ salt solution

Hidden extras
Commercial cleaning agents contain not only detergents, but mixtures of other useful chemicals in proportions that vary according to use. Detergent powders and liquids contain whiteners – compounds that absorb invisible ultraviolet radiation and reemit it as visible blue light. This overcomes any yellowish tinge that old garments may have, making colors appear brighter. So-called biological powders and liquids also contain enzymes, natural catalysts that enable the breakdown of proteins in dried blood and sweat to take place at much lower temperatures, saving energy.

The job of a washing machine is to ensure that a detergent is able to work properly. It has to mix detergent with water at the right temperature, and agitate the dirty clothes in this solution to encourage dirt particles to break free. Then it must ensure that the clothes are fully rinsed out and that most excess water is spun out. Some combined machines go one step further and are able to dry clothes still inside their drums with hot air.

Some types of fabric are damaged by soaking in water and must be dry cleaned. In this case, the word "dry" is a misnomer because the clothes are still wetted, but by a liquid organic solvent (most often trichloroethene) rather than water. The solvent is able to dissolve the organic greases and fats that have accumulated on the clothing.

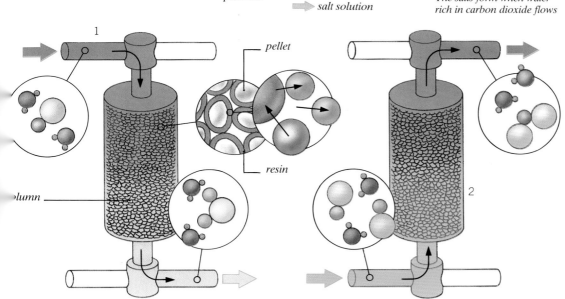

1

pellet

resin

column

2

Softening water
*In some areas, tap water contains salts of calcium and magnesium, which cause water hardness. This inhibits the action of soaps, and is responsible for deposits of limescale on the insides of hot water pipes. The salts form when water rich in carbon dioxide flows through limestone, dissolving some of the rock. When the water is heated, the carbon dioxide is driven off, and the calcium carbonate precipitates out as unwanted limescale. The problem can be avoided by passing hard water through an ion-exchange column [**C**]. This contains pellets coated in a resin that carries free sodium ions. Sodium is a more reactive metal than calcium and so sodium ions and calcium ions swap places [1]. The sodium salts that form are soluble in water at high temperatures, and so do not precipitate out when water is heated. In addition, sodium salts (in low concentrations) have no effect on the action of soaps and detergents so unsightly scum does not form. When the sodium ions on the resin have been completely used up, they are replenished simply by passing a strong solution of salt (sodium chloride) back through the column, reversing the exchange of sodium and calcium ions [2].*

Power at Home

How electricity is safely harnessed

Electricity can be a dangerous form of energy. Although domestic voltages are many times smaller than those of national grids, the supply in the home is still capable of causing severe electric shocks or starting fires. Fuses and circuit breakers protect users from some dangers, while the earth wire prevents shocks from metal-cased appliances that have become live. Other machines, such as electric drills, are double insulated, contained in two layers of plastic casing that isolate the high currents drawn by their electric motors from human contact.

A domestic electricity supply contains three cables – *live* and *neutral*, which carry currents, and a third *earth* wire for protection. The live wire is at a voltage that cycles between high positive and negative values 50 (Europe) or 60 (America) times a second. The average value of this voltage, a figure quoted on most electrical appliances, is either 230 or 110 V, depending on the country of the supply. When an appliance is switched on, current flows in both directions along a chain that runs from the live wire through the appliance to the neutral conductor, which is maintained at zero volts.

Almost every conductor of electricity – including domestic cabling – has a resistance to an electrical current, made manifest by the conductor's heating up as a current flows. The higher the current, the greater the amount of heat, so too large a current can overheat a cable and ignite its plastic cladding. Domestic wiring is protected from this by devices that limit the current flowing in each wire. The simplest of these is a fuse, which is simply a length of thin wire. If the circuit being protected draws a current larger than a certain value, the fuse wire rapidly overheats and melts, breaking the circuit and cutting off the current.

A circuit breaker is an alternative to a fuse. It acts like a switch that automatically turns itself off when an overload is detected. Once the cause of the current surge has been found, the supply can be restored by simply flipping a switch on the circuit breaker. One version of the device uses the current flowing through the live wire of a circuit to create a magnetic field. When the current is too great, the field becomes strong enough to move a lever that cuts off the supply. Other versions contain *bimetallic* strips that bend as they are heated by an excessively large current and pull two contacts apart.

Protective ground

The earth, or ground, can be thought of as an infinitely absorbent "sponge" for electricity. An electric shock occurs when a human body provides a path between a live wire and the ground. But the human body is not a good conductor of electricity. The earth or ground wire is a much better conductor along which, if given the option, a current will always travel. If the live wire accidentally comes into contact with the metal outer casing of an appliance, a very large current immediately flows to the ground, blowing the appliance's fuse.

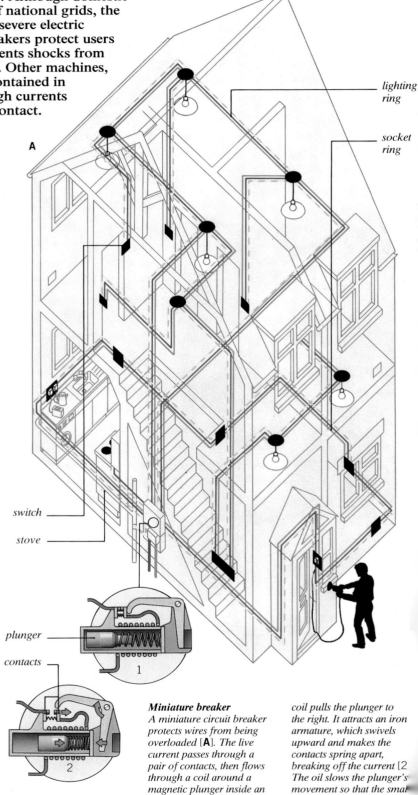

lighting ring

socket ring

switch

stove

plunger

contacts

Miniature breaker
A miniature circuit breaker protects wires from being overloaded [A]. The live current passes through a pair of contacts, then flows through a coil around a magnetic plunger inside an oil-filled capsule [1]. When current surges dangerously, the field produced by the

coil pulls the plunger to the right. It attracts an iron armature, which swivels upward and makes the contacts spring apart, breaking off the current [2]. The oil slows the plunger's movement so that the small surges produced during normal start-up of an electric motor are tolerated.

Connections: Cleaning 92 Power Stations 106 Electricity Transmission 116 Principles 246 248 254

Safety measures

Electricity, carried by live (brown) and neutral (blue) wires, enters the home via a meter that logs the amount used [**A**]. The supply then passes through an on/off switch to a distribution board, which is connected to the ground cable (yellow and green). This is bonded to metal plumbing – a safe channel to the ground in the case of a short circuit. At the distribution board, the supply divides into a number of branches. Each of these feeds several sockets or lights, or a single appliance, such as a stove, that draws a large current. Each branch is protected from carrying too great a current by either a fuse or circuit-breaker.

Sockets are connected in parallel along ring circuits running in loops that start and end at the distribution board. Each ring can carry only a limited total current. Light fixtures are wired in similar rings, but with a branch at each running to the on/off switch.

Power drill

Powering an electric drill [**B**] is a universal electric motor, which spins at high speed and is cooled by a fan. Its rotation is transferred to the bit, securely held in a chuck, by a set of gears. At the center of the motor is a spinning rotor around which a series of coils are wound, their ends connecting to two plates on a commutator. The rotor sits between two stationary outer coils. Current enters one outer coil then flows, via a sliding carbon contact, into one of the rotor coils. It exits via another sliding contact and travels around the second outer coil [1]. These currents produce magnetic fields between the stationary and rotor coils that pull one side of the rotor up while pushing the other side down, spinning the rotor. The torque, or turning force, this produces on a rotor coil is greatest when it is at right angles to the outer coils. Simplified diagrams [2,3,4] show a motor with two rotor coils (red and blue) and two outer coils (green). At first, current flows through the red coil [2], but as it turns it gives less torque [3]. The commutator then switches current to the blue coil as it approaches its position of maximum torque [4]. Real motors contain many coils to give a more constant torque.

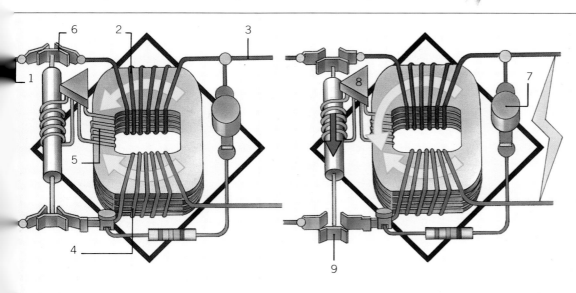

Super circuit breaker

Inside a residual current device live [1,2], and neutral [4] wires flow through contacts [6] before coiling around an iron core and passing to an appliance [3]. If all is well (left) the current in both wires is the same, and they produce magnetic fields in the core that exactly cancel one another out. But when a short circuit occurs (right), or the test button is pressed [7], the live wire's magnetic field dominates. The resulting field is sensed by a coil [5], which triggers a circuit [8] that disconnects the supply [9].

Labels: bit, chuck, gears, fan, magnetic field, coil, commutator, switch, rotor, commutator, rotor, universal motor, carbon contacts

Intelligent Building

How homes are designed for efficiency and security

Every year, huge amounts of money are spent heating and cooling homes and offices. And lighting alone accounts for 20 percent of the electricity generated in the United States. These costs are cut dramatically in a new generation of "intelligent" buildings that make maximum use of natural heat, light, and ventilation. The efficiency of existing buildings can be improved by effective insulation, the use of passive heating and cooling systems, and other energy-saving devices. At home and in the office, hi-tech devices also warn of intruders and fire.

Heat loss from a house is reduced dramatically by building in passive insulation, which adds as little as US$3000 to the cost of an average house. At its simplest, this consists of a thick layer of insulation in the loft and the use of two-layered walls, the cavity between which is filled with an insulating material, such as fiberglass. In homes not properly insulated, windows are a focus of unnecessary heat loss – a typical single-glazed window transmits eleven times more heat than the same area of well-insulated wall. Double- or triple-layers of glass, separated by pockets of xenon or argon gas (which are better insulators than air) can substantially cut these losses. The resulting *superinsulated* house is able to store the "free" heat given off by human bodies and electrical appliances so efficiently that temperature inside may rise by 30°F over external temperature on an average day. It is estimated that the best insulated houses provide fuel savings of more than 75 percent.

Built-in efficiency

In a large office building, heat is produced by photocopiers, computers, and many other pieces of equipment, as well as by the human occupants. These large buildings have a great volume relative to their outer surface area: they therefore lose heat slowly and so tend to over-heat. In addition, the large concrete mass of the building acts as a "heat sponge," soaking up the heat and releasing it back slowly. For these reasons, most office buildings require expensive air conditioning even in winter. In new build-ings this can be avoided by shaping the structure to encourage natural ventilation, allowing hot, stale air to rise out of the top while cooler, fresh air is drawn in nearer ground level. And the cost of cool air can be reduced by using the air conditioning units during the night when electricity is cheaper, to cool a large tank of water. During the day this water can in turn be used to cool circulated air.

Lighting is another area where considerable energy can be saved. Systems of blinds have been designed that reflect incident sunlight deep into a building's interior, reducing the need for artificial lighting. Working in a similar way, bundles of fiber-optic cables can channel sunlight from the roof of a building to the floors below. This light can be supplemented by illumination from highly efficient variable intensity fluorescent tubes, powered by radio-frequency electrical waves.

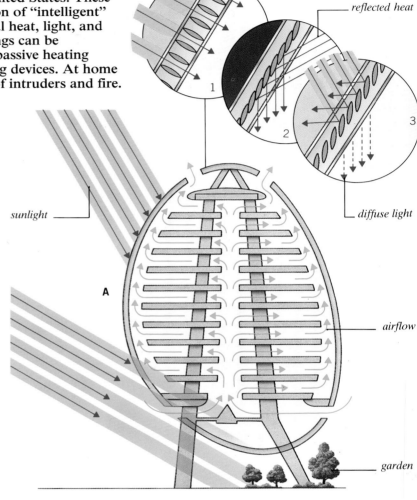

shutters

reflected heat

sunlight

diffuse light

A

airflow

garden

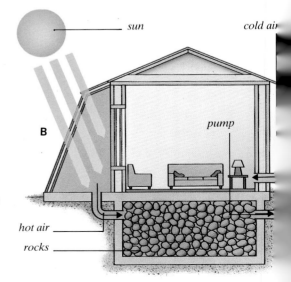

Rocks that store heat
*Some energy-efficient houses use low-tech heat storage units made from rocks [**B**]. During the summer months the sun heats air in a double- or triple-glazed greenhouse extension. A fan draws this hot air downward over rocks that fill a chamber under the house. These rocks are initially at the cooler temperature of the soil, and so absorb heat as the air flows over them. Cooled air emerges from the chamber and is pumped around the house. In winter the air flow is reversed: cold air is pumped over the warmed rocks and used to heat the house.*

sun

cold air

B

pump

hot air

rocks

Connections: Computers 84 Climate Control 88 Domestic Power 94 Renewable Energy 112 114 Principles 224 234 246 248 262

Natural ventilation

A future office block [A] may be ventilated entirely by air currents induced by its own egg shape. Fresh air enters through the bottom of the building. The air passes up the central atrium and through each floor, warming as it rises, before passing out of vents on the roof. The flow is great enough to make air conditioning unnecessary. In winter automatic shutters open fully to admit direct sunlight [1]. The shutters close at night [2], reflecting and conserving heat inside the building. They open only partially in the summer to let in diffuse, but not harsh, direct sunlight [3].

Simple security

The most basic of security devices [D] is the lock. One type [5] contains pins, split partway along their length. Each is pushed down by a spring. A key inserted into the lock pushes the pins upward against the springs. Only the correct key raises the pins to the point at which all their splits line up. This allows the lock barrel to be turned to unlock the door. A reed switch [6] is able to detect when the door has been opened. When closed, a magnet set into the door attracts an iron reed in the frame, making a circuit. This is broken when the door opens, triggering an alarm system.

Smoke signals

At the heart of a smoke detector [C] is a lump of radioactive material (purple), which emits alpha particles. These collide with gas molecules in the air, knocking electrons away and forming positively charged ions. The ions are able to carry an electric current between metal plates attached to a 9V battery. The size of this current is monitored by electronic circuits in the detector [1]. If smoke particles drift into the gap between the plates, they absorb some of the ions and reduce the size of the current that flows [2]. This current drop triggers an audible alarm.

"Smart" sensors

Any object warmer than absolute zero gives out electromagnetic radiation. In the case of a human body, this is invisible infrared radiation which, when sensed by a passive infrared detector, can trigger a burglar alarm [D]. Inside a detector is an array of photocells, electronic chips sensitive only to radiation at the wavelengths emitted by a human body. A lens at the front of the detector has a number of facets, each of which focuses radiation from a distinct narrow slice, or "finger," of a room onto a corresponding photocell. The way that the quantity of radiation sensed varies from photocell to photocell enables the sensor to distinguish between human and other sources of infrared radiation.

For example, an electric bulb emits radiation at similar infrared wavelengths to a human body. However, when a bulb is switched on the sensor does not trigger the alarm because it "sees" a constant level of infrared radiation after the initial surge [1]. But as an intruder breaks in and then moves across a room [2], the sensor will detect radiation decreasing in one finger [3] while it increases in a neighboring sector [4], a pattern that characterizes human movement and triggers the alarm.

C

battery

speaker

smoke detector

test button

smoke particle

electron

alpha particle

positive ion

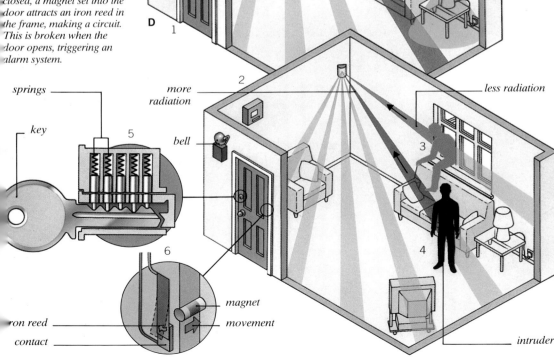

control box

sensor "fingers"

infrared radiation

lamp

D   1

springs

key

5

bell

more radiation

2

less radiation

3

4

6

magnet

movement

iron reed

contact

intruder

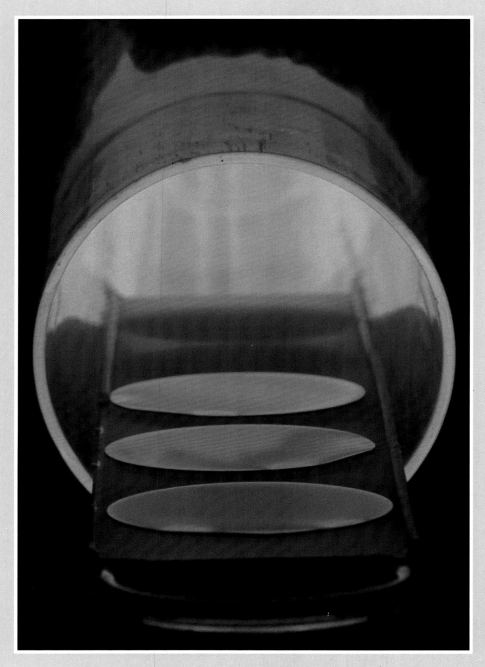

Circular silicon wafers cut from a single silicon crystal are baked at high temperature as part of the chip manufacturing process.

4
Power and Manufacturing

Mining

Oil: Exploration

Oil: Extraction

Power Stations

Nuclear Power

Nuclear Power: New Developments

Renewable Energy: Sun and Wind

Renewable Energy: The Earth

Electricity Transmission

Bridges

Tunnels

Skyscrapers

Paper Manufacture

Silicon Chip Manufacture

Robots

Waste Treatment

Buried Treasure

How the earth's mineral resources are mined

Mining is central to human civilization: its mineral products – stone, bronze, and iron – are used to identify the cultural ages of humankind. Today, mining is carried out on a larger scale than ever before: around four billion tons of coal are extracted worldwide every year, and individual mines – such as the open-pit copper mine at Bingham Canyon, Utah – can yield as much as 300,000 tons of ore and waste rock every day. In both underground and surface mines, new techniques and increasing mechanization continue to improve efficiency.

World industry depends on about 80 minerals, which range from the commonplace, such as iron, copper, and limestone, to the more exotic, such as platinum and molybdenum. Many different mining techniques have been developed to extract these minerals from the variety of locations and surroundings in which their deposits naturally occur.

Minerals are most easily and cheaply extracted from *surface mines*: these account for more than 80 percent of all minerals (excluding oil and gas) mined in the United States. In the simplest of these mines, the mineral is simply scooped out (soft ores) or blasted and carried out (hard ores) of an open pit by heavy machinery. As the pit grows in size its sides may be terraced (cut into steps) to increase stability and to allow many parts of the pit to be mined simultaneously.

Strip by strip

Whereas *open-pit mining* is suited to large and deep deposits near the surface, *strip mining* is used for relatively thin but wide surface deposits, or *seams*. It is a technique used mainly for coal, but is sometimes applied to other minerals of similar texture and strength. In strip mining, surface material, known as *overburden* or *spoil,* is removed by bulldozers strip by strip to expose the mineral. The mineral is then broken up and scooped out using large *draglines* with bucket capacities of up to 7000 cubic feet. Spoil from the newly dug strip is dumped in the space left by the one before.

Delving deep

Where the overburden is more than about 100 ft thick, surface extraction becomes uneconomic and underground mining methods are used. All these operations begin with the driving or digging of an opening into the mine. This is often a vertical opening, or *shaft,* with a diameter of around 23–26 ft, equipped with a hoisting system at the surface. Access to the buried ore may also be by *adit* – a passage that is nearly horizontal and dug into the side of a hill or mountain – or gently inclined spiral ramp. Adits and ramps are preferred to shafts as means of entry to a mine because diesel trucks can be used to carry ore out of the mine without having to transfer to hoisting systems. Horizontal passages called *levels* lead off the shaft or adit, and are used as passageways for personnel, equipment, and ore.

Sulfur extraction
The element sulfur is an important raw material for many chemical syntheses, and can be extracted from depths of more than 1000 ft by an ingenious liquid mining technique called the Frasch process [**A**], *which exploits sulfur's low melting point (250°F).*

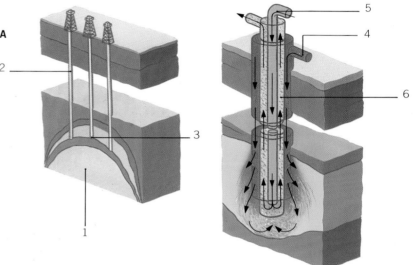

A

A common extraction technique is *room-and-pillar mining,* in which "rooms" of ore are excavated, and the roof supported by unmined pillars. This method is particularly suitable for coal, which often occurs in horizontal deposits surrounded by strong wall rock. But many ores occur as steeply dipping *veins* and demand a different extraction technique, such as *sub-level stoping.* Here, miners develop *sub-levels* – horizontal passages through the ore. The ore is then blasted and removed, forming a cavern, or *stope.* As mining continues, the stope grows in size: blasted ore fragments fall to the bottom of the stope, where they are collected.

In a situation where the walls of the mine (or the ore itself) are weak, or the workings are at great depth (and therefore pressure), the mine must be supported during ore extraction. This is done by hydraulic props or timber frames, or simply by filling stopes with waste material, such as sand or mill tailings.

Liquid assets
*The element sulfur occurs naturally [**A**], often in formations called salt domes – rocky bodies containing sodium chloride and gypsum [1]. In the Frasch process, boreholes [2] are sunk into these rocks [3], and three concentric pipes are run down into the hole. Pressurized water, heated to 310°F, is passed down the outer pipe [4]. Emerging from the pipe, it melts the sulfur. Compressed air passed down the central pipe [5] pushes the frothing mixture of molten sulfur and water to the surface up the middle pipe [6]. More than 80 percent of world sulfur is mined like this.*

Connections: Oil 102 104. Tunnels 120 Waste Treatment 130 Principles 236 240 242

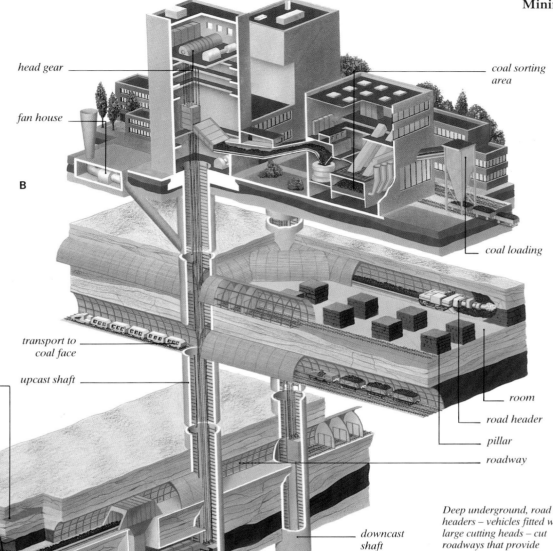
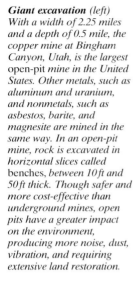

Giant excavation (left)
With a width of 2.25 miles and a depth of 0.5 mile, the copper mine at Bingham Canyon, Utah, is the largest open-pit mine in the United States. Other metals, such as aluminum and uranium, and nonmetals, such as asbestos, barite, and magnesite are mined in the same way. In an open-pit mine, rock is excavated in horizontal slices called benches, between 10 ft and 50 ft thick. Though safer and more cost-effective than underground mines, open pits have a greater impact on the environment, producing more noise, dust, vibration, and requiring extensive land restoration.

head gear

fan house

B

coal sorting area

coal loading

transport to coal face

upcast shaft

direction of coal face advance

coal face

room

road header

pillar

roadway

downcast shaft

coal skip

longwall face

conveyor

hydraulic roof supports

collapsed rock

direction of shearer advance

rotary shearer

supports move inward

Working in a coal mine
A coal mine [**B**] is joined to the surface through at least two vertical tunnels, or shafts, around 0.3 mile deep. The downcast shaft is the shaft through which coal is transported. Above its mouth is the head gear, which carries the windings that lift coal and machinery in and out of the mine. The upcast shaft is the passage that carries miners to the roadways leading to the coal face. An inclined passage leads to a fan that extracts more than 14,000 cubic feet of air from the upcast shaft every second. Fresh air enters through the downcast shaft and is drawn through the entire mine.

Deep underground, road headers – vehicles fitted with large cutting heads – cut roadways that provide access to coal seams. Today, coal is normally extracted by room-and-pillar or longwall mining. In the former, miners blast and drill the coal, leaving behind sufficient pillars to support the overlying rocks (called the hanging wall). This technique is most widely used in the United States. In the longwall system, miners use a shearer – a machine with a rotating cutting head – to cut ore from a single coal face (longwall) between 300 and 100 ft across. Hydraulic supports hold up the hanging wall above the miners. As the shearer cuts along the length of the coal face (from right to left) the supports advance to protect the miners: the hanging wall behind the supports is allowed to collapse. Conveyor belts carry the cut coal to coal skips, which are pulled to the surface.

In Search of Black Gold
How underground oil deposits are located

The industrialized world has an unquenchable thirst for oil. Ever since the first well was sunk in 1859 the techniques of prospecting and extraction have been in constant development, ensuring an uninterrupted supply of "black gold." Today, geologists armed with instruments to measure minute variations in the earth's magnetic and gravitational fields, and seismometers to analyze sounds reflected from deep within the earth's crust, can locate possible oil reserves. However, it is still only the drill that can determine whether oil is actually present below the ground.

Today's multi-billion-dollar oil industry has its roots in events that occurred tens of millions of years ago. The ancient seas teemed with microscopic plants and animals, similar to the plankton found in the oceans today. The abundant remains of these organisms settled to the oxygen-poor floors of still sedimentary basins, where they formed an organic-rich mud. Over the millennia, the mud was buried and compressed by successive layers of sediment, and the organic material was converted to a complex mixture of hydrocarbons – oil and gas.

Burial squeezed the oil and gas out of the rocks in which they formed, and toward regions of lower pressure at the edges of the sedimentary basin. Here, the hydrocarbons passed into coarse-grained sandstones, and it is these types of rock that today hold around 60 percent of the world's oil and gas reserves. Reservoirs are formed only where these oil-bearing rocks are capped with an impermeable layer of salt,

gypsum, or mudstone, which traps the oil, preventing its further migration. In some areas, the right combination of source, reservoir, and cap rock has produced oilfields holding over 500 million barrels (one barrel is 42 gallons).

Locating deposits of oil at depths of up to 3 miles is no easy matter. Aerial and satellite photographs can reveal subtle differences in the terrain that indicate the presence of an oil-trap. Conventional geological investigations can identify rock formations associated with oil-bearing areas. But before drilling starts, an accurate picture of deeply buried layers of rock, or *strata*, is needed. This is obtained by seismic survey, a technique that exploits the properties of sound.

Sound is generated at the surface, either by detonating an explosive charge or by vibrating a heavy weight. The low-frequency compression-waves travel through the rock, but are partially

Sound surveys
*Seismic surveying is used to map oil deposits not only below land but also at sea [**C**]. An air gun sends sound waves through the water, into the underlying rock. The reflected sound is detected by a string of underwater microphones or hydrophones, which trail behind the survey vessel. The hydrophones use piezoelectric crystals to convert the pressure of incoming sound waves into electrical signals. These are processed by computer to give a digital map of the oil deposit (right). Sound reflections (sonar) are also used to measure water depth at which the deposit occurs.*

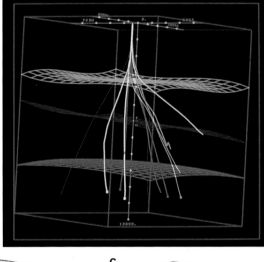

Oil-traps
*Many geological formations can act as oil reservoirs, but most of the world's oil is found in anticlinal traps [**A**]. Here, buckling of rock forms a dome capped with an impermeable layer that traps upwardly-migrating gas and oil. Anticlinal traps can be huge – hundreds of miles long and thousands of feet high. Although below the earth's surface, they can be detected by aerial or satellite survey. This is because they push up overlying rock, which is eroded unequally at the surface, leaving a pattern of concentric rings. The anticline also causes local variations in the earth's magnetic field, which can be detected on the surface using a* magnetometer.

*Oil is often found in sandstones [**B**], where it occupies the pores of the rock – tiny spaces between quartz grains. Each quartz grain is enveloped in a thin film of water, which acts as a lubricant, enabling the oil to migrate through the rock.*

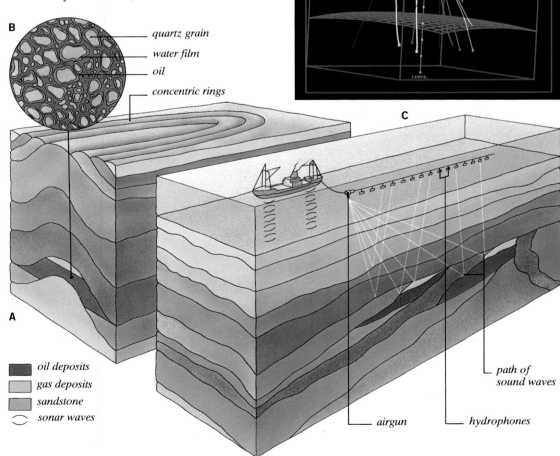

quartz grain
water film
oil
concentric rings

B

A

C

oil deposits
gas deposits
sandstone
sonar waves

path of sound waves

airgun

hydrophones

Connections: Oil: Extraction 104 Power Stations 106 Chemical Engineering 184 Oil Refining 186 Principles 218 222 228 230

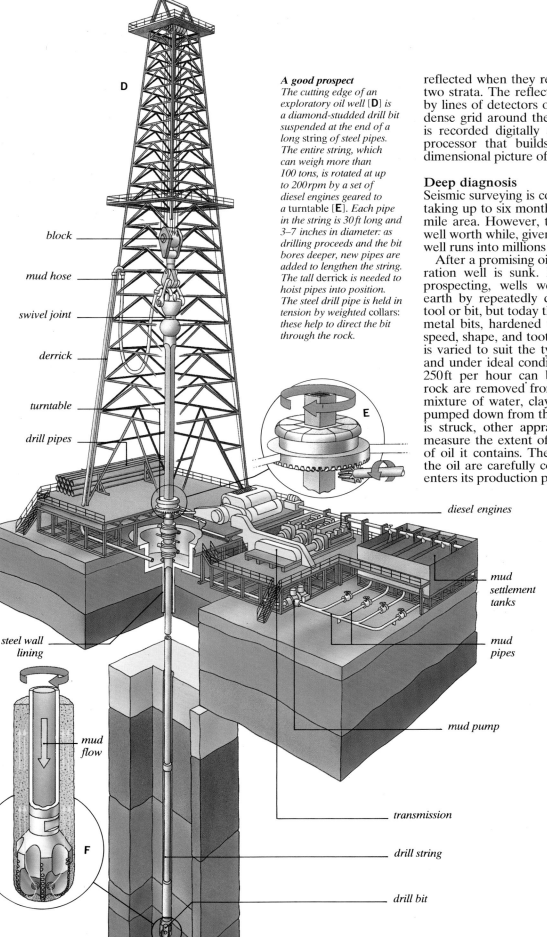

A good prospect
The cutting edge of an exploratory oil well [D] is a diamond-studded drill bit suspended at the end of a long string of steel pipes. The entire string, which can weigh more than 100 tons, is rotated at up to 200 rpm by a set of diesel engines geared to a turntable [E]. Each pipe in the string is 30 ft long and 3–7 inches in diameter: as drilling proceeds and the bit bores deeper, new pipes are added to lengthen the string. The tall derrick is needed to hoist pipes into position. The steel drill pipe is held in tension by weighted collars: these help to direct the bit through the rock.

block

mud hose

swivel joint

derrick

turntable

drill pipes

steel wall lining

mud flow

F

diesel engines

mud settlement tanks

mud pipes

mud pump

transmission

drill string

drill bit

reflected when they reach a boundary between two strata. The reflected sounds are picked up by lines of detectors or *geophones* laid out in a dense grid around the explosion site. The data is recorded digitally and fed into a powerful processor that builds up an accurate three-dimensional picture of the strata.

Deep diagnosis

Seismic surveying is costly and time consuming, taking up to six months for a typical 80-square-mile area. However, the time and expense are well worth while, given that the cost of drilling a well runs into millions of dollars.

After a promising oilfield is located, an exploration well is sunk. In the early days of oil prospecting, wells were "punched" into the earth by repeatedly dropping a heavy cutting tool or bit, but today they are drilled by rotating metal bits, hardened with diamond teeth. The speed, shape, and tooth configuration of the bit is varied to suit the type of rock being drilled, and under ideal conditions, rates of more than 250 ft per hour can be achieved. Cuttings of rock are removed from the well by "mud" – a mixture of water, clays, and other chemicals – pumped down from the surface. If and when oil is struck, other appraisal wells are drilled to measure the extent of the field and the quality of oil it contains. The economics of extracting the oil are carefully considered before the well enters its production phase.

Mud holes
For efficient drilling, broken rock must quickly be removed from around the drill tip. This is done by pumping "mud" – a mixture of water, clays, chemical additives, and suspended solids (ranging from walnut shells, to flakes of mica) down the hollow pipes of the drill string. The viscous mud emerges from holes in the bit and moves back up the well, carrying rock fragments with it all the way to the surface [F]. At the surface it is pumped into tanks, where the rock fragments settle out. The recovered fragments are examined by geologists for clues to the presence of oil, while the cleared mud is pumped back down the drill string and reused.

Mud lubricates and cools the bit, and maintains pressure in the well, preventing its sides from caving in before they can be reinforced with a steel lining. Mud also prevents oil from blowing out of the well under its natural pressure.

Plumbing the Depths

How oil is extracted and distributed

Standing 885 feet above the floor of the turbulent North Sea, the Statfjord B oil platform is one of the largest and most remarkable structures ever built. This concrete-and-steel edifice, which weighs more than both towers of New York's World Trade Center put together, holds all the equipment needed to drill wells in the sea-floor and extract from them 150,000 barrels of oil per day. Production platforms like Statfjord B and its terrestrial cousins are the first step in the long journey of oil and its products from remote underground fields to the consumer.

Most commercial oil wells, and all gas wells, start off as *flowing wells* – ones in which there is enough pressure underground to drive the hydrocarbons up to the surface. This means that on land, converting an exploration well to a production well is simple. The drill assembly, or *string,* is withdrawn, and the well sides coated with a protective lining. An array of valves, gauges, and pipe junctions known as a *Christmas tree* is installed at the top of the well to monitor and control the flow of oil. Finally, the weight of *drilling mud* – the viscous fluid pumped down the well during drilling – is reduced until the natural pressure of the oil forces it up out of the ground. After a period of extraction, natural pressure drops, and the oil must be pumped to the surface. This is often done using beam pumps – the "nodding donkeys" that dot the landscape in Texas and the Middle East – but in some cases reservoir pressure is maintained artificially by injecting water or compressed gas through adjacent *injector wells.*

Drilling beneath the waves
Around one-third of the world's oil comes from offshore fields, most of which are located in the Arabian Gulf, the North Sea, and the Gulf of Mexico. With today's advanced undersea technology, wells can be sunk in waters more than 3000 ft deep, although fixed production platforms generally operate in waters of less than 1300 ft. Modern platforms can service as many as 60 production wells, some directly beneath the platform, others remote but connected to the production platform by seabed pipelines. These *satellite wells* allow small adjacent oilfields to be tapped at minimum expense.

New directional drilling techniques have also increased the efficiency of oil extraction. Rather than just drilling vertically downward, engineers can now "steer" a drill bit to a target oilfield several miles away from the rig. Directional drilling not only requires fewer platforms but actually increases the amount of oil recovered. This is because underground oil reservoirs may be several miles wide, but only a few feet deep, so horizontal wells make far more extensive contact with the oil than vertical bores.

The flow of oil may be improved by increasing the permeability of the rock that holds it. Explosives were once used to fracture the host rock, but today this job is usually done by pumping down fluid to increase the hydraulic

Troubled waters (right)
Increases in oil traffic have been accompanied by a growing number of spills. At sea, floating booms are used to contain the oil, which is then burned, recovered, chemically dispersed, or digested. These measures help to reduce the ecological damage caused.

Up and coming
When the natural pressure of a reservoir falls, oil must be pumped to the surface to maintain production. On land, this is most simply done using a beam pump [A], in which a "nodding" beam [1] is attached to a "plunger" [2] submerged in the fluid of the well.

Heavy platforms
Gravity platforms [D] operate in seas up to 100 ft deep. These enormous structures are all the more remarkable because they are built in sheltered waters and then towed out to sea, their ballast tanks filled with air. Moving these huge masses is a precision operation, for inertial forces are so great that the smallest collision could prove disastrous. Once in the correct position, their ballast tanks are filled with water, and sink to the seabed. Around the bottom of the concrete base, a steel skirt cuts into the seabed as the massive platform settles.

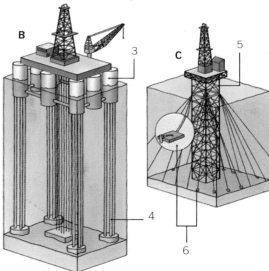

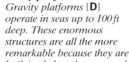

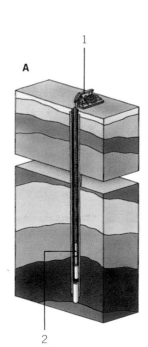

In deep water
Today, oil is extracted from reservoirs lying more than 6500 ft below the seafloor. Production platforms in deep water are compliant – designed to "give" a little with the movement of wind and water – making them lighter and cheaper than rigid platforms. In a tension-leg platform [B], buoyancy tanks [3] hold the drilling platform aloft. The buoyancy applies tension to steel tubes [4] anchored in the seabed. A guyed tower platform [C] consists of a narrow steel jacket [5] that rests on the seafloor. Guy wires attached to weights [6] allow some movement.

Connections: Ships 36 Underwater Technology 38 Mining 100 Oil: Exploration 102 Oil Refining 186 Principles 218 222 228 230 236

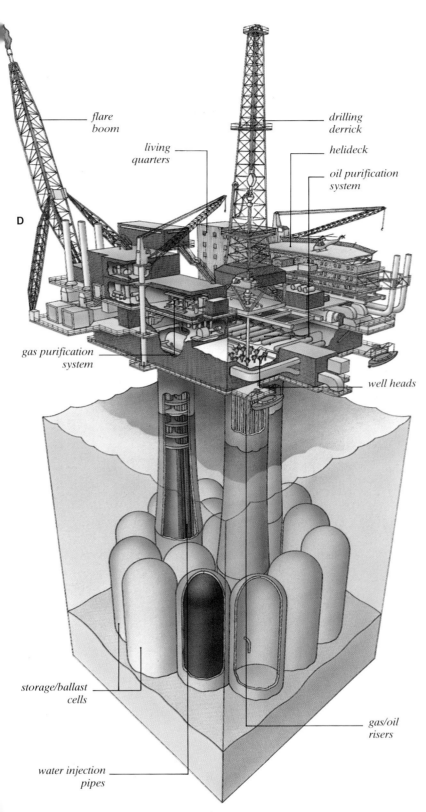

flare boom

drilling derrick

living quarters

helideck

oil purification system

gas purification system

well heads

storage/ballast cells

gas/oil risers

water injection pipes

pressure underground. Once the pressure is released, the rock closes up again, so *propping agents*, such as glass or plastic beads, are pumped down to hold the rock fractures open.

Oil on the move

On the production platform, oil is separated from water and gas before transportation to a refinery. In the early days of the industry, crude oil was refined close to the oilfield and its products distributed, but as demand grew for a wider range of oil products, it became more practical to transport the crude by pipeline or tanker to a refinery closer to centers of population.

Pipelines are used over land and for short distances at sea. The most impressive is the Trans-Alaska Pipeline which links the Prudhoe Bay oilfield, 250 miles north of the Arctic Circle, to the ice-free port of Valdez, 600 miles away on the south coast of Alaska. The pipeline, capable of transporting 2,000,000 barrels of crude oil a day, crosses three mountain ranges, more than 800 rivers and streams, and passes through three major earthquake zones. For transporting oil over longer distances, oil tankers prove more efficient than pipelines. In fact, tankers are so widely used that they make up over one-third of the world's merchant shipping. Very large crude oil carriers (VLCCs) are capable of carrying 300,000 tons, and some ultra-large crude oil carriers (ULCCs) have a 500,000-ton capacity.

Centralized production
A single oil platform is at the center of a complex production system [E]. It can receive oil from up to 60 satellite wells [1], channeled through seabed pipelines [2]. And a technique called directional drilling allows wells to be drilled at many angles [3] *other than straight down, so that many oil fields can be reached from a single platform, and geological obstructions can be avoided.*

Oil is transported to the shore by pipeline, or by tankers [4], which draw alongside a loading buoy [5] situated a safe distance away from the platform.

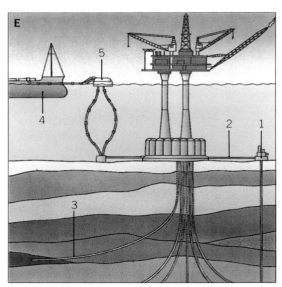

E

Offshore industry
A gravity platform [D] rests on the seabed, stabilized by its own immense weight (up to 800,000 tons). Its base consists of reinforced concrete cells, from which rise concrete legs filled with a network of pipes carrying oil, gas, and water. On top is a separate steel deck,

which supports all the equipment for drilling and exploiting oil wells. Oil (red) and gas (green) flow from the production wells through risers to the well heads. Oil and gas are separated from impurities and dried. Gas is exported to the mainland by pipeline: during shutdown periods it

must be flared off. Purified oil is pumped down to the ballast cells at the base of the platform, where it is stored until collected by a tanker. The cells are flooded with sea water when empty of oil. Water (blue) can be injected into the production wells to maintain pressure in the oil reservoir.

Power to the People
How power stations produce electricity

Day and night, power stations keep running to meet the electricity demands of modern society. Producing over 9000 megawatts of electricity – enough to light a hundred million lightbulbs – the largest plants eat through 20,000 tons of coal a day. At the heart of a station are its generators, 10-foot-long dynamos that weigh 400 tons and spin 50 or 60 times a second. Unfortunately, power also brings pollution – acid rain and greenhouse gases. Clean-coal technologies and gas-burning stations aim to cut the huge volumes of these harmful emissions.

A power station generator differs from a bicycle dynamo only in scale. In each case, a spinning magnet *induces* a current in coils of copper wire – known collectively as the *armature*. The powerful electromagnets in an industrial generator spin at 3000 rpm to induce a current of more than 10,000 amps in an armature of thick copper conducting rods.

The energy needed to turn these generators comes ultimately from burning fuel, usually coal, in a giant boiler. A maze of tubing, many miles long, carries water through the boiler. As it absorbs the heat of combustion, the water turns into high-pressure steam with a temperature of more than 930°F. This rushes through a series of windmill-like turbines, connected by a shaft to the generator.

Clean coal technology
Coal is used in the majority of conventional power stations because it is both abundant – reserves could last more than 350 years – and relatively cheap. However, burning coal extracts a high price from the environment. Most coals contain about 3 percent sulfur. In the combustion process this forms sulfur dioxide (SO_2), a gas that reacts with water vapor in the atmosphere to produce acid rain.

Older power stations can be fitted with a device called a *scrubber*, which sprays a slurry of limestone and water into the waste gases of combustion. Most of the SO_2 combines with the limestone to form gypsum, which can be sold on as a building material. Scrubbers have disadvantages – they are very costly and produce a useless sludge in huge quantities. A large power station produces enough of this a year to cover over half a square mile to a depth of one foot.

Sulfur can also be removed at the burning stage. In *fluidized-bed combustion*, coal and limestone are crushed into a fine powder, which is fed to the boiler. Forcing compressed air through this mixture as it burns creates a bubbling "liquid" in which the sulfur forms a harmless slag. Steam pipes run directly through the combustion bed, picking up heat much more effectively than in a normal boiler.

An alternative way of using coal is *gasification*. In this process, coal and water are heated with a small amount of oxygen, producing clean-burning hydrogen and carbon monoxide. Power stations that burn this mixture of gases are far less polluting, but are expensive and consume huge quantities of water.

The best of both worlds
A combined cycle station [**A**] extracts as much energy as possible from burning gas, using the hot waste gases to produce steam as well as to turn a turbine connected to a generator. Stations using this design are clean – the exhaust is mostly steam and carbon dioxide – and convert as much as 50 percent of the energy released by the gas into electricity. A coal-fired station, with pollution filters, has an efficiency of only around 35 percent.

Air is drawn into the compressor [1] where it is put under high pressure before passing to the combustion chamber [2]. Here gas is burnt in the compressed air, producing hot exhaust gases that expand through turbine blades [3], spinning the compressor and a 200 MW generator [4]. Also connected is an exciter, a smaller dynamo that provides the d.c. (direct current) electricity that powers the electromagnet in the generator.

Once the hot gases have passed out of the turbine, they are ducted through a flue [5]. Here their excess heat turns water into steam in a stack of pipework loops, before passing through a chimney into the atmosphere.

Second generation
Water is pumped from a cold water tank [6] into the highest loop in the flue. Here it is heated almost to boiling point as it snakes through the hot exhaust gases. The water is then fed round a second loop, the evaporator [7], where it turns from liquid to gas (steam). A collection tank [8] separates out any residual water from the steam before it passes to the superheating *loops* [9,10].

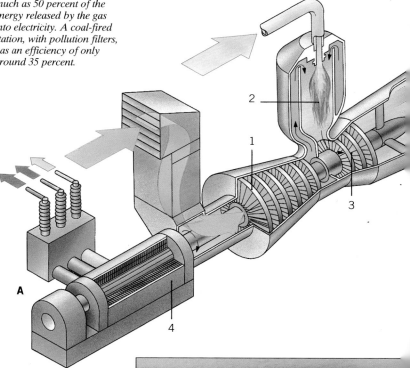

Steam stacks
The warm water from the condenser loses its heat in cooling towers *(right)*. It is sprayed over gravel and cooled by air, which enters the towers through holes at their bases. Despite the clouds of steam only a fraction of the water is lost.

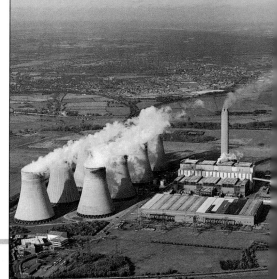

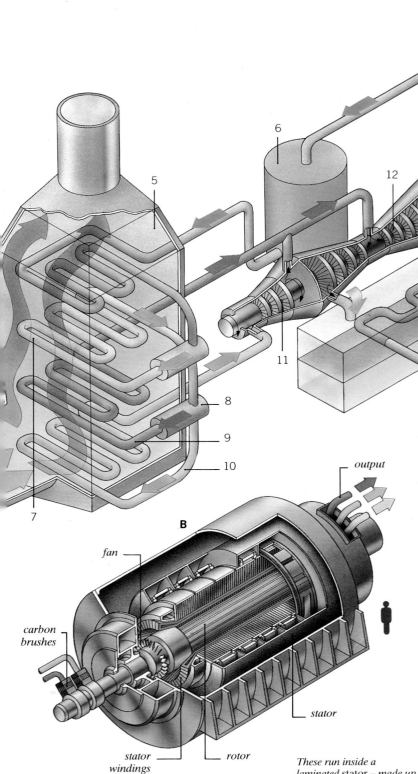

Steam engines
Two superheating coils *feed the steam turbines. The lowest coil* [10] *superheats the steam to over 930°F before it expands through a high-pressure turbine* [11]. *The steam in the second superheating loop is not quite as hot and is passed* directly to the low-pressure turbine [12], *where it is joined by the partially expanded steam from the high-pressure turbine. After expanding outward through the twin low-pressure turbines, the steam has had all of its useful energy removed. It is then condensed by passing over pipes carrying cold water* [13]: *this water in turn loses its heat in cooling towers.*

The blades of the two steam turbines turn at 3600rpm, spinning a shaft connected to a generator [14] *producing 22kV of three-phase electricity supplied to a national grid.*

fan

carbon brushes

output

B

stator

stator windings **rotor**

Turning power
When a magnet turns inside a coil of wire, an alternating electric voltage is produced. A car alternator and an industrial generator [B] *both work on this principle – but at widely different voltages. The rotor is a huge electromagnet, powered by* direct current from the exciter *and spinning at 3600rpm. Its power supply comes through sliding contacts or* brushes *made from graphite, a form of carbon that conducts electricity as well as being a low-friction material. A 22kV alternating current is induced in the three sets of solid copper windings.*

These run inside a laminated stator *– made up of thin sheets of iron with an insulating coating of lacquer to stop dangerous* eddy currents *building inside it. 1000amp currents flow in the windings, producing a lot of heat. Cooling water pumps through the cores, removing most of the heat; in addition, fans at each end blow hydrogen gas through the rotor to carry off its excess thermal energy.*

More energy from less fuel

Most old-technology power stations convert only around 32 percent of the energy in the fuel they burn into useful electricity. It is never possible to build a station that is 100 percent efficient – this is one way of expressing the second law of thermodynamics – but there are losses that are avoidable. The newer ways of burning coal not only reduce emissions but extract more energy from the fuel at the same time. Fluidized-bed and coal-gasification stations convert around 42 percent of the heat from the burning coal to electricity, compared with only 32 percent in a traditional station.

Other fuels can be used even more efficiently. Natural-gas-burning, *combined cycle* stations convert 50 percent of the heat of combustion into electricity through two extraction processes. The gas is first burnt in a turbine that drives its own generator. The waste gases from this are then used to produce steam, which passes through a set of turbines, driving a second generator.

A similar approach is used in combined heat and power (CHP) schemes. These are practical when a power station is situated near a town. Instead of producing steam, the waste gases are used to supply hot water to heat nearby houses and factories. Schemes like this can have an overall efficiency of 60 percent, and have been used successfully in Russia, Scandinavia, Japan, and Germany.

Energy from Atoms

How a nuclear power station generates electricity

Electricity was first generated by a nuclear reactor in 1951, when the EBR-1 test reactor in the USA lit up four lightbulbs. Nuclear power promised to provide "electricity too cheap to meter" and heralded a new age of prosperity. Today, some 400 nuclear power stations worldwide satisfy 17 percent of global electricity demand. Nuclear reactors are powerful (one pound of the fuel uranium produces as much heat energy as 5350 barrels of oil) and relatively "clean," but questions of safety and cost have led many countries to reevaluate their nuclear programs.

Nuclear power stations operate in much the same way as oil- and gas-fired plants. In all cases, heat energy is produced and used to pressurize a gas, which in turn drives turbines linked to electrical generators. In a nuclear power station the heat energy is derived from the fission of unstable *isotopes* of heavy metals, most commonly uranium-235.

Uranium occurs naturally in a variety of minerals. Its most important ore, uraninite, also known as pitchblende, is mined in many countries. The ore is ground to the consistency of fine sand and treated with chemical solvents, releasing a mixture of uranium oxides called yellowcake. Only 0.7 percent of the uranium present in yellowcake is U-235; most of the remainder is the isotope U-238, which is unsuitable as fuel. U-235 content can be increased by converting the uranium into gas and spinning it at high speed in a *centrifuge*. The heavier U-238 atoms collect at the walls of the centrifuge and

are removed: the remaining *enriched* uranium contains up to 3 percent U-235. It is compacted into pellets, which are sealed into cylinders, or *fuel rods,* and introduced into the reactor.

Uranium-235 is radioactive: its nuclei split spontaneously, creating two smaller atoms, heat in the form of infrared radiation, and two or three high-speed neutrons. If these neutrons then collide with other U-235 nuclei, they induce them to split, liberating still more heat and neutrons. When the amount of U-235 present exceeds a critical mass (around 8.8lb), a *chain reaction* starts, and the rate of nuclear fission grows rapidly, releasing vast amounts of energy. Such uncontrolled fission gives nuclear weapons their immense destructive power.

In a nuclear reactor, the rate of fission is carefully controlled by ensuring that every neutron that causes fission of a nucleus is replaced by *just one* new neutron. This is done by inserting neutron-absorbing material, typically boron, in

from the reactor to the sets of turbines.

First a set of high-pressure turbines, then sets of lower-pressure turbines extract the steam's useful energy, turning a generator connected to a local or national electricity supply. Residual steam is condensed by the third water coolant loop (green) and returned to the steam generators.

The reactor building is made of reinforced concrete which absorbs radiation; its steel inner lining prevents gas leakage. If the primary coolant circuit fails, the reactor core is flooded with cold, boron-rich water, which slows the fission reaction to a safe level.

Pressurized water reactor
*Inside a PWR [**A**], the heat of nuclear fission is used to produce steam, which in turn drives sets of turbines. Heat is relayed from the reactor core to the turbines via three separate coolant water loops.*

Water in the primary loop (light blue) is cycled through the hot reactor core by pumps. Pressurizers keep it at 150 atmospheres to prevent it from boiling as

it is superheated to 572 °F. This water then passes into a set of four heat exchangers known as steam generators. *Here it flows through thousands of metal tubes immersed in the cool water of the secondary coolant loop (dark blue). This water boils, forming high-pressure steam (pink), which passes*

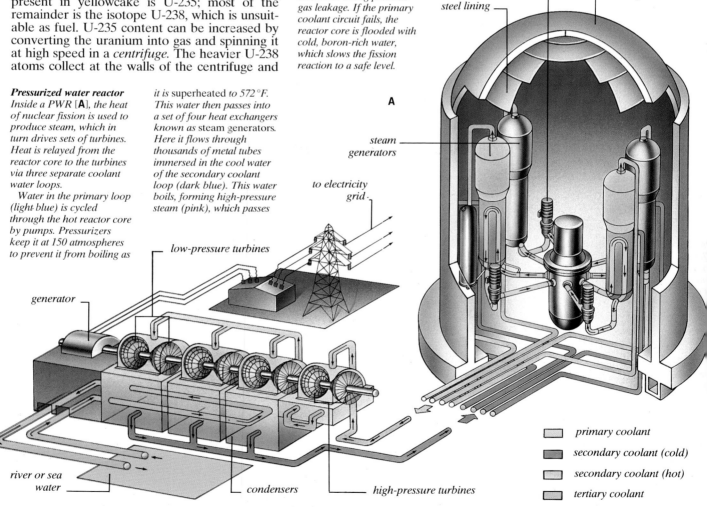

A

coolant pump

concrete containment building

steel lining

steam generators

to electricity grid

generator

low-pressure turbines

river or sea water

condensers

high-pressure turbines

	primary coolant
	secondary coolant (cold)
	secondary coolant (hot)
	tertiary coolant

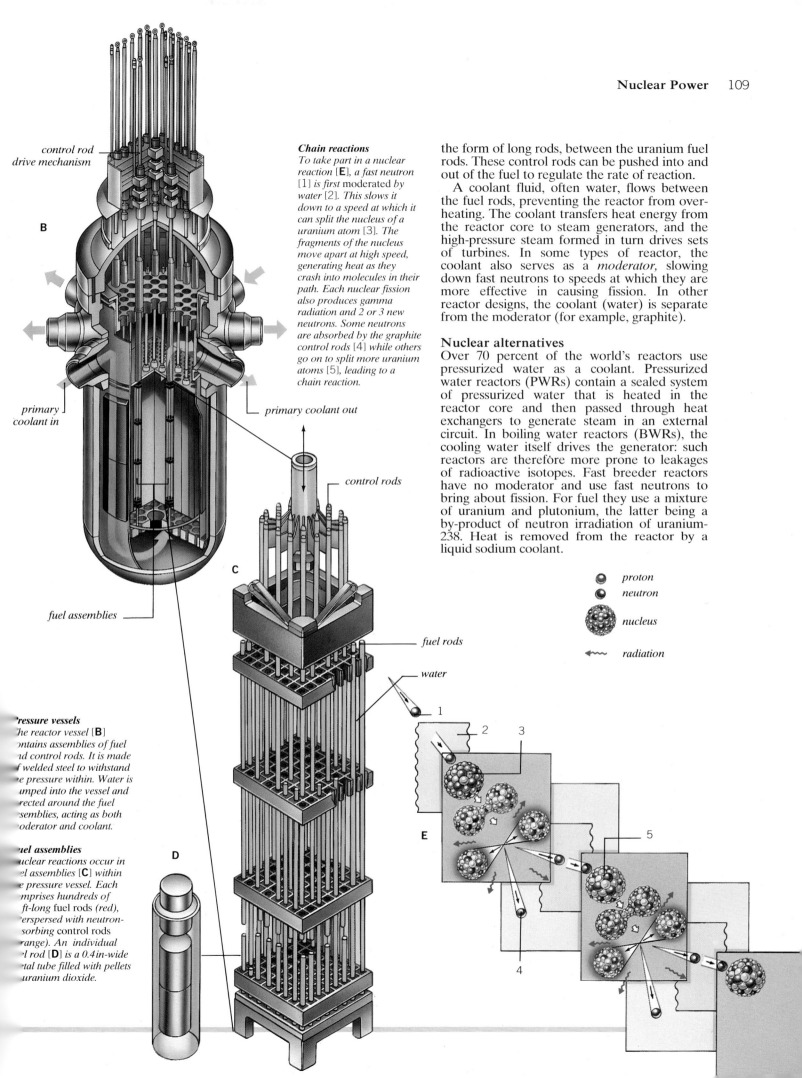

control rod drive mechanism

B

primary coolant in

primary coolant out

fuel assemblies

control rods

C

fuel rods

water

D

E

Chain reactions

To take part in a nuclear reaction [E], a fast neutron [1] is first moderated by water [2]. This slows it down to a speed at which it can split the nucleus of a uranium atom [3]. The fragments of the nucleus move apart at high speed, generating heat as they crash into molecules in their path. Each nuclear fission also produces gamma radiation and 2 or 3 new neutrons. Some neutrons are absorbed by the graphite control rods [4] while others go on to split more uranium atoms [5], leading to a chain reaction.

the form of long rods, between the uranium fuel rods. These control rods can be pushed into and out of the fuel to regulate the rate of reaction.

A coolant fluid, often water, flows between the fuel rods, preventing the reactor from overheating. The coolant transfers heat energy from the reactor core to steam generators, and the high-pressure steam formed in turn drives sets of turbines. In some types of reactor, the coolant also serves as a *moderator*, slowing down fast neutrons to speeds at which they are more effective in causing fission. In other reactor designs, the coolant (water) is separate from the moderator (for example, graphite).

Nuclear alternatives

Over 70 percent of the world's reactors use pressurized water as a coolant. Pressurized water reactors (PWRs) contain a sealed system of pressurized water that is heated in the reactor core and then passed through heat exchangers to generate steam in an external circuit. In boiling water reactors (BWRs), the cooling water itself drives the generator: such reactors are therefore more prone to leakages of radioactive isotopes. Fast breeder reactors have no moderator and use fast neutrons to bring about fission. For fuel they use a mixture of uranium and plutonium, the latter being a by-product of neutron irradiation of uranium-238. Heat is removed from the reactor by a liquid sodium coolant.

proton
neutron
nucleus
radiation

Pressure vessels

The reactor vessel [B] contains assemblies of fuel and control rods. It is made of welded steel to withstand the pressure within. Water is pumped into the vessel and directed around the fuel assemblies, acting as both moderator and coolant.

Fuel assemblies

Nuclear reactions occur in fuel assemblies [C] within the pressure vessel. Each comprises hundreds of ft-long fuel rods (red), interspersed with neutron-absorbing control rods (orange). An individual fuel rod [D] is a 0.4in-wide metal tube filled with pellets of uranium dioxide.

Atomic Alternatives

What the future holds for nuclear power

Under normal conditions, nuclear reactors are safer and far less polluting than comparable coal-fired stations. But the prospect of nuclear disaster – brought sharply into focus by the accident at Chernobyl in 1986 – and the dilemma of how and where to store radioactive waste, have fueled public opposition to nuclear power. Rising costs have deterred investment in new reactors, and the future of nuclear power is uncertain. However, the development of safer fission reactors and ultimately fusion power stations may help the nuclear industry to fulfill its potential.

At present, there are around 400 nuclear power stations in operation around the world. This number is unlikely to rise significantly in the near future because many nations have called a halt to new reactor construction. In the United States, for example, no new reactors have been ordered since 1978, and Sweden plans to abandon nuclear power generation altogether by the year 2010. The waning enthusiasm for nuclear power can be attributed mainly to the growing problem of radioactive waste, which is generated at almost every stage of production and use of nuclear fuel.

A persistent problem

Nuclear waste is classified as *high-level* or *low-level* depending both on its *radioactivity* (the rate at which it emits high-energy particles and radiation) and its *half-life* (an indication of the time taken to decay to a safe level). Low-level waste includes the vast quantities of mill tailings (formed when uranium ore is crushed) and contaminated tools, building materials and glassware etc. In the past these materials were poorly managed: mill tailings were abandoned in piles near uranium mines; other low-level wastes were packaged in canisters and dumped at sea. Today, low-level wastes are generally disposed of more carefully – usually by burial in a sanitary landfill.

High-level waste poses a greater problem. Consisting mainly of spent fuel rods from power plants, and by-products of nuclear weapon manufacture, this waste contains highly radioactive isotopes (some of which are also toxic) that have extremely long half lives, requiring storage for at least 10,000 years before their activity falls to a harmless level. At present, most high-level waste is temporarily stored in water-cooled tanks at nuclear power plants or fuel reprocessing facilities, awaiting a decision about its ultimate fate. Disposal of this waste will most likely involve its fixation in an inert glass or ceramic matrix, followed by encapsulation in a canister made of inert metal. Burial of the canisters at a depth of at least 650 ft in a geologically stable area will minimize the risk of leakage. However, every proposed site for long-term disposal facilities has met with huge public and legal opposition, and to date no long-term sites exist anywhere in the world.

Although accidents involving nuclear power stations are extremely uncommon, when accidents do occur their effects are long-lasting and

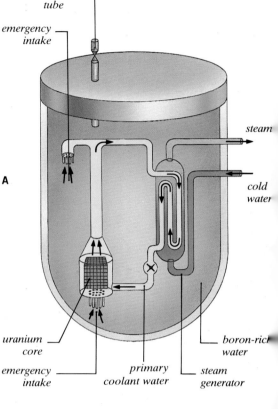

A

refill tube

emergency intake

steam

cold water

uranium core

emergency intake

primary coolant water

boron-rich water

steam generator

Safety first
Radical new reactor designs could make fission power far safer in the future. One such design is known as the Process Inherent Ultimately Safe (PIUS) reactor [A]. As in a conventional fission reactor, heat from its uranium core is transferred via primary coolant water to a steam generator, which drives a steam turbine. Significantly, the core and coolant loop are submerged in boron-saturated water (the element boron "poisons" the nuclear fission reaction). If coolant pressure falls, the higher pressure "poison" floods the core through emergency intakes, and stops fission.

Fusion energy
Hydrogen nuclei fuse with one another at temperatures of 200 million °F and pressures of several billion atmospheres. Under these conditions, electrons are stripped away from atoms, and the forces that hold the nuclei apart are overcome. To harness fusion energy

these extremes must be reproduced on earth. So far, the most promising results have come from ring-shaped reactors called Tokamaks [B]. Inside the Tokamak, a mixture of the hydrogen isotopes tritium [1] (the nucleus of which has one proton and two

neutrons) and deuterium [2] (one proton and one neutron) is bombarded with beams of ions, intense radio emissions, and 700 MW electrical pulses. These heat the hydrogen, whose atoms dissociate into a plasma – a hot mixture of positively charged nuclei and negatively charged electrons [3]. The plasma is too hot to be contained by any known material, and is confined by powerful magnets [4], which set up an intense field [5]. The charged particles of the plasma spiral along the magnetic field [6], but cannot stray away from it.

The "naked" nuclei in the plasma fuse [7], producing fast-moving helium nuclei (two protons, two neutrons [8] and single neutrons [9] These products collide with a "blanket" [10] around the plasma, generating heat, which is collected by a heat exchanger [11]. This heat is used to make steam from water [12]: the steam drives turbines coupled to electricity generators.

far-reaching. Three Mile Island and Chernobyl did massive damage to the reputation of the nuclear industry, and efforts are being made to find new, safer reactor designs. However, the real solution to the pollution and safety problems of nuclear power is to minimize the amounts of long-lived waste and radiation generated. Increases in efficiency are made with each new reactor design, but the ultimate dream of nuclear scientists is to harness nuclear fusion – the energy that powers the sun. In this process, instead of being split apart, atoms are joined together, releasing vast amounts of energy in a self-sustaining reaction. The raw ingredients most suitable for fusion are the hydrogen isotopes *deuterium* (abundant in water) and *tritium* (which can be derived from the element lithium). At present, fusion has been achieved for only a matter of seconds, and commercial fusion power is thought to be several decades away.

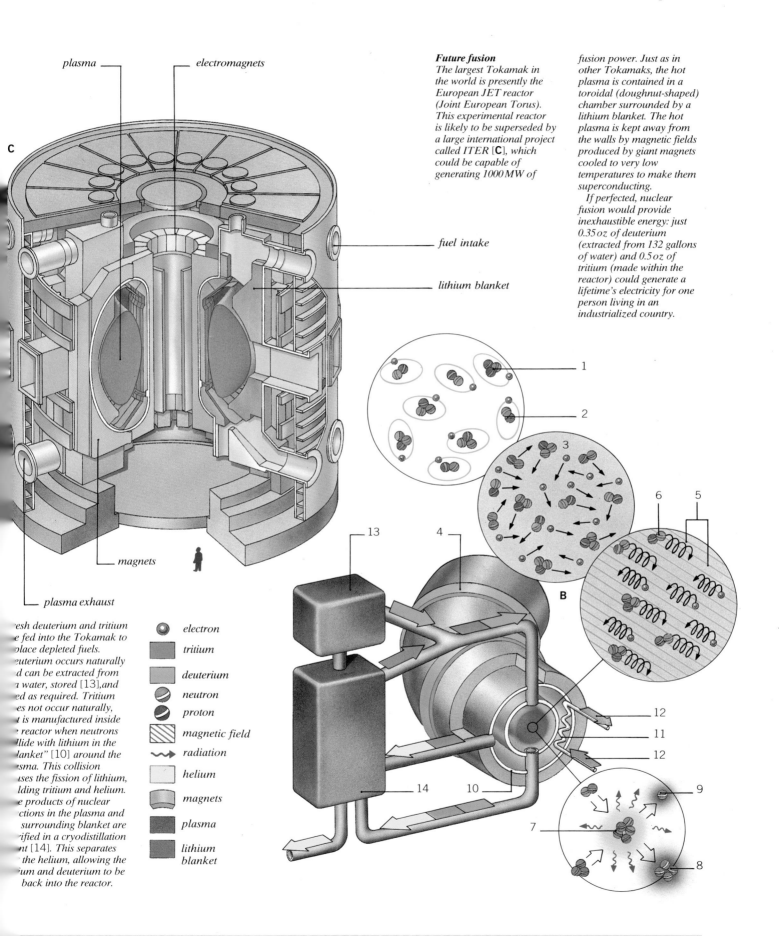

plasma

electromagnets

C

fuel intake

lithium blanket

magnets

plasma exhaust

esh deuterium and tritium
e fed into the Tokamak to
place depleted fuels.
euterium occurs naturally
d can be extracted from
water, stored [13], and
ed as required. Tritium
es not occur naturally,
t is manufactured inside
e reactor when neutrons
lide with lithium in the
lanket" [10] around the
sma. This collision
ses the fission of lithium,
lding tritium and helium.
e products of nuclear
ctions in the plasma and
surrounding blanket are
ified in a cryodistillation
nt [14]. This separates
the helium, allowing the
um and deuterium to be
back into the reactor.

○ electron

▬ tritium

▬ deuterium

◑ neutron

◐ proton

▨ magnetic field

〜 radiation

▭ helium

▬ magnets

▬ plasma

▬ lithium blanket

Future fusion
The largest Tokamak in
the world is presently the
European JET reactor
(Joint European Torus).
This experimental reactor
is likely to be superseded by
a large international project
called ITER [C], which
could be capable of
generating 1000 MW of
fusion power. Just as in
other Tokamaks, the hot
plasma is contained in a
toroidal (doughnut-shaped)
chamber surrounded by a
lithium blanket. The hot
plasma is kept away from
the walls by magnetic fields
produced by giant magnets
cooled to very low
temperatures to make them
superconducting.
  If perfected, nuclear
fusion would provide
inexhaustible energy: just
0.35 oz of deuterium
(extracted from 132 gallons
of water) and 0.5 oz of
tritium (made within the
reactor) could generate a
lifetime's electricity for one
person living in an
industrialized country.

1
2
3
6   5
B
13
4
12
11
12
14   10
7
9
8

Harnessing the Sun
How solar energy is collected and used

In 1981, *Solar Challenger,* an aircraft powered solely by sunlight, successfully flew across the English Channel. Covered with more than 16,000 solar cells, each one converting light directly into electricity, this unwieldy craft graphically demonstrated the potential of solar energy. Megawatt power stations based on solar cells could be in service within two decades; but more conventional methods of trapping sunlight already provide significant amounts of energy in the southwestern United States and many Mediterranean countries.

In the space of one second, the sun gives out enough energy to meet the annual electricity demand of the United States 13 million times over. Most of this energy radiates out into space; some is absorbed by the earth's atmosphere; and only a tiny fraction reaches our planet's surface in the form of visible and infrared light. When it reaches the surface, this energy has a low intensity – around 200 watts per square meter – and is spread out over a huge area, making it difficult to collect. Solar energy is, however, an enormous resource, and the technology to exploit it will inevitably develop as fossil fuel reserves run down.

Energy in focus
Methods of collecting solar energy fall into three broad categories – *photovoltaic, passive,* and *active.* Photovoltaic systems use *solar cells* made of a semiconductor material to convert light directly into an electrical current. The earliest solar cells, developed by space scientists as a power source for satellites, were expensive to manufacture, costing more than US$2000 per watt of output. Today's versions cost less than US$5 per watt, comparing favorably with the costs of commissioning nuclear power stations.

Passive solar energy systems rely on the absorption of heat by structures with no moving parts. These can provide a low-tech, low-cost way of trapping the sun's energy. The solar oven, for example, is no more than an insulated

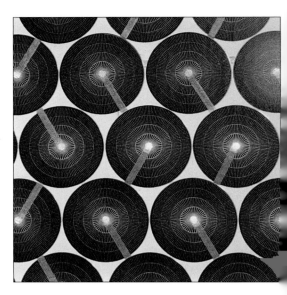

Solar cells (right)
Individual photovoltaic cells are combined in panels, and are put to work in everyday devices such as calculators, watches, and portable radios. They also provide electrical power to settlements in remote areas, and to satellites in orbit around the earth.

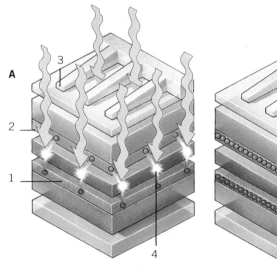

Photon traps
*A photovoltaic (solar) cell [**A**] consists of two silicon semiconductors, one of which "tends" to collect positive charge [1], the other negative [2]. The silicon layers are sandwiched between two metal contacts, and covered with a protective grid [3]. When a photon of light strikes the junction between the semiconductors [4], it dislodges an electron from the region, leaving an area of positive charge, known as a "hole." Electrons and holes are attracted to their respective semiconductors. When linked to an electrical load [5], a current flows between them.*

Sun farm (left)
This solar "farm" consists of 1818 mirrors, each measuring 23×23 ft, which focus sunlight onto a huge collector atop a 300 ft tower. Water pumped through the collector is heated to more than 930°F. The resultant steam drives a turbine, generating up to 10 MW of power for eight hours a day.

Connections: Power Stations 106 Renewable Energy: The Earth 114 Electricity Transmission 116 Satellites 172 174 176 Principles 224 234 240 252

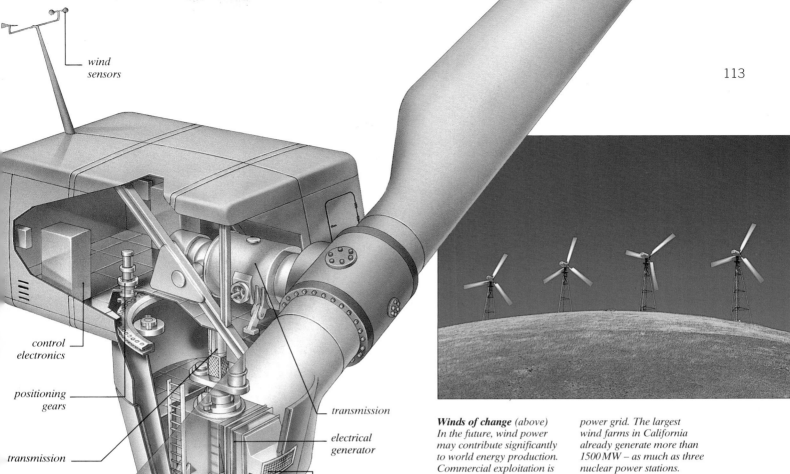

wind
sensors

control
electronics

positioning
gears

transmission

B

transmission

electrical
generator

cooling air
in/out

*Modern wind turbines have
sensors to detect wind speed
and direction. This
information is processed
and used to control a drive
gear that swivels the entire
assembly so as to position
the rotor into the wind. The
pitch of the blades may also
be changed to optimize the
speed of rotation.*

Winds of change *(above)
In the future, wind power
may contribute significantly
to world energy production.
Commercial exploitation is
likely to center on "wind
farms" – arrays of
thousands of turbines sited
on high ridges, sea coasts, or
open plains, and feeding
electricity directly into the*
*power grid. The largest
wind farms in California
already generate more than
1500 MW – as much as three
nuclear power stations.
    Wind farms, however, are
not a panacea and they do
have environmental
impacts, causing noise and
"visual" pollution, and
endangering bird life.*

Modern windmills
*Wind turbines [**B**] have
paired airfoil-shaped blades,
spanning up to 200 ft.
Commercial turbines can
produce several hundred
kilowatts of electrical power.
The blades are made from a
weather-resistant, strong,
light material, such as wood
impregnated with epoxy
resin. The rotors turn at
around 20 to 50 rpm and are
coupled to a generator
through a set of gears.*

box with a black interior and glass lid. In constant sunlight, it can reach 266°F – hot enough
to cook food. In developing countries, the use
of this simple device gives considerable savings
in fuel wood, and in time spent searching for it.
    Active solar energy systems are more versa-
tile. A fluid – often water – passes through a
solar collector, where it is heated, and is then
channeled to wherever it is needed. Small-scale
systems, comprising arrays of black water-con-
ducting pipes sandwiched between two double
layers of glass, already provide 70 percent of
hot water in some Mediterranean countries.
    Large-scale active solar energy systems use
thousands of parabolic mirrors to focus light on
to a central collector. Water pumped through
the collector is vaporized, producing pressurized
steam, which can then be made to drive turbines. Such solar power stations could generate
100 MW of power – enough for 30,000 homes –
from a mirror area of 100 acres.

Ancient technologies revived
Sunlight is naturally converted into mechanical
energy in the form of wind currents, which
result from the uneven heating of air above
land and sea. Windmills capable of extracting
this energy were among the first machines ever
built, and have been used to grind grain and
pump water for more than 5000 years. Today,
the windmill (or wind turbine) is being revived
and refined, and promises to play an important
role in future power generation. In contrast to
traditional multisail windmills, modern wind
turbines have just two or three airfoil-shaped
blades, and computerized control systems,
allowing them to operate at high efficiency in
strong winds.

Meeting demand
Although solar energy systems produce abundant "clean" electricity, they do not always
generate the power where or when it is needed.
Storing electrical energy in batteries is ineffi-
cient and expensive, so other solutions to this
problem have been sought. Perhaps the most
promising is the use of the electricity generated
to split, or *electrolyze*, water into hydrogen and
oxygen. These products can be liquefied for
easier storage and transportation, and recom-
bined in a *fuel cell* to produce electrical power
on demand. Alternatively, they can be burned
in an internal combustion engine without producing toxic or greenhouse gases.

Renewable Resources

How energy is generated from the environment

The industrialized world demands plentiful energy in a highly concentrated form. But the fossil fuels on which we rely so heavily today are a finite resource: global coal reserves are likely to last no more than 350 years, and oil extraction will become uneconomic by the year 2035. In the next millennium, the world's thirst for power will increasingly be satisfied by harvesting the energy that flows naturally through the environment. Power stations that exploit these renewable sources emit little waste heat and few pollutants, but are not without environmental cost.

The sun is a giant nuclear furnace that bathes our planet in free energy. Most methods of generating renewable power depend ultimately on solar energy, collected directly in the form of heat and light, or indirectly, by tapping into the weather cycle, which is driven by solar radiation. Renewable energy can also be generated by harnessing the gravitational force of the moon conducted through the twice-daily rise and fall of the tides; or by utilizing the heat of the earth's core, which breaks through to the planet's surface at certain locations.

Watts from water

Hydroelectric power is a form of recycled solar energy. The sun evaporates surface waters, which rise, cool, and fall again as rain and snow, filling streams and rivers. When river water flows down a steep gradient it is channeled through a series of pipes and made to drive sets of turbines connected to electrical generators.

The power generated by a hydroelectric station depends both on the volume of water flowing through the turbines and on the vertical distance through which the water drops, known as the *head*. High-head stations, in which water falls through 500ft or more can generate large amounts of power from small volumes of water. For this reason, mountainous countries, such as Norway and Switzerland, can rely heavily on hydroelectric power (99 and 75 percent of total electrical power generation respectively).

Low-head sites, in which a large volume of water falls a short distance (65ft or less), can also provide abundant power. At such locations, concrete dams are built to contain the flow, thereby increasing the head and consequently the power output.

In the past, hydroelectric dams have been built on a massive scale: the Aswan High Dam on the Nile River in Egypt is 365ft high, 2.4 miles long, and forms a reservoir (Lake Nasser) with an area of 1930 square miles. It generates 2100MW of power – as much as two nuclear power stations – and provides many additional benefits to the economy. Large-scale dam projects, however, do have environmental costs: they displace people and disrupt traditional patterns of life. And in some cases, silt accumulation in the reservoir drastically reduces the working life of the plant. To resolve these problems, new hydroelectric stations are commonly built on a smaller scale, supplying just a local village or small town with electricity.

Oceans of energy
Energy stored in warm sea water can be extracted in an Ocean Thermal Energy Conversion (OTEC) plant [A]. Warm surface water [1] is drawn into a vacuum chamber [2] and pumped through nozzles [3] that convert it into fine droplets. The droplets quickly evaporate, forming steam, which drives a low-pressure turbine [4] linked to an electrical generator [5]. The steam then passes over two sets of condensers, each cooled by water pumped up from the ocean depths [6]. In the first condenser [7], steam comes into contact with pipes carrying cold water. Water condensing on the surface of these pipes is free of salt – a valuable by-product. The remaining steam is condensed in the second condenser [8], and pumped into the ocean [9].

Cold water is pumped from depths of half a mile or more through a long pipe supported by buoys; warm water is drawn in from the surface; recondensed water is discharged into the ocean through an outlet pipe. Future OTEC stations may be built alongside bottling plants for desalinated water: and large tanks containing cold, nutrient-rich water could support commercial fish farms.

Water wheels (right)
In a high-head hydroelectric plant, water falls through a height of at least 500ft. It is forced through a nozzle, emerging as a high-speed jet, which is directed at a series of paddles, uniformly arranged on the outer rim of a wheel – a configuration known as an impulse *turbine. The turbine rotates, driving a shaft linked to an electrical generator.*

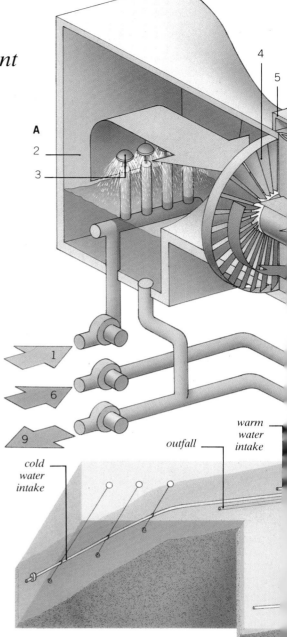

Connections: Power Stations 106 Renewable Energy: Sun and Wind 112 Electricity Transmission 116 Principles 224 234 240 252

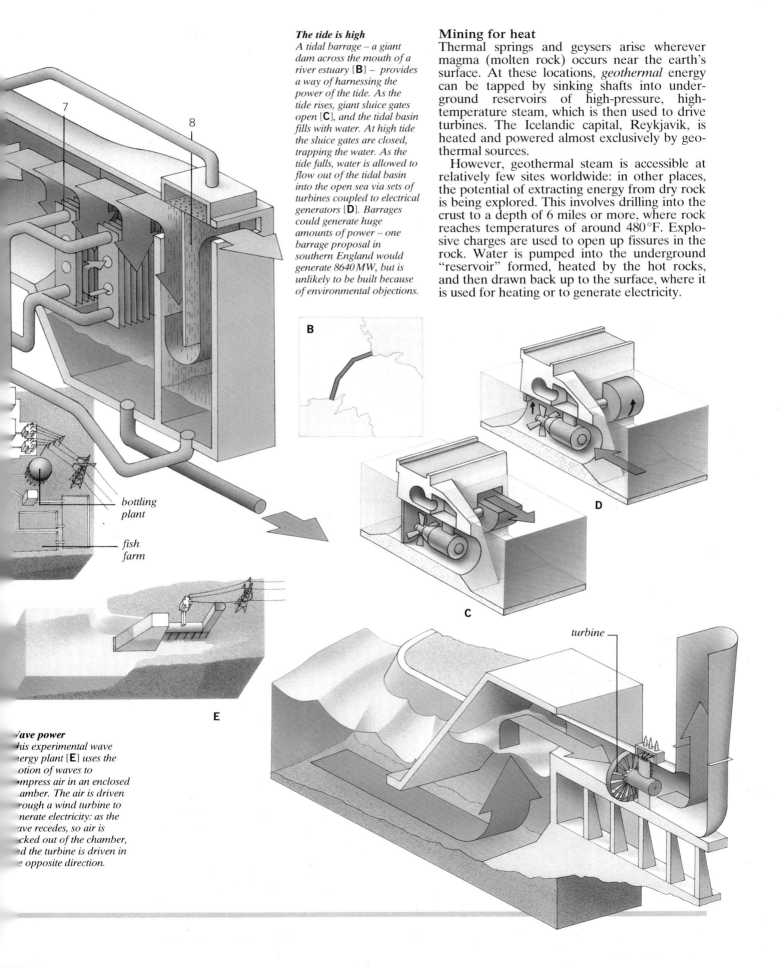

7

8

bottling plant

fish farm

The tide is high
A tidal barrage – a giant dam across the mouth of a river estuary [B] – provides a way of harnessing the power of the tide. As the tide rises, giant sluice gates open [C], and the tidal basin fills with water. At high tide the sluice gates are closed, trapping the water. As the tide falls, water is allowed to flow out of the tidal basin into the open sea via sets of turbines coupled to electrical generators [D]. Barrages could generate huge amounts of power – one barrage proposal in southern England would generate 8640 MW, but is unlikely to be built because of environmental objections.

Mining for heat

Thermal springs and geysers arise wherever magma (molten rock) occurs near the earth's surface. At these locations, *geothermal* energy can be tapped by sinking shafts into underground reservoirs of high-pressure, high-temperature steam, which is then used to drive turbines. The Icelandic capital, Reykjavik, is heated and powered almost exclusively by geothermal sources.

However, geothermal steam is accessible at relatively few sites worldwide: in other places, the potential of extracting energy from dry rock is being explored. This involves drilling into the crust to a depth of 6 miles or more, where rock reaches temperatures of around 480°F. Explosive charges are used to open up fissures in the rock. Water is pumped into the underground "reservoir" formed, heated by the hot rocks, and then drawn back up to the surface, where it is used for heating or to generate electricity.

B

D

C

turbine

E

Wave power
This experimental wave energy plant [E] uses the motion of waves to compress air in an enclosed chamber. The air is driven through a wind turbine to generate electricity: as the wave recedes, so air is sucked out of the chamber, and the turbine is driven in the opposite direction.

Corridors of Power
How electricity travels from station to socket

Like surging tides, the ebb and flow of electricity demand has daily highs and lows. From the early hours of the morning, when most people are asleep and factories are closed, demand can leap to the peak of the early evening when lighting, heating, and cooking power is needed. Meeting these needs is a grid of interconnected power stations, being brought on-line as and when they are needed. The electricity they generate is transported across hundreds of miles at 400,000 volts, before reaching homes and workplaces.

Electricity is in greatest demand in cities, where it is used for lighting, heating, and industrial machinery. However, almost all power stations are in remote, sparsely populated areas. Carrying the power from the stations to the users is a network, or *grid,* of high voltage electricity cables, connecting the many different users in a way that allows their changing demands to be met from a variety of sources.

People use more power in the cold winter months than in the summer. Within any 24-hour period, more electricity is used in the early evening than in the small hours of the morning. And power demands can vary from minute to minute, for instance when stoves are switched on at the end of a major televised event. Grid managers have to plan for these different cycles, making sure that extra capacity is available at short notice.

The largest power stations can take hours or even days to build up to peak operation, and so are kept running continuously. When extra power is needed, gas-turbine-powered generators can be called upon which deliver 200 MW a few minutes after starting. No batteries exist that are capable of storing the surplus energy generated at periods of low demand in the early morning. *Pumped-storage stations* are hydroelectric plants that use this unwanted electricity to pump water from one reservoir to another higher up a hill. The water is released through hydroelectric generators when power is needed once again.

A multitude of voltages

An electric current heats up a conductor as it passes along it, wasting energy. Electrical power (the amount of energy transmitted each second) is equal to the current in a wire times the voltage across it. So to minimize losses in the long stretches of cable between power stations and towns, transformers are used to convert the high-current low-voltage output of generators into low-current high-voltage 400 kV electricity for distribution. Once the supply nears its users it is stepped down to 33 kV for local distribution. Further transformers cut this down to the 110 V (US) or 230 V (Europe) of the domestic supply.

The voltages used in the grid are so high that when they are disconnected a highly dangerous arc is formed that would weld a normal switch together. Special circuit breakers avoid this using powerful blasts of gas to blow the white-hot plasma out from between the contacts.

large heat losses in the wires transmitting power and thus wastes electricity. To avoid these losses, a transformer just outside the station steps up this voltage to 400 kV for transmission across country. Ceramic insulators are needed (right) to insulate the supporting pylons from this high voltage.

Large circuit breakers [B] are used to switch the high-voltage supply on and off. At the moment that the two switch terminals are separated, a blast of gas is fired up the spine of the circuit breaker, safely blowing away the electric arc that inevitably forms between the terminals.

Electrical grid
A typical electricity supply grid [A] links a variety of power sources with many different consumers. Power stations, whether coal-, oil-, or gas-burning, nuclear, or hydroelectric, produce electricity at a low voltage and high current. However, a high current results in

terminals

air storage

pump

power station

B

20 kV

circuit breakers

A

transformer

high voltage terminals

low voltage terminals

radiator pipes

cooling oil

outer coil

inner coil

laminated iron core

C

Step up, step down
Electricity is generated as three sets, or phases, of alternating current (a.c.), the voltage changing from positive to negative 50 or 60 times a second. This form of electricity can easily be converted from high to low voltage and back using a transformer [C]. Inside is a

core made from sheets of iron. Each of the three legs carries two nested coils of wire. When an a.c. supply is fed through the inner coil, it sets up a magnetic field in the iron core. This in turn induces a second alternating current in the outer coils. The ratio of the voltages in the two coils is that of the

number of turns in each. Near towns, the supply is stepped down to 275 kV, then again to 33 kV. At this voltage it is sent to large factories and railroad systems. Another step down reduces the voltage to the domestic 230 or 110 V, each of the three phases sent to every third house.

Connections: Power Stations 106 Nuclear Power 108 110 Renewable Energy 112 114 Principles 226 244 246 248 250

Pumped storage station

At times, more electricity is generated than there is demand for it. No batteries are big enough to store this surplus power, but it can be used to pump water from a low-level reservoir to one farther up a hillside [D]. The demand for electricity can suddenly surge, especially during evenings. Unlike a normal coal-powered station, which can take up to an hour to warm up, the stored water can be released instantly. It flows downhill along the penstock and through the turbine. This drives the motor in reverse, making it behave like a generator and produce electricity.

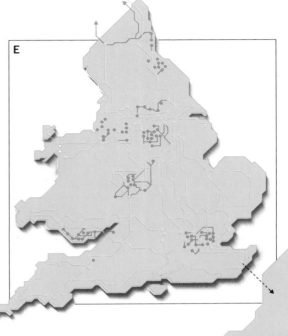

400 kV lines

275 kV lines

- - - - - - - - - - -

d.c. link to France

● ● stations and substations

○ pumped-storage stations

National Grid

The varying power demands of a country can be met most efficiently by linking all the power stations and urban centers in a national grid [E]. This particular map shows that of England and Wales, with 400 kV lines running between transformer substations. In cities 275 kV lines proliferate, offering a variety of routes so that power cuts are unlikely. A link with France carries direct current (d.c.) to avoid having to synchronize the two systems.

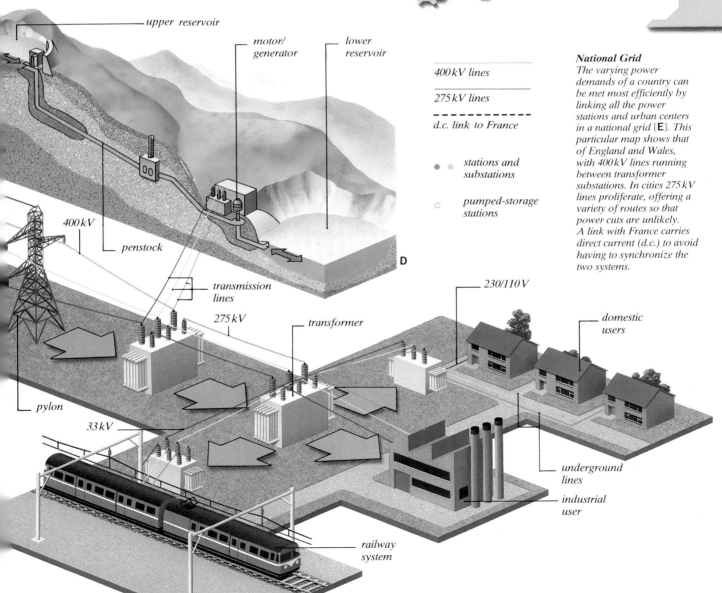

upper reservoir

motor/generator

lower reservoir

400 kV

penstock

pylon

transmission lines

275 kV

transformer

33 kV

railway system

230/110 V

domestic users

underground lines

industrial user

D

Spanning the Abyss

How bridges are constructed

The Ponte Vecchio in Florence has carried traffic over the Arno River since 1345, and has become one of the world's great aesthetic landmarks. Today, architects strive to match its elegance and longevity while providing for an ever-increasing flow of traffic. By applying the principles of physics and using stronger, lighter materials, they have built structures remarkable in their own right. The Golden Gate Bridge in San Francisco covers 1.7 miles, with a center span of 4200 feet. Its two support cables are more than three feet thick and contain 80,000 miles of steel wire.

There is no "ideal" design for a modern bridge. In each case, the architect considers numerous factors – the length of span needed, type of traffic to be carried, geology of the area, cost of the materials, and aesthetic suitability – before arriving at an acceptable compromise. Although every bridge is unique, designs fall into four main categories – beam, arch, cantilever, or suspension – which differ in the way that they bear the combined weight of the bridge itself and the vehicles using it (known as the *load*).

In a beam bridge, the load is carried by a horizontal concrete or steel beam resting on two pillars or piers. The main drawback of this design is that the beam bends under its own weight, compressing its upper surface and stretching its lower surface, which makes it prone to cracking. The weight of the beam and the traffic it carries also tends to to pull the piers toward each other, and this pull is stronger if the beam bends. The bridge can be

Cable construction (below)
Though at first sight resembling a suspension bridge, the cable-stayed bridge is in fact a modification of the cantilever design. The deck, which comprises horizontal beams that cantilever out from piers, is supported by inclined steel cables fanning out from the top of the towers.

The cables prevent the deck from bending, and cable-stayed bridges can therefore be built with relatively thin decks, creating a sense of airiness and space. Cable-stayed designs have become increasingly popular since the 1950s.

Design alternatives
Modern beam bridges [A] are often braced with steel trusses to increase stiffness. In some variants, the roadway sits on a "beam" made of long steel girders with a square or rectangular cross section. Such box girder bridges can span more than 82 ft.

Arch bridges [B], are made of steel or reinforced concrete. The deck sits on the arch, supported by spandrel columns, or is suspended from it by cables.

Cantilever bridges [C] are often built from box girders or steel trusses. One of their main advantages is that they can be erected from the shore outward without any temporary supports.

Suspension bridges [D] allow the longest spans of all. On completion in 1998, the Akashi Strait Bridge in Japan will have a center span of 1.25 miles.

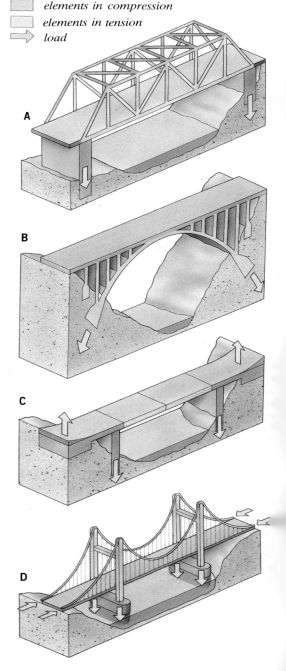

☐ elements in compression
☐ elements in tension
⇨ load

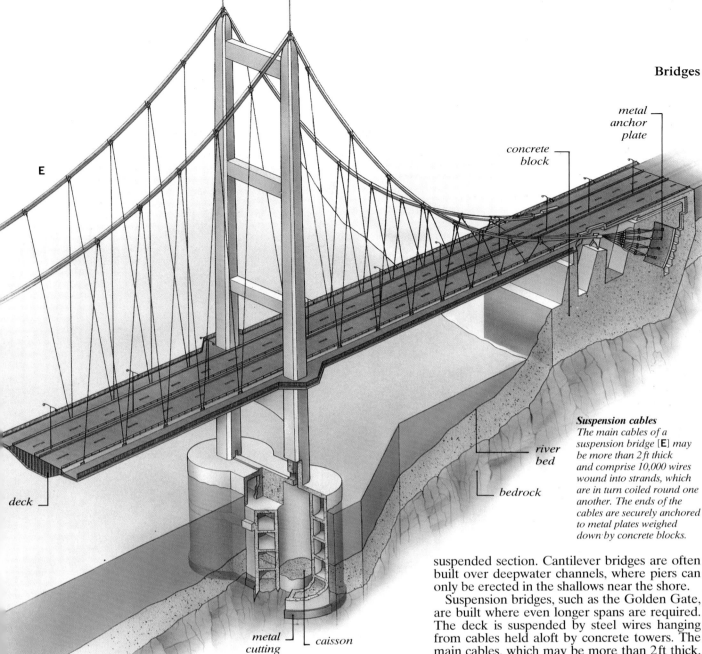

E

metal
anchor
plate

concrete
block

deck

metal
cutting
edge

caisson

river
bed

bedrock

Suspension cables
The main cables of a
suspension bridge [E] may
be more than 2 ft thick
and comprise 10,000 wires
wound into strands, which
are in turn coiled round one
another. The ends of the
cables are securely anchored
to metal plates weighed
down by concrete blocks.

Firm foundations
The towers of a typical
suspension bridge [E] are
hollow reinforced-concrete
structures, sometimes more
than 330 ft in height.
Because they are anchored
to the riverbed, their
foundations must be laid in
a special way. A caisson –
a large concrete chamber,
open at the bottom – is first
sunk. Compressed air is
pumped in, allowing
engineers to remove rock
and sediment from within.
The caisson is weighted
down and gradually sinks to
the required depth. Its base
is then sealed and its interior
chambers filled with
concrete to provide a solid
foundation for the tower.

strengthened by increasing the number of piers
and the stiffness of the beam. Beams of modern
bridges are made from concrete, or from frame-
works of steel girders called *trusses*, which add
strength without increasing weight. The bridge
can then be built with longer spans and fewer
piers, reducing the cost of construction.

Arch bridges are built where it is difficult to
construct piers – for example, over deep gorges
or fast-flowing rivers. The arch is strong
because it is always in compression – the forces
acting on the arch squeeze it together rather
than pulling it apart. The weight of the load is
converted into a sideways force, which is transmitted via the ends of the arch to abutments.

A cantilever bridge is designed like two
springboards facing one another. Each "board"
is firmly anchored to the shore and supported a
short distance away from the shore by a pier.
The two "boards," which carry the roadway or
deck, are usually linked in the middle by a short

suspended section. Cantilever bridges are often
built over deepwater channels, where piers can
only be erected in the shallows near the shore.

Suspension bridges, such as the Golden Gate,
are built where even longer spans are required.
The deck is suspended by steel wires hanging
from cables held aloft by concrete towers. The
main cables, which may be more than 2 ft thick,
are firmly embedded in concrete on the shores.

Material considerations
The maximum span of any type of bridge
depends in part on the strength-to-weight ratio
of the materials from which it is built. Cantilever bridges made from high-tensile steel
trusses can economically cover 700 yd in a single
span and steel arches 900 yd. Suspension bridges
span distances of more than 1500 yd, because
steel cables are stronger in tension than beams.

Concrete is an inexpensive and convenient
material. It is strong in compression but weak in
tension, making it suitable for arches and piers,
but less so for beams and cantilevers. However,
since the 1950s, the technique of *prestressing*
has increased the versatility of concrete. Prestressed concrete is cast around a series of steel
wires held at high tension by hydraulic jacks.
When the jacks are released the wires pull the
concrete together, greatly increasing its strength
under tension. Many modern beam and cantilever bridges use prestressed concrete and are
surprisingly slender in their proportions.

Going Underground

How a tunnel is excavated

On December 1, 1990, Great Britain ceased to be an island. The joining of the two ends of the Channel Tunnel connected England to France for the first time in 10,000 years, and opened up a new era in rail transportation. Running 165 feet below the seabed, the 30-mile-long tunnel is actually made up of three separate shafts, excavated by giant boring machines weighing 500 tons. A total of 13,000 workers toiled beneath the sea for 33 months to complete the giant project, removing enough rock to fill the Great Pyramid at Giza three times.

Since the early 1800s engineers have dreamed of driving a tunnel beneath the English Channel. Past proposals foundered due to a lack of technical ability, funding, or political will (and the fear of a French invasion), and it was not until May 1994 that the tunnel, the longest undersea passage in the world, was finally opened to traffic. It had cost more than US$15 billion to build.

Before excavation began, alternative routes were assessed using *seismic* surveying techniques to map the layers of rock, or *strata*, beneath the channel, and temporary drilling platforms to drill core samples from depths up to 260 ft below the seabed. These surveys revealed one particular stratum around 165 ft down that was an ideal tunneling medium. The layer was composed of chalk marl, a mixture of chalk and clay soft enough to cut through easily, yet strong enough to avoid collapse. Importantly, marl does not usually contain one of the tunneler's enemies, *aquifers* – pockets of highly porous rock containing large quantities of water. When a tunnel breaks into one of these, almost instant flooding can occur.

Journey to the center of the earth

Of the three parallel tunnels, the 15.7 ft-diameter service tunnel, which runs between the two main rail tunnels, was drilled first. This enabled the state of the chalk marl layer to be assessed before the main bores went through it. The giant tunnel-boring machines were preceded by narrow drills that probed 30 ft ahead, searching for unexpected aquifers and other hazards.

Tunnel-boring machines, or TBMs, are 655 ft-long mechanical earthworms. At the front are cutting heads, drums more than 26 ft wide. Their faces, which are fitted with cutting picks (or diamond-studded wheels), turn 2 to 3 times a minute against the chalk, chipping out *spoil*. Just behind the cutting head are hydraulic gripping pads that press against the sides of the newly dug tunnel, providing a firm grip against which the cutting head can push forward. Slightly farther back, huge robot arms swing curved concrete segments into place, building up the tunnel lining. Other sections *grout* the lining (filling in gaps between the chalk and concrete) and lay the temporary narrow-gauge railway used to transport spoil to the surface. Lasers keep the vast machines on course: even the slightest deviation is corrected by using the eight hydraulic rams to steer the cutting head.

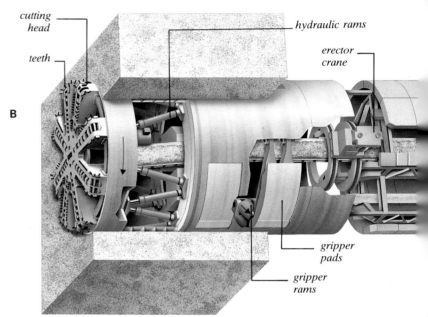

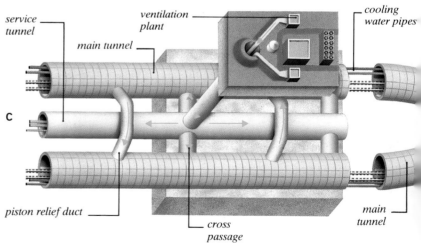

Continental link
The channel tunnel [C] *is in fact three separate bores – two* running tunnels *with railroad tracks, and a third* service tunnel *for access and evacuation. Around 2500 cubic feet of ventilation air are pumped into the service tunnel each second through a shaft from the surface. This flows into the railroad tunnels through cross passages, which link all three bores every 1230 ft. In addition, piston relief ducts every 820 ft balance out the pressure wave that a fast-moving train creates. A few miles from each coast is a crossover cavern more than 60 ft wide and 560 ft long, where trains can cross from one track to another. Two main types of train use the tunnel: the passenger-carrying Eurostar and automobile-transporting Le Shuttle. The service tunnel has its own emergency vehicles, automatically steered by following wires buried beneath the road.*

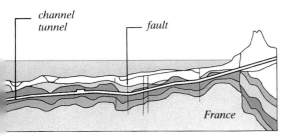

channel tunnel — fault

France

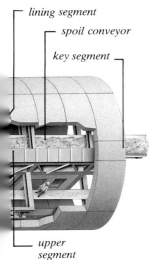

lining segment

spoil conveyor

key segment

upper segment conveyor

Extending the land

More than two hundred million cubic feet of spoil were removed by TBMs in the excavation of the Channel Tunnels. On the French side, the spoil was first mixed with water to form a slurry before being pumped to fill in lagoons that were later grassed over. In England, the waste was used to extend the land at the foot of the famous white cliffs of Dover, providing a site for the air-conditioning and pumping plants needed for the tunnel. The ground on the French side of the Channel is heavily faulted, and thus more likely to let water flood into a newly dug tunnel. To guard against this, the cutting heads of the TBMs had watertight seals fitted just behind them, and the tunnels on the French side were lined with stronger cast iron.

At one-third and two-thirds of the way under the sea, the two rail tunnels join for a short stretch in vast *crossover caverns*. In these, trains are able to cross from track to track, so that only one-sixth of the tunnel system has to be closed down for maintenance at any one time. The English crossover was excavated using the *New Austrian Tunneling Method*, in which *roadheaders* – huge drills on adjustable booms – clawed out the chamber a section at a time. The newly exposed rock was sprayed immediately with *shotcrete*, a type of quick-setting concrete. Where the rock was more likely to fracture, rock bolts were inserted, and glued in place with a strong epoxy resin grout.

TBM in place while other rams push the cutting head forwards and keep it on course. Every 5 ft, an erector crane *puts a ring of six concrete lining segments in place, locked in by a seventh key segment. The segments arrive on two conveyor belts at the top and bottom of the TBM.*

Inching forward
The main bores of the channel tunnel were excavated by Tunnel-Boring Machines (TBMs). These are huge machines, almost 30 ft high and dragging a service train 850 ft long. Behind this is a railroad system that carries waste and supplies between the cutting face and surface. At the front of the TBM [B] is the rotating cutting head, the hard-tipped teeth of which cut into the chalk marl stratum that the tunnel runs through. The spoil – freshly cut chalk – falls to be carried away on a conveyor belt. Gripper pads, pushed against the walls by hydraulic rams, hold the

The perfect level
The rock strata beneath the English Channel [A] are ideal for tunneling. One layer in particular, of chalk marl, is soft enough to be cut through quickly, but importantly it is also impervious to water. In addition, it is strong enough not to collapse before the tunnel is lined. After running all the way across the Channel, at the French end the marl is cut by several geological faults before diving downward. This meant that the tunnels dug from France progressed slowly at first because they had to be lined with bolted-together sections of waterproof cast iron.

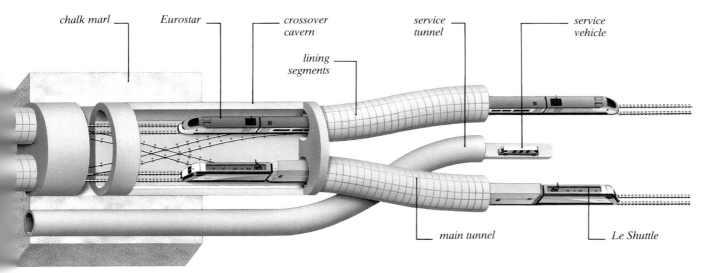

chalk marl   Eurostar   crossover cavern   service tunnel   service vehicle

lining segments

main tunnel   Le Shuttle

Vanishing point (left)
This dramatic view of the inside of a main railroad tunnel clearly shows the large pipes at the right that carry cooling water to remove the heat generated by passing trains. Concrete ledges on both sides can be used in an emergency to walk from a stranded train.

The High Life

How skyscrapers are built

Dominating the Chicago skyline, the 1454-foot Sears Tower is the tallest building in the world. But how long the 20-year-old, 110-story skyscraper will continue to hold the record is uncertain. By the end of the century, it is estimated that 300 million people will live in 21 *megacities,* in which demand for space will be met by building even higher upward. A number of Japanese construction companies have developed plans for futuristic buildings that will top one mile, but a skyscraper of 2625 feet could be built today using tried and tested technologies.

For many years, fabricated steelwork has provided the structural backbone for the tallest buildings in the world. In the last decade, high-strength, *prestressed* concrete has increasingly been used to make skyscraper frames, but the tallest of the concrete edifices – the 65-story, 945 ft tower at 311 South Wacker in Chicago – is still no match for its steel-framed neighbor, the Sears Tower.

A sure footing

The foundations of most skyscrapers are laid in a similar way. First, the site is excavated – a depth of soil beneath the proposed building is scooped or blasted out, and the walls of the excavation are supported by boards pinned into place by steel piles known as *soldiers.* Some deep excavations may reach soil or rock capable of sustaining the weight of the building directly, but more often numerous broad piles must be hammered into the ground by pile drivers in order to reach layers of stronger soil. Occasionally, the soil is so weak that it must be bound together by chemical treatment: the twin towers of the World Trade Center in New York stand on soil *consolidated* in this way.

The piles are usually tied to an underground concrete *raft* in which the vertical columns of the skyscraper are anchored. The raft and piles act as anchors not only supporting the weight of the building and its contents, but also reacting against the lateral forces on the building, which are caused by wind and seismic activity. Not surprisingly, therefore, a skyscraper's foundations must be massive – 311 South Wacker, for example, rests on an 8 ft-thick concrete raft tied to 100 piles, ranging in diameter from 3 ft to 9 ft and up to 112 ft in length.

Tall orders

When high-rise construction was in its infancy, most buildings were of simple beam-and-column design. The structural frame was a forest of steel columns linked together by beams. This limited the building's height to around 20 stories, above which height the beams and columns would have had to be uneconomically thick to withstand any swaying movement of the building.

Heights of between 30 and 40 stories became possible with the advent of the core wall system. In this design, the building has a stiff, central core, often made from reinforced concrete walls arranged in box sections. Structurally, the core

Soaring Sears (right)
The 110-story Sears Tower in Chicago is, in structural terms, a "bundle" of steel-framed tubes (see below). At the base, nine tubes in a 3 x 3 grid occupy a total area of 226 x 226 ft. The tubes terminate at different heights; two terminate at the 15th floor; two at the 66th; three more at the 90th; and only two tubes reach the full 1454 ft-height of the building. This arrangement has the advantage of providing a variety of floor plans, as well as good resistance to side winds. The greatly reduced mass at the top of the skyscraper also makes it less prone to earthquake damage.

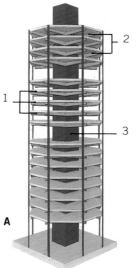
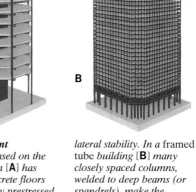
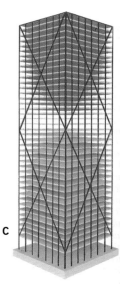

A   B   C

Structured ascent
A skyscraper based on the core wall system [A] has lightweight concrete floors [1] supported by prestressed concrete cantilevers [2]. These in turn are attached to a concrete box section at the building's core [3]. In taller buildings, peripheral columns provide additional

lateral stability. In a framed tube building [B] many closely spaced columns, welded to deep beams (or spandrels), make the structure equivalent to a hollow cantilever. When subjected to lateral loads, such as high winds, the tube bends. Compression of the columns on the leeward side

and extension of the windward columns resists further bending. To build higher without increasing the number and width of columns (and therefore retaining maximum windo space), framed tubes are trussed [C] with diagonal beams. Main structural elements are shown in red.

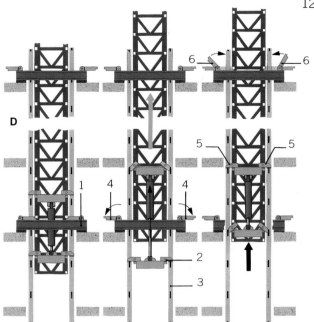

Climbing cranes

When building a skyscraper, cranes "climb up" the inside of the tower [**D**] to keep pace with the work. At first, the crane is firmly clamped to a beam on one of the floors [1]. When it needs to move up , a dog clamp attached to the crane via a hydraulic jack [2] engages with a "ladder" [3] running up the inside of the tower. The main clamp is released [4] and the jack extends, lifting the crane upward. Another dog clamp engages [5], and the jack is retracted. This is repeated until the right level is reached. A main clamp then engages to secure the crane [6].

resists bending forces on the building, and bears some or all of its weight, but the core also provides a convenient location for elevators and ducts that run services up and down. The floors are attached to the core itself, and cantilever away from it: this gives the architect freedom of design for the areas surrounding the core.

Next in design ascendancy are *framed tube* buildings. These are designed in the opposite way from core wall buildings because all the major structural components are located around the periphery, making what amounts to a huge, stiff tube. The tube carries the building's weight and provides resistance to lateral loading. Many modern skyscrapers are of this design. They are characterized by their dense structural exteriors – numerous peripheral columns linked by deep beams – but internally they give the architect a completely free rein in the layout of the large floor areas.

The upward growth of buildings has forced structural engineers to adopt new techniques to prevent or minimize seismic damage. There are two main designs of "earthquake-proof" building. In the first, a heavy concrete block on top of the building, activated by computer-controlled dampers, is shifted across the roof to balance the forces of the earthquake. In the second, the building's foundations incorporate giant rubber "shock absorbers," which allow the building to move with the seismic shocks, thus minimizing structural damage.

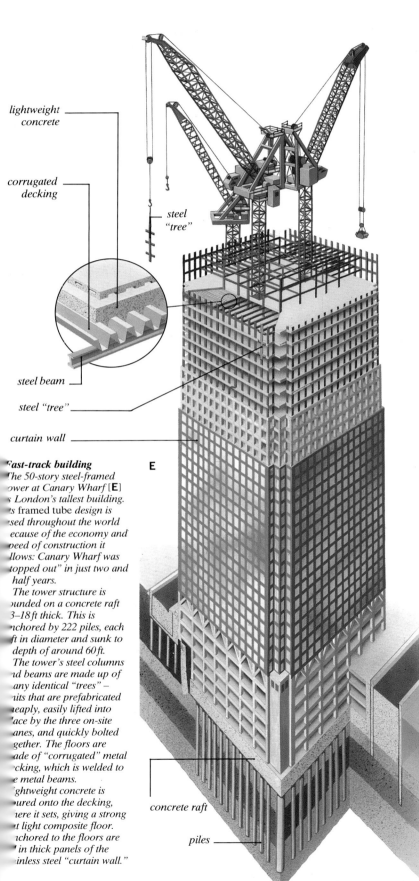

lightweight concrete

corrugated decking

steel "tree"

steel beam

steel "tree"

curtain wall

E

concrete raft

piles

Fast-track building

The 50-story steel-framed tower at Canary Wharf [**E**] is London's tallest building. Its framed tube *design* is used throughout the world because of the economy and speed of construction it allows: Canary Wharf was "topped out" in just two and a half years.

The tower structure is founded on a concrete raft 3–18 ft thick. This is anchored by 222 piles, each ft in diameter and sunk to a depth of around 60 ft.

The tower's steel columns and beams are made up of many identical "trees" – units that are prefabricated cheaply, easily lifted into place by the three on-site cranes, and quickly bolted together. The floors are made of "corrugated" metal decking, which is welded to the metal beams.

Lightweight concrete is poured onto the decking, where it sets, giving a strong but light composite floor. Anchored to the floors are in thick panels of the stainless steel "curtain wall."

A New Leaf

How paper is manufactured

The development of papermaking in China some two thousand years ago was one of the seminal events in the history of civilization. Paired with the printing press, it constituted an enormously powerful means for the storage and dissemination of knowledge. The basic process of papermaking has changed little in two millennia, although automation now allows paper to be produced in the huge quantities demanded by the printing and packaging industries. Today, a single machine is able to produce more than 330 tons of paper in a day.

A sheet of paper consists of a mat of randomly interwoven *cellulose* fibers, each no more than 4mm in length. Cellulose is a *polymer* made up of thousands of glucose molecules linked together in long, unbranched chains. It is the most abundant organic molecule on earth, being the main structural element of every plant cell wall. Cellulose fibers give plant cell walls a combination of strength and flexibility, and it is these properties that make cellulose ideal for paper manufacture.

In theory, paper can be made from almost any plant material, but the best source of cellulose fibers is wood, which contains more than 60 percent of the polymer. At present, around 50 percent of the world's forestry output goes to supply the papermaking industry. But virgin timber is not the only source of cellulose fiber for papermaking. Most paper mills use more than 50 percent recycled paper and board, as well as rags, sawdust, and woodchips.

Mechanical pulping
*Logs arriving at a paper mill [**A**] have their bark stripped and are wetted before passing to either chemical or mechanical pulping machinery. In mechanical pulping, the logs are loaded into a mill where they are pressed against a pulpstone spinning at 360rpm. The surface of the stone is covered with grains of abrasive silicon carbide. The pulp formed is first bleached, then washed.*

Chemical pulp
Debarked logs destined for chemical pulping pass to a chipper. The woodchips formed are then fed into the top of a chemical digester – a cylindrical tower up to 200ft tall. Caustic chemicals – usually sodium hydroxide and sodium sulfide – are mixed with the chips, and temperature and pressure within the digester are raised. Under these extreme conditions, the woodchips are converted into a pulp, which is then washed to remove the caustic chemicals. Modern digesters can produce 660 tons of pulp per day. Both chemical and mechanical pulps are

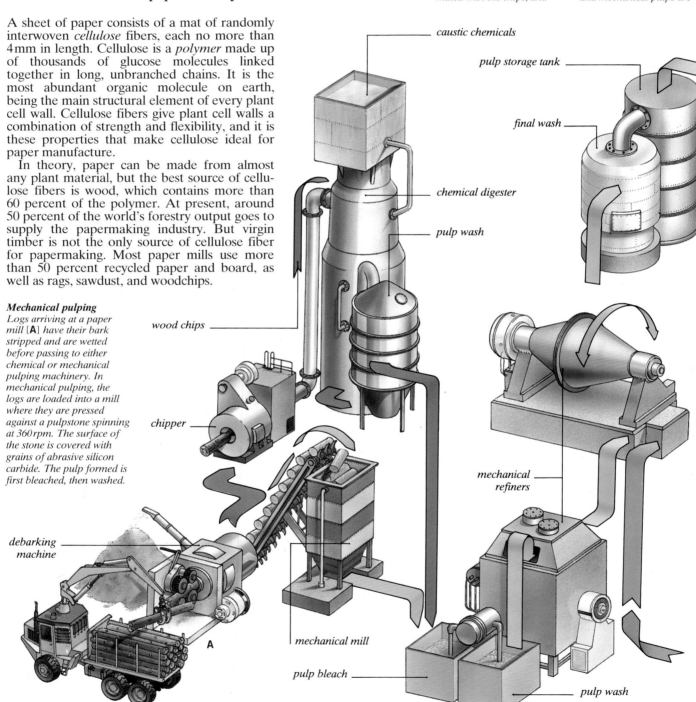

caustic chemicals

pulp storage tank

final wash

chemical digester

pulp wash

wood chips

chipper

mechanical refiners

debarking machine

A

mechanical mill

pulp bleach

pulp wash

Connections: Printing 76 Waste Treatment 130 Polymers 192 Principles 226 228

bleached with chlorine dioxide to improve brightness, and the bleaching agent is then washed out. Two mechanical refiners then beat the pulp to increase fibrillation. The pulp is given a final wash and then stored in a large vat ready for the paper machine.

At the mill, these raw materials are treated chemically and mechanically to separate and thoroughly soak their constituent fibers. The resulting *pulp* is then fed to a papermaking machine. A thin layer of pulp is spread onto a moving wire mesh to form a matted sheet of fiber. The wet sheet is compressed and dried into the finished product – a roll of paper.

The properties of different types of paper – from highly absorbent tissue to glossy "art" paper – depend on the type of pulp used and the way in which it is processed. Many writing papers use a mixture of pulp derived from softwood and hardwood trees. The former contain long cellulose fibers that knit together well to form a strong but coarse sheet: the latter provide shorter fibers that "fill in" the sheet, giving it opacity and a smooth surface.

Pulp can be made by the mechanical abrasion of wood or by digesting woodchips with caustic chemicals. Mechanical pulp is cheaper to make, but also relatively weak because individual fibers are broken into small fragments. It is used for short-life products, such as newsprint. Chemical pulping preserves the integrity of the wood fibers, making a stronger paper.

Before passing to the papermaking machine, the pulp is beaten and cleaned. Each fiber is made up of many cellulose molecules bound together. Beating the fibers partly unravels the molecules – a process called *fibrillation*. The unraveled fibers are able to form more bonds with adjacent fibers, giving a strong, dense paper, such as tracing paper. Low fibrillation gives a weaker product, such as blotting paper.

The final product is also influenced by additives. For high-quality printing and writing papers, fillers, such as clay (kaolin) or chalk, and whiteners such as titanium dioxide, are added to the pulp to make the paper smoother, brighter and more receptive to ink.

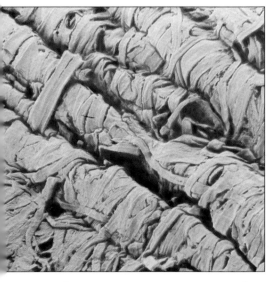

Fourdrinier machine — B — pulp — headbox — suction boxes — water out — rollers

Papermaking
Papermaking [**B**] starts as wood fibers are fed into a headbox. *This forms a wide belt of pulp, which is fed onto the moving mesh belt of a Fourdrinier machine. The wet mat of fibers moves over suction boxes that draw out water, and is then* compressed by a dandy roller. Other rollers squeeze out more water. The damp paper web is dried by passing over a series of heated cylinders. Felt belts hold the paper tightly against the cylinders. The dry paper then passes over stacks of steel rollers (calender stacks), which impart a smooth finish.

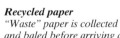

paper web — linear cellulose molecule — heated cylinders — felt belts — calender stacks — hydrogen bonds — glucose subunits — D

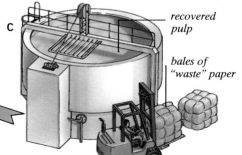

C — recovered pulp — bales of "waste" paper

Recycled paper
"Waste" paper is collected and baled before arriving at the mill. Here it is pulped and de-inked [**C**] before being mixed with virgin wood pulp. Paper cannot be recycled indefinitely because repeated pulping fragments its fibers, weakening the paper formed.

Fiber structure
A molecule of cellulose is a long chain of glucose subunits. Adjacent chains are held together by hydrogen bonds to form microfibrils [**D**]: these coil around one another to form a single fiber. Many fibers, joined by hydrogen bonds, form a sheet of paper (left).

Silicon Architecture

How silicon chips are made

Each silicon chip is a minutely etched end product of an intricate production process. Beginning with the commonest of raw materials – sand – silicon is processed and purified to produce a three-foot-long single crystal. A diamond-tipped saw slices this giant crystal into ultrathin wafers, onto which hundreds of individual chips will be etched, each containing over a million tiny electronic devices. Finally, the completed integrated circuits are separated to produce the end product – all the elements of a computer in a square smaller than a fingernail.

A pure crystal of silicon does not conduct electricity unless traces of other substances (or *dopants*) are diffused into it. Different dopants give silicon special properties, which are used to make electronic devices such as transistors and *integrated circuits* or chips.

The pure silicon needed for chip manufacture results from a lengthy refining process. It starts with silica, the main component of sand, which is heated with carbon to make 98 percent pure silicon – still too contaminated to be useful. This silicon is then dissolved in hydrochloric acid and the resulting liquid fractionally distilled, like crude oil, to separate out almost all the impurities. Heating the remaining liquid in a hydrogen atmosphere yields ultrapure silicon.

However, this silicon is in the form of a jumble of many crystals of different sizes and orientations. In the *Czochralski pulled crystal* process the silicon is melted in a large crucible into which a probe, tipped with a small *seed crystal*, is immersed. Silicon atoms in the melt attach themselves to the seed in perfect alignment with its structure while it is rotated and pulled slowly upward. In this way the seed grows into a 3 ft-long, cylindrical, single crystal.

Into the clean room

All manufacturing steps take place in what is known as the "clean room," an area where each 1 ft cube of air must contain fewer than 1000 tiny specks of dust and zero humidity. The temperature is maintained at a constant 68°F and all workers have to wear coats, gloves, masks, and overshoes. All this is necessary because even one dust particle or water droplet can ruin a batch of chip production.

The giant ingot of silicon is ground into a perfect cylinder, which is sliced by a diamond-tipped saw into *wafers* 1 mm thick. Using a slurry of particles only one-tenth of a micrometer wide, the faces of these wafers are polished to give a mirror-smooth base onto which up to two hundred identical chips can simultaneously be photoetched.

Each chip consists of more than a million interconnected transistors, diodes, and resistors. These are built up from differently doped layers of silicon, with insulating silica and metal connections. The completed chips, encapsulated in a protective plastic package and connected to spiderlike metal legs by fine gold wires, find their way into devices as diverse as wristwatches and interplanetary probes.

Crystal grower
The next step in chip manufacture is to make a single large silicon crystal. [**B**]. Pure silicon from a zone refiner [7] is broken up and melted in a crucible [8]. The crucible spins in the inert gas atmosphere inside a pressure vessel [9]. A "seed" crystal, orientated at a precise angle, is lowered into the melt on the end of a revolving metal rod [10]. The temperature of the melt is lowered so that as the crystal is slowly withdrawn from the crucible, silicon atoms attach to it in exact alignment, solidifying to form a perfect cylindrical crystal three feet long [11].

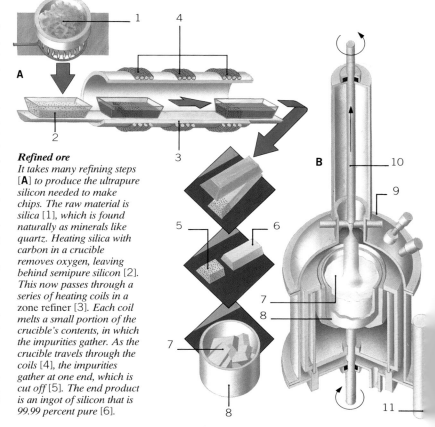

Refined ore
It takes many refining steps [**A**] to produce the ultrapure silicon needed to make chips. The raw material is silica [1], which is found naturally as minerals like quartz. Heating silica with carbon in a crucible removes oxygen, leaving behind semipure silicon [2]. This now passes through a series of heating coils in a zone refiner [3]. Each coil melts a small portion of the crucible's contents, in which the impurities gather. As the crucible travels through the coils [4], the impurities gather at one end, which is cut off [5]. The end product is an ingot of silicon that is 99.99 percent pure [6].

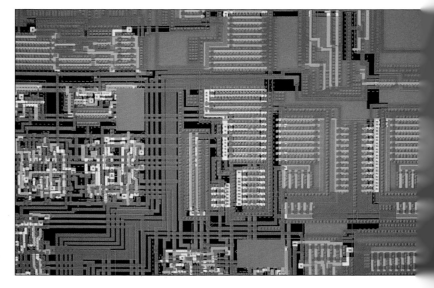

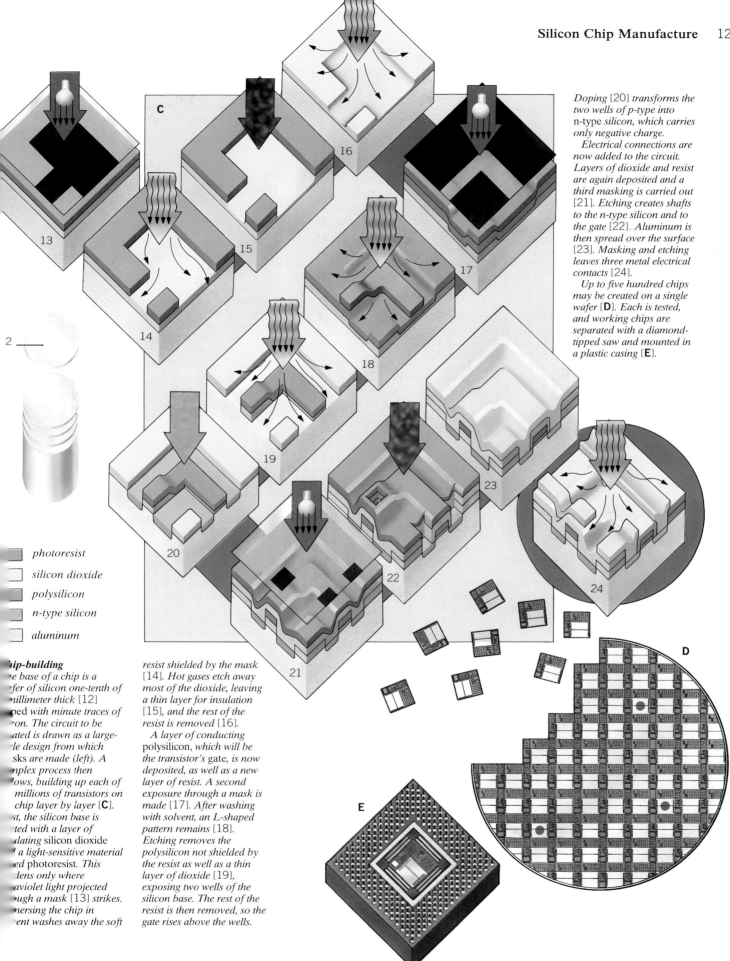

Doping [20] transforms the two wells of p-type into n-type silicon, which carries only negative charge.

Electrical connections are now added to the circuit. Layers of dioxide and resist are again deposited and a third masking is carried out [21]. Etching creates shafts to the n-type silicon and to the gate [22]. Aluminum is then spread over the surface [23]. Masking and etching leaves three metal electrical contacts [24].

Up to five hundred chips may be created on a single wafer [D]. Each is tested, and working chips are separated with a diamond-tipped saw and mounted in a plastic casing [E].

photoresist

silicon dioxide

polysilicon

n-type silicon

aluminum

hip-building

e base of a chip is a fer of silicon one-tenth of millimeter thick [12] ped with minute traces of ron. The circuit to be ated is drawn as a large-le design from which sks are made (left). A nplex process then lows, building up each of millions of transistors on chip layer by layer [C]. st, the silicon base is ted with a layer of ulating silicon dioxide a light-sensitive material ed photoresist. This dens only where aviolet light projected ugh a mask [13] strikes. nersing the chip in ent washes away the soft

resist shielded by the mask [14]. Hot gases etch away most of the dioxide, leaving a thin layer for insulation [15], and the rest of the resist is removed [16].

A layer of conducting polysilicon, which will be the transistor's gate, is now deposited, as well as a new layer of resist. A second exposure through a mask is made [17]. After washing with solvent, an L-shaped pattern remains [18]. Etching removes the polysilicon not shielded by the resist as well as a thin layer of dioxide [19], exposing two wells of the silicon base. The rest of the resist is then removed, so the gate rises above the wells.

Willing Workers

How robots manipulate and sense their surroundings

The term "robot," from the Czech word *robota* meaning "labor," was first coined in the 1920s by the writer Karel Capek. But the first practical steps in robotics did not come until the 1950s, when the "static industrial manipulator" – an early form of the robot arm – was patented. Its sophisticated successors are now commonplace, performing tasks in complex manufacturing processes. Others have been developed that mow lawns, administer drugs, and even shear sheep. However, self-contained, truly intelligent robots are still only distant dreams.

The science-fiction vision of the android – a robot in human form – has not yet become reality. However, most of today's industrial robots do in fact resemble at least a small part of the human anatomy, the arm. They feature joints that work like waists, shoulders, elbows, and wrists, and a tool-handling *end effector* for a hand. The resemblance is misleading, because robots have distinctly unhuman abilities. They are strong – able to support heavy loads without tiring – and are also highly and repeatedly accurate. This makes them ideal for performing the arduous and boring tasks involved in many production processes.

The number of movable joints or *degrees of freedom* that a robot has determines its abilities. Three degrees are needed to position the end effector anywhere in the arm's reach, and another three are required for the end effector to grasp an object or

Artificial stupidity (above)
American researchers have produced several robot insects capable of searching for and hiding in shadows – a simple task referred to as "artificial stupidity." Future robot insects may be used in industry to make checks and repairs on otherwise inaccessible machinery.

Industrial strength

*Automobile production involves repetitive tasks that must be performed accurately and at high speed. Far stronger than human workers, robots carry out paint-spraying, and component installation, as well as welding (right). A typical robot arm [**A**] turns about several joints, powered by electrical* actuators [1], *to move itself into position. Another joint at the "hand" controls the* end effector. *The robot is controlled by a computer* [2], *which is programmed via a* teaching pendant [3].

Robots commonly determine their position with incremental encoders [4]. *As a joint turns, a light cell* [5] *gives a pulse each time a line on a disk passes it. By counting the pulses the computer "knows" the joint's position at all times.*

The same robot may be used for different tasks by changing its end effector. Vacuum grippers [6] *have a number of suction cups, and are used to lift and position*

smooth-surfaced objects, such as automobile windshields. The general-purpose two-fingered claw [7] *often has sensors on its tips so that an object is not crushed. Delicate objects need a ring gripper* [8]. *This is placed around the object and a rubber tube inflates, gripping with even pressure.*

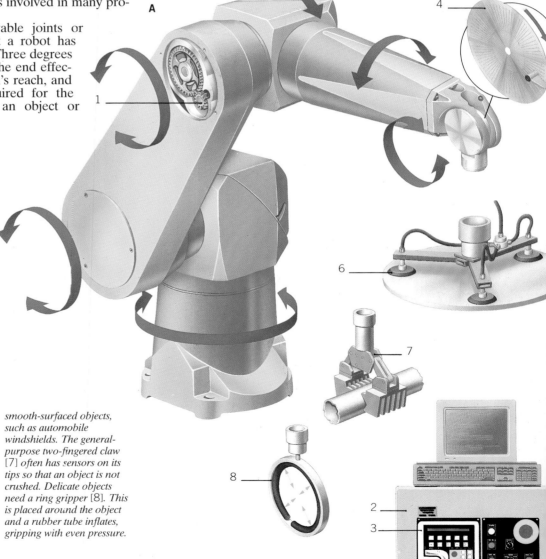

Connections: Cars: Systems 14 Miniaturization 146 Principles 228 236 238 240 250 254 256

workpiece from any direction. Although a robot needs six degrees (plus a seventh – the opening and closing of the end effector) to mimic the versatility of a human arm, each joint introduces some play, making accurate positioning of the arm more difficult. For this reason, and also to save expense, many industrial robots make do with four or five degrees of freedom.

Arms need to be highly rigid, and hence tend to be very heavy. So the motors that move them need to be powerful yet capable of working very accurately. Smaller robots sometimes use *stepper* motors, which move through a preset distance each time they receive an electrical pulse. The largest arms are too heavy for electric power on its own. Instead, hydraulic motors and cylinders are used, under the control of a central computer.

Sensors

In order to carry out its tasks, a robot needs to call upon equivalents of at least some of the five senses. Touch can be emulated by simple microswitches. Their contacts close under very slight pressure, giving a signal that can tell when the end effector is gripping a workpiece. If a simple electric motor is being used to control the actuator, contact can be detected even more simply, through the increase in current drawn as the closing end effector encounters resistance. Proximity sensors can sense nearby objects without having to touch them, sometimes by

Do the hop
A hopping robot is constantly falling off balance, and so is a difficult machine to control successfully. One experimental design [B] can hop on the spot as well as move in all directions. The 3-foot-high machine has two main parts: a body

carrying valves, sensors, and electronics, and a leg to provide the motion. A cable connects both to a nearby computer. The body's tilt is measured by a gyroscope, while other sensors measure the length of the leg, its position in relation to the body, and whether it is touching the ground.

When the machine starts to hop, hydraulic actuators push the foot to the left [1] so that the body begins to tip to the right. A piston, powered by compressed air, pushes the leg off the floor [2]. In the air, the leg retracts and pivots around to the right to balance the machine on landing [3].

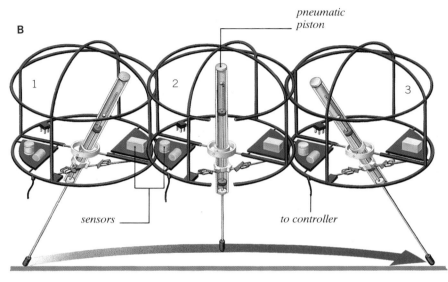

B

pneumatic piston

sensors

to controller

generating an electromagnetic field. A metal object in the vicinity alters the value of this field, and the change can be used to alert the robot's controller. Other types monitor the pattern of a reflected light, or the time taken to reflect an infrared pulse.

Much more sophisticated sensing is achieved with video-camera-based vision systems. The seemingly simple task of being able to recognize objects from a video image in fact takes large amounts of computing power and complex *artificial intelligence* programs. Even smell and tasting devices, often dependent on the conductivity of certain substances, have been developed – a robot winetaster already exists.

Remote control

One of the major advantages robots have is their ability to function in dangerous situations where human beings could not work – or at least not for very long. Strictly speaking, these are not robots but *teleoperators* because they are not controlled by a computer but mimic the movements of a human controller. Bomb-disposal machines already exist and similar devices have been used to clean up the nuclear contamination at Chernobyl. The development of virtual reality technology, and improvements in the mechanics of robots, may eventually enable humans to carry out dangerous and intricate tasks – even surgery – from the other side of the world.

Trash Culture

How society deals with the waste it produces

Every year, each man, woman, and child in the United States produces around twenty times his or her own weight in garbage – a total of more than 330 million tons annually. Although a small proportion of this waste is recycled, the bulk is disposed of by incineration or by burial in giant excavations, such as the Fresh Kills Landfill on Staten Island, New York City. With an estimated mass of 100 million tons, and a volume of 350 million cubic feet, Fresh Kills ranks among the largest human-made structures in North America.

Around 50 percent of all the solid waste produced by a modern society is agricultural in origin, composed of animal manure and crop residues. Most of this is recycled into the soil at the farms where it is made and is of great value as a fertilizer and soil stabilizer. Another 40 percent comes from metal processing – mine tailings and smelter slag, for example – which are typically stored at their site of production. Five percent is industrial waste which, depending on its toxicity, is buried, recycled, or treated to make it less harmful. But it is the remaining five percent – household and commercial refuse – that attracts the most public attention, because it impinges on our lives most directly, and we have most control over its fate.

Green cleaning
The three main methods for dealing with municipal waste are recycling, burial, and burning. Of these, recycling is usually the best alternative. It can cut the volume of waste for disposal by up to 50 percent; it reduces the volume of pollutants released into the atmosphere; and it lessens demand on raw materials and energy. For example, making aluminum from scrap rather than its ore, bauxite, gives an energy saving of more than 90 percent; and recycling plastic to make bottles can cut the manufacturer's fuel bill in half. In spite of these advantages, recycling still plays a relatively small role in waste management, accounting for just 11 percent of waste in the US, and 15 percent in Germany.

Out of sight, out of mind
In most countries burial constitutes the main disposal method. Giant landfill sites around most American and European cities swallow up to 80 percent of all municipal waste. Today's landfills are more than just holes in the ground: they are carefully engineered to prevent the seepage of toxic *leachates*, such as ammonia, organic matter, and other polluting chemicals, into the soil and groundwater. Buried pipes carry methane gas, which is given off by the decomposing garbage, to the surface, where it can be burnt off safely.

Over the last 50 years, landfills have been a convenient and inexpensive option. However, rising land prices and ever-stricter environmental legislation have driven up the cost of landfill disposal, and waste authorities have begun to rely increasingly on incineration.

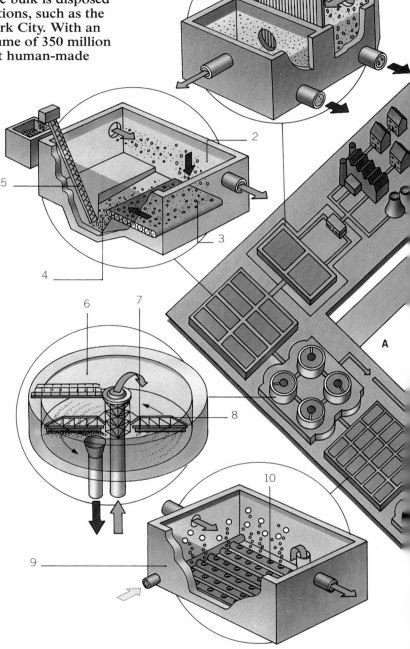

Fit to drink
An average European household produces more than 100 gallons of waste water every day. This water drains into the sewage system, running along underground pipes to a sewage treatment plant [A], where it is purified before being discharged back into rivers and streams. As sewage enters the plant, a series of mechanical devices removes from it progressively smaller solid impurities. Coarse screens [1] remove the largest solids, such as rags. The water then passes slowly through settling tanks [2]. Moving at low speed (12 ft/s), particles of grit and sand larger than 1/125 in sink, falling down onto a conveyor belt [3]: this carries the particles into a trough [4] from which they are removed by an elevator [5]. The water is then pumped into sedimentation tanks [6], where it remains for around 90 minutes.

Connections: Mining 100 Power Stations 106 Paper Manufacture 124 Principles 222 224 228 232

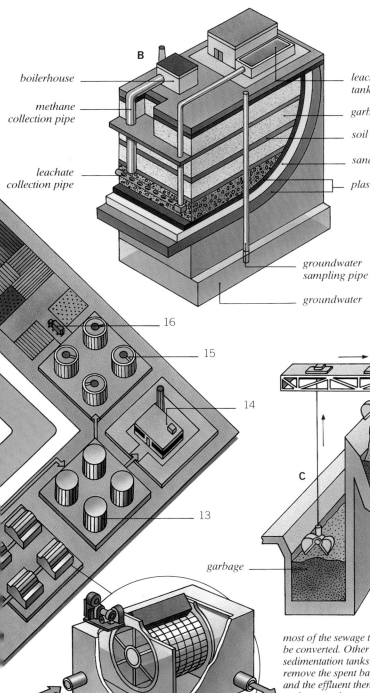

B

boilerhouse

methane
collection pipe

leachate
collection pipe

leachate
tank

garbage

soil

sand

plastic lining

groundwater
sampling pipe

groundwater

16

15

14

13

garbage

2

C

4

3

2

5

6

1

ash

Buried rubbish

*Garbage emptied into a landfill [**B**] is daily compacted by bulldozers and covered with a layer of soil to control odor and infestation. A lining composed of two layers of thick plastic prevents leakage of toxic leachate. The leachate collects in a layer of gravel and is pumped out of the landfill for treatment. Other buried pipes conduct methane gas, produced by the decomposing garbage, to the surface where it is burned off to produce useful energy. A deep bore allows groundwater beneath the landfill to be monitored for contaminants.*

Up in flames

*A mass-burn incinerator [**C**] can turn 1100 tons of waste into 275 tons of ash every day, thereby reducing the amount of solid waste that needs to be landfilled.*

Incinerators also provide energy: as the garbage is burned in a furnace [1] the heat produced is transferred to a boiler [2]. Water pumped into the boiler [3] is converted into steam [4], which is used to generate power. The smoke produced is cleaned by a scrubber [5], which sprays calcium compounds into the smoke: these react with toxic compounds. Any remaining particles are removed by an electrostatic precipitator [6].

ere, even the finest
spended solids (known as
ude sludge) have time to
her sink to the bottom of
e tank to be removed by
ctrically driven scrapers
, or float and be removed
surface skimmers [8].
The water is now much
arer, but must be treated
ologically to remove any

potentially dangerous organic compounds. In the aeration tank [9], sewage is mixed with a mixture of aerobic bacteria. As oxygen is bubbled through the mixture [10], these bacteria break down organic matter and convert it into harmless minerals, gases, and water. It takes about 8 hours for

most of the sewage to be converted. Other sedimentation tanks [11] remove the spent bacteria, and the effluent then undergoes microscreening [12]. The water enters a rotating drum made of an extremely fine screen, which filters out any remaining solids. The purified water can now be discharged into a river, lake, or the sea.

Crude sludge from the sedimentation tanks [6,11] is treated in sludge digestion tanks [13]. Here, bacteria "eat" the sludge, producing methane gas, which may be burnt to generate electricity [14]. After a second digestion [15], sludge can be used as fertilizer [16].

A burning issue

Waste incinerators can burn the diverse range of materials that make up household waste. The heat energy released may be used to generate power or heat local buildings. Inside the incinerator, rotating kilns or agitating grates are used to promote burning: air is injected above the fire to maintain temperatures of at least 1470°F for 15 seconds to destroy the products of combustion, some of which are toxic. Burning destroys most of the garbage, but about 20 percent of the original volume remains as ash, which is usually sent to landfill. The *bottom ash*, charred remains of large objects and unburned material, is usually landfilled in the same way as household waste. Fine-grained *fly ash*, which contains higher concentrations of heavy metals and other pollutants, should be disposed of more carefully in secure landfill, though in practice most incinerators mix bottom ash and fly ash together for disposal.

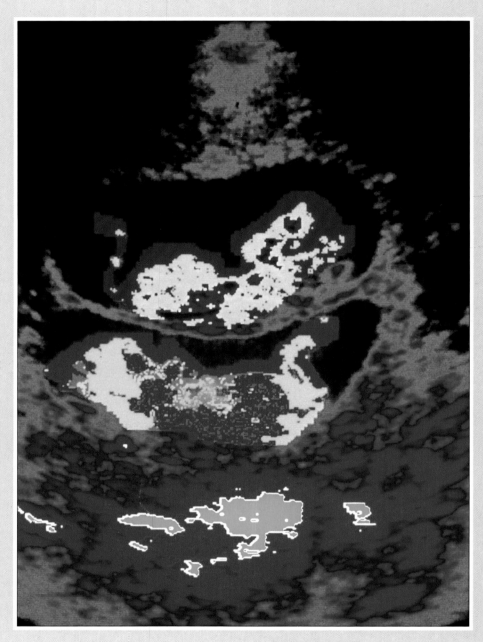

An obstetric ultrasound scan, artificially colored by computer, clearly shows twins in the womb of an expectant mother.

5
Research and Medicine

Making Light Work

How systems of lenses magnify and focus images

The inventors of the first microscopes must have felt a great sense of wonder as they gazed upon a whole new world of unseen complexity. The simple glass beads used as lenses in these early instruments produced badly distorted images. In contrast, a modern microscope contains a large number of different glass elements, all working together to produce a highly magnified image free from aberrations. Similarly, a high-quality camera zoom lens is made of as many as thirty different elements, assembled so as to cancel out one another's distorting effects.

A single lens, such as a magnifying glass, cannot produce perfectly focused images, no matter how accurately its surfaces are ground. It will inevitably introduce *aberrations* – distorted shapes or colored fringes around the image. A camera lens avoids these problems by using combinations of lens *elements* made from different types of glass. Some of these elements have concave as well as convex surfaces, while others are irregularly curved. Only with this degree of complexity can the lens produce a sharply focused image on the film.

A lens works by *refracting* light, bending the path of light rays to bring them to a focus. White light is a mixture of different wavelengths, or colors, from red to violet. Unfortunately, the degree to which a ray is refracted depends on its wavelength. This means that a beam of white light is split into its component colors as it passes through a lens, and the different colors are focused at slightly different points. A fringe of color forms around the focused image. This is called *chromatic aberration*. To avoid this problem, a compound lens is used. This contains two separate elements, each made from a different type of glass. The combination has the same focusing power as the lens it replaces, but much reduced chromatic aberration. Such lenses are still a compromise because they do not eliminate error across the entire spectrum. But because color photographic film and video cameras are only sensitive to red, blue, and green light, it is sufficient to reduce chromatic aberration around these wavelenths. Astronomers cannot afford any aberrations when viewing distant stars. For this reason, astronomical telescopes use huge curved mirrors that are immune to chromatic aberration.

Distorted reality

Chromatic aberration is not the only form of lens distortion. Most lenses, for simplicity of manufacture, have spherically ground surfaces. In such a design, light rays passing through the center of the lens are focused at a slightly different point to those passing through its edges. This effect is strongest in powerful (strongly curved) lenses. It is reduced by using a compound lens made of several weaker components in series. It is also minimized by blocking out rays from all but the central parts of the lens. This is why reducing camera aperture usually gives a crisper image on a photographic film.

A closer look
A professional microscope [B] contains prisms that split and direct light to its binocular eyepiece as well as to several other openings where a variety of stills and TV cameras can be attached. The double eyepiece allows a researcher to view a specimen at a slightly different angle through each eye, and thus to perceive a sense of depth in the magnified image. Using a special polarized light source and filters, the microscope is able to produce images that reveal minute fossil structures found in rock 37–54 million years old (below).

Color separation
A camera zoom lens [A] contains as many as 30 separate glass elements fixed together in three movable groups. The focal length of the lens (its "power") is changed by turning the zooming ring, which makes the three groups move relative to one another. The movements are calculated to change the field of view while keeping the image in sharp focus.

White light is made up of many different colors. A single glass element [1] acts just like a prism, splitting a beam of white light up so that the blue parts are bent more sharply than the red components. The result is chromatic aberration, an effect marked by colored fringes that appear around an image. It can be avoided by using an achromatic doublet made of two elements, one concave, the other convex, ground from different glasses [2]. Their chromatic aberrations work in opposite directions and cancel one another out.

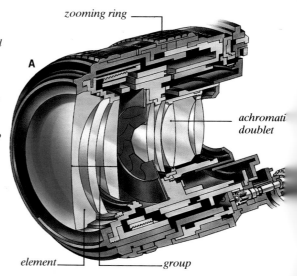

zooming ring

achromatic doublet

element   *group*

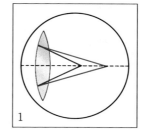
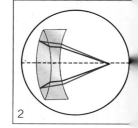

1   2

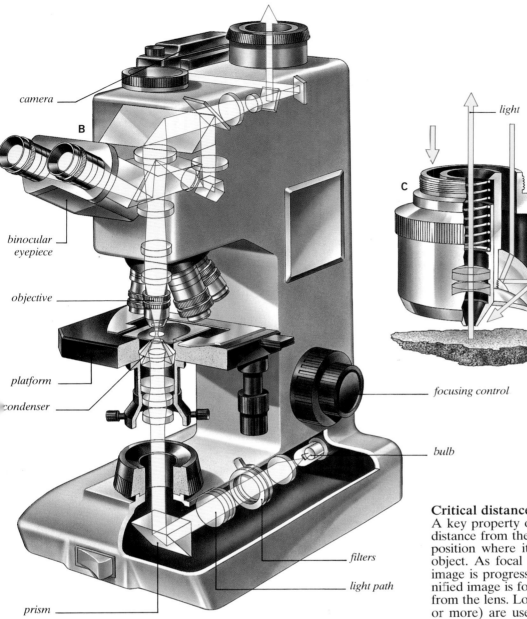

camera

B

binocular
eyepiece

objective

platform

condenser

prism

1

2

3

4

5

light

C

focusing control

bulb

filters

light path

Powers of concentration
*A professional microscope [**B**] usually illuminates a specimen from below. The light from a bulb is concentrated into a beam and passes through a series of filters. These select particular wavelengths of light from the broad spectrum produced by the bulb. Next, a prism reflects the beam upward into the condenser. This is a series of lenses that concentrate the light into a small spot on the specimen. Light that passes through the specimen is gathered by one of several objective lenses that can be rotated into place to give different magnifications. The chosen lens directs light upward into a complex of optics that channels the magnified image to the binocular eyepiece or camera. Some objectives [**C**] contain an outer ring condenser that focuses light downward onto the specimen. Reflected light passes into the objective lens, providing an image of the specimen's surface.*

Critical distances
A key property of a lens is its *focal length,* the distance from the central axis of the lens to the position where it forms an image of a distant object. As focal length increases, the resulting image is progressively magnified, but this magnified image is formed farther and farther away from the lens. Long focal length lenses (500mm or more) are used by sports and wildlife photographers to bring their subjects into close view. When attached to a 35mm camera, such lenses would need to be separated from the film by a distance of at least 500mm – a bulky and impractical arrangement. Instead, these lenses are cleverly designed so that the distance between lens and film is shortened while maintaining the power of a long focal length. They are known as *telephoto* lenses.

Many stills cameras, and practically all video and movie cameras, are now fitted with zoom lenses. These have variable focal lengths, adjusted either by twisting a ring on the lens assembly or pressing a button to operate an electric servo motor. This action causes groups of lens elements to move together or apart in complicated ways, altering the focal length and changing the size of the image accordingly. When this happens, the position of the image also changes – it goes out of focus. Other elements in the lens move to refocus the image automatically, making zooming from one focal length to another a simple process.

Cutting losses
*The surface of a lens reflects a small proportion of the light falling onto it [**D**]. Such reflections occur at each surface of the many elements in a typical zoom lens, dissipating a significant percentage of the light that enters the lens [1]. This loss is reduced by blooming the elements. In this process, a thin coating of lacquer is applied to each surface [3]. This has exactly the right thickness to ensure that light waves reflecting from the outer surface of the lacquer [4] are exactly half a wavelength out of phase with those coming back from the glass/lacquer interface [5]. The two waves interfere destructively, canceling each other out, so that very little light is reflected and most passes through the lens [2]. The coating is least successful for the colors at either end of the visible spectrum, which reflect and give bloomed lenses a characteristic red/violet appearance.*

Miniature Vision

How electron microscopes can see minute objects

Some electron microscopes are capable of magnifying objects one million times – equivalent to enlarging a postage stamp to the size of a small country. But the value of the electron microscope depends as much on its resolution – its ability to reveal detail in the smallest of objects – as on its magnifying power. Using a high-speed beam of electrons, rather than light, to "see" with, the electron microscope can distinguish objects separated by less than one nanometer – one-billionth of a meter or the length of ten atoms – and produce images of viruses and individual molecules.

Atomic images *(above) Using an instrument known as a tunneling microscope, individual atoms may be viewed. The microscope works by moving a probe, the tip of which is just one or two atoms wide, over the sample surface. The height of the probe's tip reflects the contours of the sample. Shown here is a clump of gold, around three atoms thick (yellow and red), on a base of regularly spaced carbon atoms (green).*

A keen-eyed observer can separate, or *resolve*, two dots one-tenth of a millimeter apart. Using an optical microscope that magnifies 1000 times, the viewer is able to distinguish two objects separated by one-thousandth of a millimeter. Although higher magnifications can be achieved by adding still more powerful optical lenses, they are of little value because the larger image formed carries no extra *detail*. This is because the resolving power of the optical microscope is ultimately limited by the properties of light itself: objects smaller than its wavelength (around 500nm) are bypassed by light waves and are thus impossible to detect.

Electron microscopes make use of electron, rather than light, waves. Electrons are usually thought of as particles of electricity, but when moving at high speed, they also show wavelike characteristics. Electron waves have a wavelength of around 0.005nm and so are capable of picking out much finer details than light waves.

Microscopic components

The electron microscope consists of a tall, sealed column containing a vacuum. At the top is an electron gun composed of a tungsten filament, held at tens or hundreds of thousands of volts relative to the rest of the instrument. When the filament is heated, electrons "boil off" its surface and pass into the vacuum, where the high voltage accelerates them down the column. The high-speed electron beam is focused and projected onto a thin slice of specimen, held on a *stage* midway up the column. Electrons pass straight through "empty" parts of the specimen, but are absorbed or deflected by denser areas. The undeflected electrons strike a fluorescent screen or photographic plate at the base of the column. In the image formed, the brightest areas correspond to the least dense parts of the specimen. Microscopes of this type are called Transmission Electron Microscopes (TEMs) because images are formed by electrons passing through the specimen. The Scanning Electron Microscope (SEM) works in a different way, building up a three-dimensional image by detecting electrons reflected from the surface of a specimen.

The electron beam in the TEM is controlled by magnetic fields produced by electric coils. These "electron lenses" are similar in function to the glass lenses of an optical microscope. A condenser controls the diameter and brightness of the beam before it strikes the specimen.

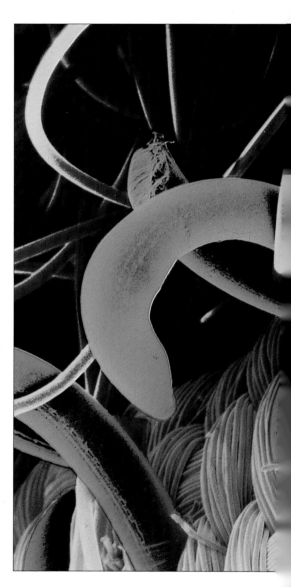

Another electron lens, called the objective focuses the beam on the specimen, and magnifies the image around 50 times. Electron lenses below the specimen then further magnify the image. Most TEMs cover the magnification range of x100 to x500,000.

Slices of life

Before examination in a TEM, a specimen goes through a number of preparatory stages. First is cut into slices thin enough to be penetrated by an electron beam (no more than 100nm, one-thousandth of the thickness of a sheet of paper). Specimens may be sliced in a precision cutter called an ultramicrotome, or thinned by *sputtering* – bombardment with high-energy ions. The slices are then mounted on copper grids and placed onto the specimen stage of the TEM. Biological specimens may also be immersed in solutions containing heavy (electron-dense) metals to improve image contrast.

Connections: Lenses and Microscopes 134 Spectrographs 140 Particle Accelerators 144 Miniaturization 146 Principles 212 214 244 248 258 262 264

Surface scanner

*Whereas the TEM is used to examine the internal structure of thin specimens, the SEM [**A**] provides three-dimensional images of the surfaces of objects. Though it shares many features with the TEM, it works in a basically different way. Like the TEM, it possesses an* electron gun – a heated tungsten filament held at a voltage of around 20,000 V, – and electromagnetic lenses that project the beam of electrons on to the specimen. But unlike the TEM, the electron beam does not form an image directly. Instead, the beam is focused into a spot, no more than 10 nm across, and scanned over the surface of the specimen.

Controlled beam

At the top of the SEM column, electrons "boil off" the surface of the tungsten electrode and accelerate toward the positively charged anode. They then continue down the column, which is evacuated by powerful pumps, toward the specimen. Three sets of electromagnetic lenses focus the electron beam into a tiny spot. An aperture ring allows the spot diameter to be finely controlled. A fourth set of electromagnetic coils – the scan coils – deflect the beam in a precise pattern, causing it to "paint" a series of parallel lines on the surface of the specimen.

The scan coils are linked to a computer, which synchronizes the movement of the beam with the lines scanned in a grid pattern on a video monitor.

The high-speed electron beam knocks electrons out from atoms on the surface of the specimen. These secondary electrons are in turn accelerated toward a target that emits a flash of light when struck. The flashes of light (which correspond to bumps on the specimen's surface) are detected and converted to pulses of electricity. The computer displays this stream of pulses as bright or dark areas on the monitor, building up an image line by line. Shown here is a diatom – a microscopic marine plant – magnified 300 times.

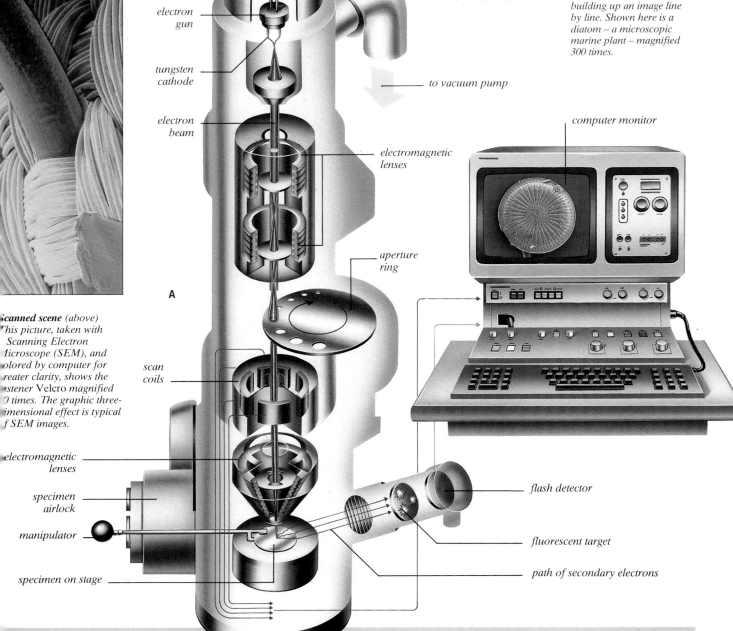

Scanned scene (above) This picture, taken with a Scanning Electron Microscope (SEM), and colored by computer for greater clarity, shows the fastener Velcro magnified 0 times. The graphic three-dimensional effect is typical of SEM images.

high voltage supply

electron gun

tungsten cathode

electron beam

to vacuum pump

electromagnetic lenses

A

aperture ring

computer monitor

scan coils

electromagnetic lenses

specimen airlock

manipulator

specimen on stage

flash detector

fluorescent target

path of secondary electrons

Seeing Stars
How telescopes probe ever deeper into space

Perched two and a half miles up Mauna Kea on the island of Hawaii is the most powerful telescope on earth. The Keck telescope permits astronomers to see into space to the farthest reaches of the universe. Its 33-foot mirror grasps ten billion times more light than the naked eye, and its sensitive instruments can reveal the detailed structure of stars and galaxies. But it is not only light that carries information about these distant bodies to the earth. Radio waves from space are detected by huge parabolic dishes – the largest, built in a natural crater, is 1000 feet across.

Small telescopes are *refractors* – like binoculars and microscopes, they use lenses to bend light and form a magnified image. Lenses, however, have their drawbacks. They are subject to *chromatic aberrations,* colored fringes that appear around the image, and above a certain size cannot support their own weight. The largest astronomical telescopes instead use a different design in which light is collected and focused by a large curved mirror, or *reflector.* Mirrors are relatively light and easy to grind to the exact *hyperbolic* shape required, allowing ever bigger telescopes to be built. Wider reflectors gather more light and give better *resolution,* letting astronomers see deeper still into the universe.

Bigger is better
Traditionally mirrors were ground very slowly from a thick slab of glass. The largest telescope built this way is the 16 ft Hale reflector on Mount Palomar in California. However, any reflector wider than this would sag under its own weight, losing the fine balance of its optics. The Keck telescope avoids this problem by using a lighter mirror made up of 36 hexagonal sections combined in a 33 ft-wide mosaic.

Large mirrors can also be made by spinning molten glass while it solidifies. The resulting concave block can be ground to a hyperbolic shape relatively easily. Mirrors made this way are very thin and light – the Japanese Subaru telescope has a mirror 27 ft across, but only 8 in thick. To allow for the flexing of such a slim reflector, the telescope has 250 sensors and motors built into its mounting. A computer constantly monitors the shape of the mirror and can correct any distortions caused by heat or movement, as well as compensating for errors introduced by atmospheric disturbance.

Tuning in to the universe
A telescope dish has to be much wider than the wavelength of the radiation it examines in order to pinpoint where a signal is coming from. This is not a problem for optical instruments, for the wavelength of light is roughly a millionth of a meter. However, radio waves that reach the earth's surface can be tens of meters long, and consequently radio telescopes need dishes at least 82 ft (25 m) wide. These focus the waves onto a *detector horn* – an aerial that turns radio waves of a particular wavelength into an electrical signal. Telescopes often have a number of horns, each tuned to a different wavelength.

Radio pictures
Radio telescopes operate on the same principle as optical instruments but on a larger scale, to match the longer radio wavelengths. Radiation is collected by a parabolic dish and reflected to a second surface which focuses it onto a detector.

The Very Large Array (VLA) in New Mexico (below) consists of 27 dishes. They are mounted on a Y-shaped railroad track, with arms 12.5 miles long. Using the principle of interferometry, the signals from all 27 dishes are combined to make up images with the clarity of a single 13-mile telescope.

Images are built up by measuring radio intensities and frequencies from different parts of the sky. Color is added by a computer to make the images easier to read. This image (right), assembled from 3 cm radio waves, shows clouds of hydrogen gas (yellow and orange), which are thought to be a stage in star formation.

On its own, even the largest radio dish cannot obtain an image comparable with the best optical results. Sharper results are obtained by using a technique called *interferometry* by which the signals from two telescopes pointing at the same object are combined electronically by a computer. Using paired telescopes in this manner gives a resolution equivalent to that of a single dish as large as the *baseline* – the distance between the two reflectors.

More than two entire observatories can work together this way – one array combines observations from Australia and Japan with data from an unmanned orbiting observatory, creating a radio telescope with the resolution of one twice as wide as the earth.

Interferometry can also be used to improve the performance of optical telescopes. Keck 2 is now being built on Hawaii, a twin to the nearby Keck 1. The combination will in effect produce images from a telescope 280 ft wide.

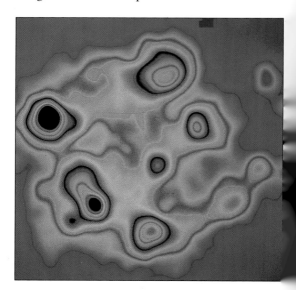

Connections: Video Technology 58 Photographic Cameras 62 Lenses and Microscopes 134 Spectrographs 140 Principles 258 262 264 266

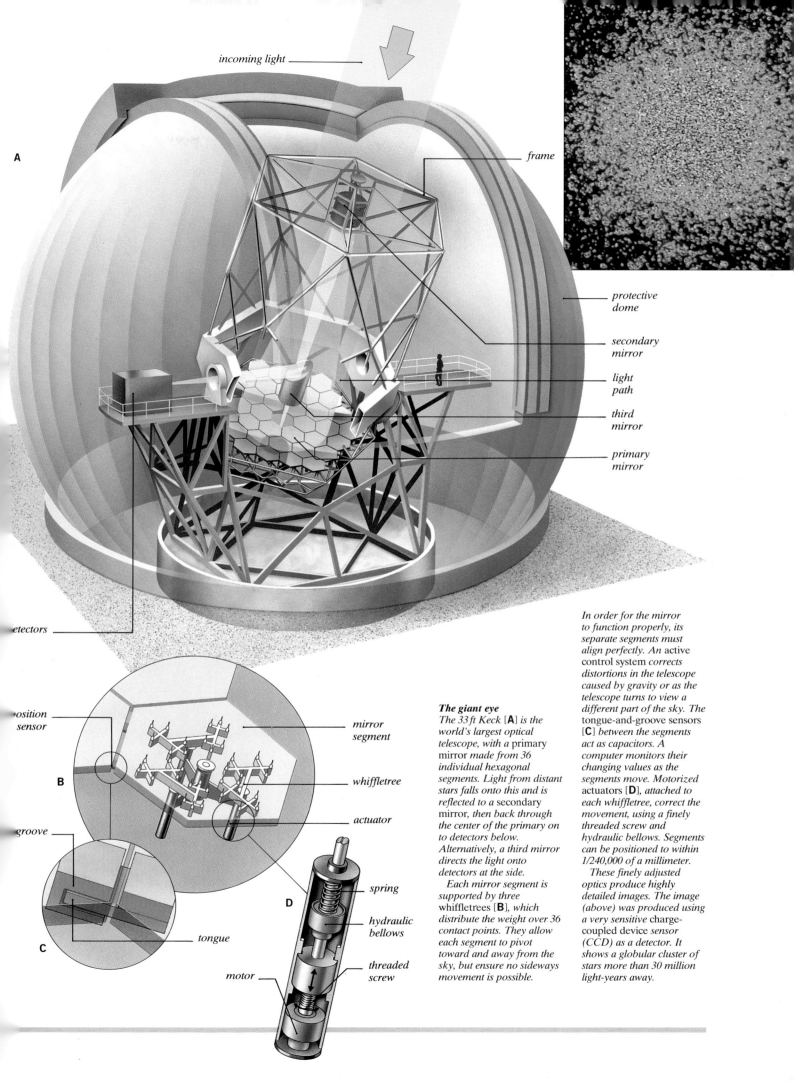

A

incoming light

frame

protective dome

secondary mirror

light path

third mirror

primary mirror

detectors

position sensor

B

mirror segment

whiffletree

actuator

groove

C

tongue

D

spring

hydraulic bellows

threaded screw

motor

The giant eye

The 33 ft Keck [**A**] is the world's largest optical telescope, with a primary mirror *made from 36 individual hexagonal segments. Light from distant stars falls onto this and is reflected to a secondary mirror, then back through the center of the primary on to detectors below. Alternatively, a third mirror directs the light onto detectors at the side.*

Each mirror segment is supported by three whiffletrees [**B**], *which distribute the weight over 36 contact points. They allow each segment to pivot toward and away from the sky, but ensure no sideways movement is possible.*

In order for the mirror to function properly, its separate segments must align perfectly. An active control system *corrects distortions in the telescope caused by gravity or as the telescope turns to view a different part of the sky. The tongue-and-groove sensors [**C**] between the segments act as capacitors. A computer monitors their changing values as the segments move. Motorized* actuators [**D**], *attached to each whiffletree, correct the movement, using a finely threaded screw and hydraulic bellows. Segments can be positioned to within 1/240,000 of a millimeter.*

These finely adjusted optics produce highly detailed images. The image (above) was produced using a very sensitive charge-coupled device *sensor (CCD) as a detector. It shows a globular cluster of stars more than 30 million light-years away.*

Atom Detectives

How chemists identify different elements and molecules

In 1989, scientists were able to establish the true age of the Turin Shroud, a holy relic previously thought to be 2000 years old. Using carbon dating, they found that the cloth was in fact made only 800 years ago. Carbon dating is one of a battery of powerful spectroscopic techniques that allow scientists to look directly at the properties of atoms. Spectroscopy has diverse applications: it is used by chemists to identify unknown substances and to determine the purity of drugs, and by astronomers to investigate the structure and composition of distant galaxies.

The identity of an unknown substance can be investigated in a number of ways. It can be treated with chemical reagents to reveal the distinctive way in which it forms bonds. Alternatively, the reaction of its atoms to external forces such as light or magnetism can be used to reveal its unique physical properties. Both components of an atom – the heavy nucleus and the light electrons orbiting it – produce telltale signs that can be used to identify a sample.

Atoms and molecules have a characteristic mass, which can be measured in a sophisticated instrument called a *mass spectrometer*. The atoms to be investigated are first turned into charged *ions* by being bombarded with high-energy electrons. These ions are then accelerated toward a strong magnetic field, which bends their paths, causing the ions to travel in an arc. Significantly, the paths of heavy ions, which have a lot of *inertia,* are bent less than the paths of light ions. Each atom or molecule is therefore deflected through a known angle. By measuring the numbers of atoms traveling through each angle, the precise composition of a very small sample can be determined.

Reading the rainbow

An electron can only occupy certain orbits around its nucleus, each corresponding to a specific level of energy. When an electron jumps between two of these levels, it releases or absorbs a *photon* (a "packet" of light) of a wavelength directly related to the energy difference between the two levels. Because the electron orbits of different elements have different energies, the wavelengths – and hence colors – of light given out by each element are distinct and can be used for identification.

When a voltage is placed across a sample of gas in a discharge tube, electrons jump to higher energy orbits. As they return to lower orbits, the electrons release photons of characteristic wavelengths. In a *spectrometer*, this emitted light is passed through a prism or diffraction grating to split it into its component colors. The intensity of light at each part of the spectrum is measured, giving a unique fingerprint of the element concerned. Conversely, if white light (a mixture of all colors) is shone through a sample, some wavelengths are absorbed, boosting the electrons of the elements present to higher energies, while others pass through. In this case, the absence of certain parts of the spectrum is used to determine composition.

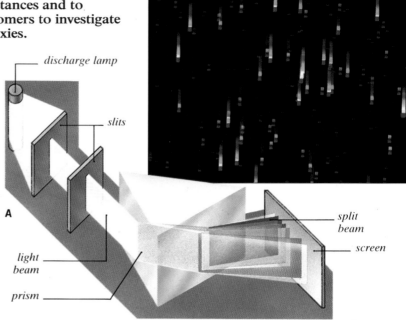

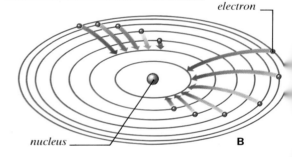

Splitting light
The electrons orbiting an atom can only occupy specific energy levels. Photons – particles of light – with specific wavelengths are emitted or absorbed when electrons jump from level to level. Electrons jumping from the outer to the innermost levels of hydrogen atoms give out ultraviolet photons. Jumps from the outer levels to the second orbit result in the emission of visible light [B], which can be split into individual colors, each with its own wavelength, by a spectrometer [A]. A current passing through a hydrogen-filled discharge lamp stimulates electron jumps. As the light this produces passes through a prism, blue light bends more than red, so that the colors separate to form the hydrogen spectrum. Each element has its own spectral "fingerprint," so passing the light from a telescope through a spectrometer reveals the composition of the stars (top right).

Carbon Dating

Carbon occurs naturally in three atomic weights or *isotopes*. Carbon-12 and carbon-13 are stable, but carbon-14 (^{14}C) is radioactive, decaying to nitrogen with a *half-life* – the time taken for half the atoms in a sample to decay – of 5730 years.

A living organism is composed mostly of carbon compounds. It possesses all three isotopes in a constant ratio because the carbon is constantly replenished as the organism nourishes itself. However, when the organism dies, the amount of carbon-14 steadily reduces as individual atoms decay, while the quantities of the two stable isotopes remain constant. Therefore, finding the ratio of the masses of the three isotopes in an archeological sample – such as bone, wood, or cloth – is a way of determining the sample's age.

Connections: Telescopes 138 Chemical Analysis 142 Particle Accelerators 144 Principles 212 214 216 222 262 264 266 268

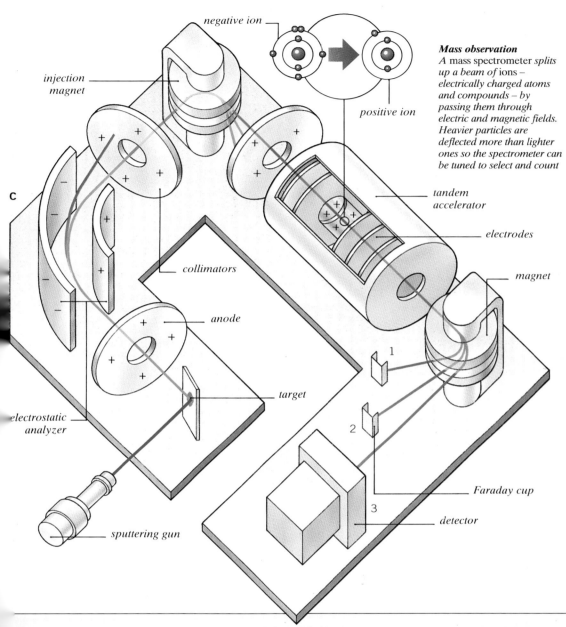

negative ion

injection
magnet

positive ion

C

collimators

anode

electrostatic
analyzer

sputtering gun

target

tandem
accelerator

electrodes

magnet

1

2

Faraday cup

3

detector

Mass observation
A mass spectrometer *splits up a beam of ions – electrically charged atoms and compounds – by passing them through electric and magnetic fields. Heavier particles are deflected more than lighter ones so the spectrometer can be tuned to select and count* particles of a particular mass. A much more sensitive version of this instrument is the accelerator mass spectrometer [**C**]. This is used in carbon dating, *in which measuring the relative amounts of carbon atoms with different masses gives the age of a sample.*

Cesium ions from a sputtering gun bombard the sample target, which gives off a spray of negatively charged ions. These are accelerated toward a positive anode, passing as a beam through an electrostatic analyzer made up of large positive and negative plates. The electric field between the plates bends the beam so that particles at only one energy level pass through a plate and into the injection magnet. This has a magnetic field with just the right strength to let only carbon ions pass into a tandem accelerator. Highly positively-charged electrodes strip electrons from the ions as they pass through so they emerge carrying a triple positive charge. They emerge at high speed to pass between the poles of a second magnet, which splits the beam so that only the rare carbon-14 atoms are counted by a solid-state detector [3]. The more common carbon-13 and carbon-12 are collected in Faraday cups [1,2].

Carbon dating is traditionally done by counting the number of ^{14}C atoms decaying in a given period of time. Because the half-life is so long that there are only a very few decays per minute, the technique only works for specimens less than 0,000 years old. Above this age, the amount of C in the sample is almost impossible to detect. A second, more recent, technique is to separate the different isotopes by their masses, using an accelerator mass spectrometer (AMS). This directly counts the number of ^{14}C and ^{12}C atoms a sample and is much more sensitive than counting very rare decay events. The AMS has extended the range of ages that can be accurately found up to 75,000 years. The development of this technique has also led to the detection of other radioactive isotopes that may be used to date much older specimens from a few thousand to millions of years old.

Dating a relic (right)
The Turin Shroud is a linen cloth that apparently bears the image of a crucified man. It is believed by many to be the cloth in which Christ's body was wrapped and was first displayed in the 1350s. It is now kept in Turin Cathedral, Italy.

Doubts about its origins would only be dispelled if it was shown to be 2000 years old. Carbon dating enabled a scientific estimate of the age of the cloth to be made, but early techniques required the destruction of a large portion of the Shroud to use as a sample.

The much smaller sample needed by Accelerator Mass Spectrometry (AMS) – a few square inches – made dating feasible.

In 1988, a sample was cut from the cloth, divided into three and sent to three separate laboratories in the United Kingdom, the United States, and Switzerland. Each carbon-dated the cloth using AMS, and compared the result with three other samples, two medieval and one from the 1st century AD. All three laboratories concluded with 95 percent certainty that the shroud dates from the period AD 1260-1390.

Analytical Thinking

How chemical compounds are identified and examined

Today's chemists rely on a battery of sophisticated analytical tools to separate, identify, and characterize complex compounds. Most of these procedures are confined to the research laboratory: X-ray diffraction, for example, is used to unravel the structures of organic molecules like hemoglobin and DNA. But some techniques, most notably chromatography, have wider uses. Chromatography forms the basis of alcohol and drug testing; it is used to monitor levels of air and water pollution and to determine the purity of foods and pharmaceuticals.

More than one million chemical compounds are known to exist. Whether they arise naturally or as products of laboratory manipulation, they often occur in *mixtures* with other compounds. Before the individual components of a mixture can be identified and put to use, they must first be separated and purified. There are many ways to do this, but the most widely used are *distillation* and *chromatography*.

Distillation is a process in which a mixture of liquids is evaporated and then recondensed: liquids with low boiling points condense first and are collected; others distill later, or not at all. A variation on this technique, known as *fractional distillation,* is used to separate liquids in which the boiling points are very close. In this process, liquids are repeatedly evaporated and recondensed in a tall column and the different *fractions* collected at different heights. Distillation is used to separate chemicals on a large scale, particularly in the petrochemical industry,

Microanalysis
*Gas-liquid chromatography (or GLC) is used to analyze samples of a few millionths of a liter [**A**]. The sample – an unknown mixture of compounds [1] – is injected from a microsyringe [2] through a self-sealing membrane [3] into a stream of inert carrier gas, such as*

helium [4]. The carrier gas first passes through a "scrubber" [5] to remove trace impurities that could confuse the results. Heating elements [6] vaporize the sample, which mixes with the helium and is carried into the chromatography column [7]. This is tightly coiled, to accommodate as

great a length as possible in a small space. The column, which is around 0.2in wide, is packed with granules of silicon [8], each covered by a film of a high-boiling-point liquid [9]. The whole column is heated in an oven [10] to prevent the sample from cooling down and condensing in the column.

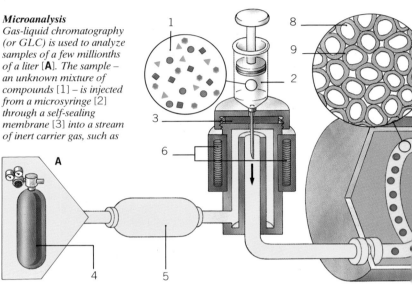

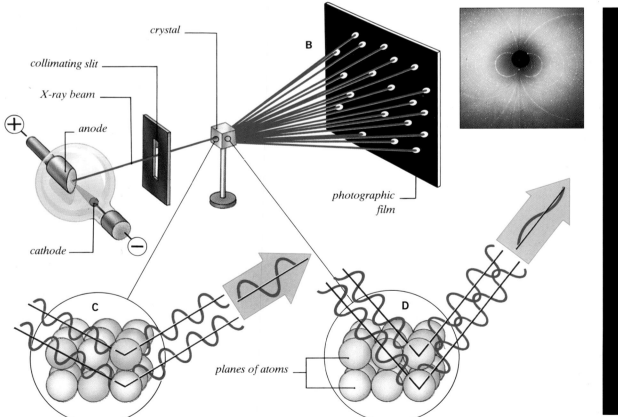

crystal

collimating slit

X-ray beam

anode

cathode

B

photographic film

C

D

planes of atoms

Connections: Spectrographs 140 Monitoring the Body 156 Oil Refining 186 Biotechnology 196 Principles 212 214 216 218 224 226

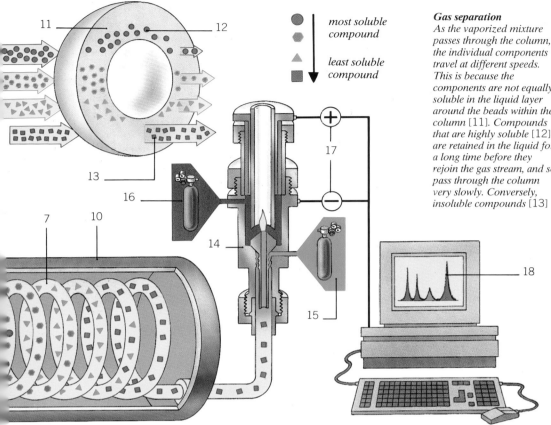

most soluble
compound

least soluble
compound

Gas separation
As the vaporized mixture passes through the column, the individual components travel at different speeds. This is because the components are not equally soluble in the liquid layer around the beads within the column [11]. Compounds that are highly soluble [12] are retained in the liquid for a long time before they rejoin the gas stream, and so pass through the column very slowly. Conversely, insoluble compounds [13]

are not significantly retarded and pass rapidly through the column.

The components of the mixture emerge from the column one by one and are carried into the detector [14]. Here, the helium gas carrying the separated components is mixed with hydrogen [15] and oxygen [16] before passing over a flame in a burner.

Whenever one of the separated components arrives at the flame, some of its molecules become ionized by the flame's energy. The charged ions move between two electrodes within the detector [17], constituting an electric current. This current is registered by a computer and displayed as a series of peaks [18]. The size of each peak corresponds to the amount of each component present, and the position of the peak (compared to the peaks produced by a known set of "reference" compounds) gives a clue to the identity of the unknown component.

-ray vision
*powerful technique called -ray crystallography [B] an be used to investigate e structure of a solid. A eam of X-rays is generated bombarding a tungsten rget with high-speed ectrons in a vacuum. The eam passes through a slit d is directed onto a crystal the substance being amined. X-rays are flected from the planes of oms in the crystal: at a rtain angle [C], the stance between any two jacent planes in the crystal such that the reflected rays reinforce one other: this constructive erference is recorded on heet of film as a bright ot. At other angles, X-rays erfere destructively, and spot is recorded [D]. amination of the crystal's ray "fingerprint" (far *), allows distances ween atoms to be culated, and the structure the molecule can be luced. X-ray studies ped reveal the structure he DNA molecule (left).*

but is not ideal for the small-scale separation of biological molecules. These compounds may break down or be inactivated at temperatures near their boiling point, and may be present in minute quantities only. They are more efficiently separated by *chromatography.*

Color signatures
The term chromatography, derived from the Greek words for "color writing," was coined early in the 20th century by the Russian botanist Mikhail Tsvet. While investigating the chemicals responsible for photosynthesis in green plants, Tsvet passed solutions of plant extract through glass tubes packed with powdered chalk. Individual pigments in the mixture had different affinities for the particles of chalk in the column. Those with low affinities traveled down the column rapidly, while those with higher affinities passed down slowly. As a result, distinct colored bands developed along the length of the tube, each band corresponding to one of the pigments involved in photosynthesis. Today, chromatography is applied not just to colored compounds but to colorless mixtures, in which the individual components are identified by a variety of detectors.

Variations on Tsvet's simple column are still in use today, but more rapid and accurate chromatographic techniques have been developed. All rely on the same basic principles. The mixture to be separated is dissolved in a liquid

or a gas, known as the *mobile phase*. This is passed over a *stationary phase*, which may be solid or a liquid, held on an inert support. The molecules in the sample mixture interact with the stationary phase: some "stick" to it for a long time before redissolving in the mobile phase, while others pass through quickly. This means that components of the mixture can be identified by the speed at which they move through this molecular "obstacle course."

In *gas-liquid chromatography*, the mobile phase is an inert gas, which carries the sample gases over silica beads impregnated with a liquid, the stationary phase. In *thin-layer chromatography*, the stationary phase is an alumina gel coated on to a glass plate. A spot of the sample is placed at one end of the plate, which is then suspended vertically with the sample end dipped in a solvent, the mobile phase. The whole apparatus is held within an enclosed tank so that the solvent vapor surrounds the plate. The solvent rises through the gel by capillary action, carrying components of the mixture different distances up the plate.

Chromatography tells a chemist not only which chemicals are present, but in what quantities. It is used by scientists in many disciplines: food technologists use chromatography to isolate a desired flavor (the taste of coffee is influenced by over 300 separate compounds); and police scientists use it to measure the quantity of alcohol in a driver's bloodstream.

Probing the Nucleus
How particle accelerators are put to use

The CERN particle accelerator near Geneva is the largest scientific instrument ever built. Serviced by a staff of more than 3000, it draws as much electrical power as a city of 200,000 inhabitants. This vast amount of energy is used to propel electrons and their *antiparticles,* or positrons, around a 17-mile ring at velocities approaching the speed of light, and to steer them into high-speed collisions. The fragments produced in the collisions are analyzed, giving scientists vital clues about the structure of matter and the origins of the universe.

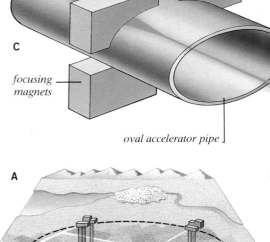

C

focusing magnets

oval accelerator pipe

Any charged particle, such as an electron, positron, proton, or ion, accelerates when placed in an electric field. Devices that make use of this property are called *particle accelerators*. Not all are exotic research instruments. For example, the tube inside every television set is a particle accelerator in which electrons are "boiled off" a filament and accelerated through a vacuum by an electric field set up between a positive and a negative electrode. Steered by electromagnets they strike a fluorescent screen in a pattern that forms a visible image.

The energy given to a particle by an electric field is measured in electron volts (eV). By accelerating from one electrode to another in an electric field produced by 1 volt (V), an electron is said to gain 1eV. Inside the tube of a television set, electrodes carry a charge of around 10,000V (10kV), and therefore accelerate electrons to energies of 10keV.

Large research accelerators work in much the same way as television tubes. Charged particles are produced and accelerated in a vacuum by an electric field, with powerful magnets controlling their direction. However, the charged particles are given colossal energies, measured in thousands of millions of electron volts (GeV), and are accelerated to more than 99.999999 percent of the speed of light. By smashing fast-moving particles into stationary targets or into other particles, they can create subatomic fragments, many of which are highly unstable and not normally found in the natural world. It is thought that some of these fragments are the fundamental building blocks of matter itself.

Energy to mass
In large accelerators, the energy of collision is changed into matter in accordance with Einstein's famous equation, $E = mc^2$. The faster the particles before the collision, the more energy there is, and the more chance that heavy fragments will result. Ever more powerful accelerators are being built in the search for bigger collisions and the new particles that will form.

Linear particle accelerators, or *linacs*, accelerate particles down a long, straight tube. In the 2-mile-long linac at the Stanford Linear Accelerator Center in California, electrons are accelerated to energies of 50GeV. Their targets are either protons or neutrons, and the collisions reveal that these particles consist of smaller particles called *quarks*.

Accelerator ring
CERN, the European laboratory for particle physics, operates the Large Electron–Positron Collider (LEP), situated on the French/Swiss border [**A**]. *The huge accelerator is used to study the fragments produced when electrons collide with their antiparticles – positrons.*

The accelerator ring is 330ft below ground. Particles move within an oval tube with a diameter of 6in. The interior of the tube is evacuated by powerful pumps. Positrons and electrons move around the ring at identical speeds, but in opposite directions because they carry an equal but opposite electric charge. Situated around the ring are magnets to control the particle beams, accelerator sections to supply energy to the speeding particles, and vast detectors, which track the minute fragments that result from the collisions.

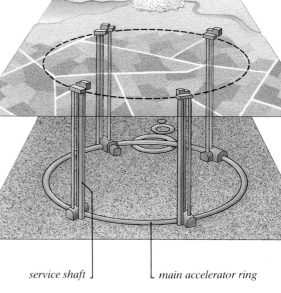

A

service shaft

main accelerator ring

Small linacs, able to reach energies of around 10MeV (millions of electron volts), are produced commercially. Such instruments are used in hospitals to make radioactive isotopes needed for the diagnosis and treatment of cancer. These isotopes have a short half-life and must be manufactured and used on site.

Circling the target
To reach higher energies, particles are accelerated in a *cyclotron*, where they spiral out from a central source before being steered in the direction of a target. Cyclotrons can accelerate protons up to an energy of around 25MeV. For still higher energies, a slightly different machine called a *synchrotron* is used. Here, particles race around a circle of constant radius. The accelerating force is provided by radio waves that set up an electric field inside hollow metal structures called cavities, and sets of giant magnets keep the particles on course (see main picture).

LEP layout
Before they are fed into the main ring of the LEP [**B**], *electrons and positrons are accelerated by a linac and three smaller rings* [1] *that raise their energy to 20GeV. The main ring is broken up into sections, each of which has a particular function. Focusing sections contain magnets that concentrate the beam* [**C**], *which would otherwise tend to spread out. Bending magnets* [**D**] *steer the beam around the curves of the accelerator ring. The particles are accelerated in radio frequency cavities* [**E**]. *A pulse of radio waves* [2], *moving at the speed of light, is fired down the*

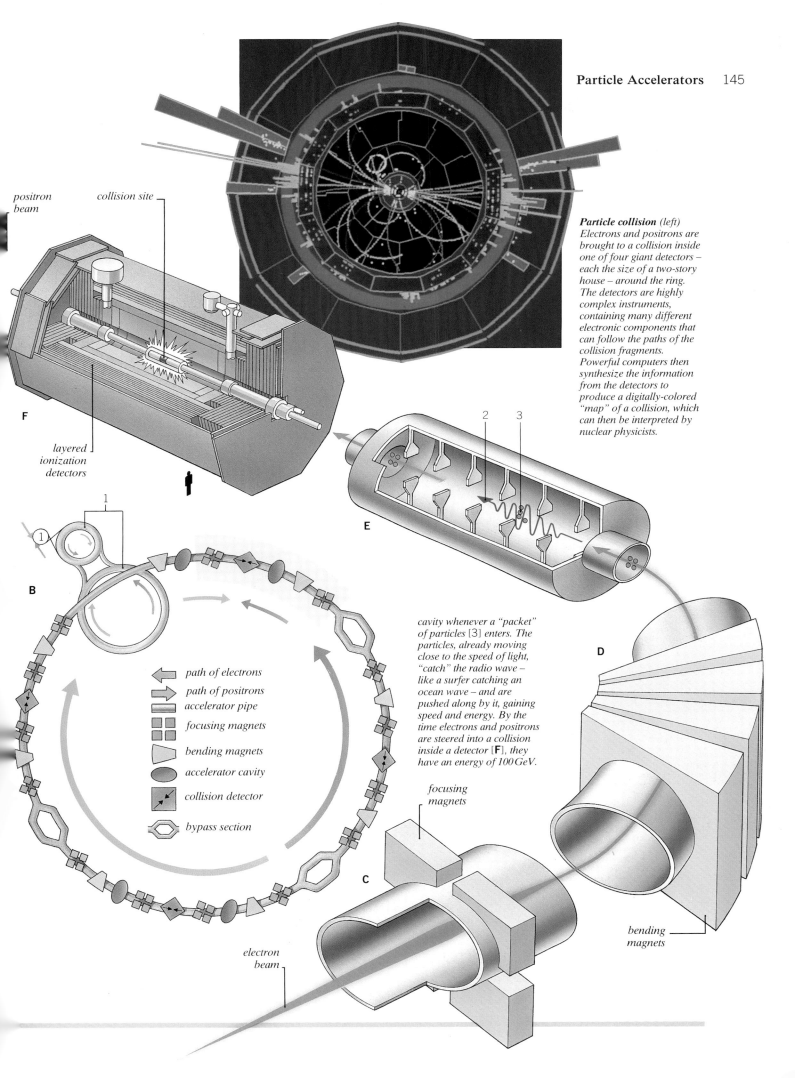

positron beam

collision site

F

layered ionization detectors

Particle collision (left)
Electrons and positrons are
brought to a collision inside
one of four giant detectors –
each the size of a two-story
house – around the ring.
The detectors are highly
complex instruments,
containing many different
electronic components that
can follow the paths of the
collision fragments.
Powerful computers then
synthesize the information
from the detectors to
produce a digitally-colored
"map" of a collision, which
can then be interpreted by
nuclear physicists.

2   3

E

B

1

1

path of electrons

path of positrons

accelerator pipe

focusing magnets

bending magnets

accelerator cavity

collision detector

bypass section

cavity whenever a "packet"
of particles [3] enters. The
particles, already moving
close to the speed of light,
"catch" the radio wave –
like a surfer catching an
ocean wave – and are
pushed along by it, gaining
speed and energy. By the
time electrons and positrons
are steered into a collision
inside a detector [**F**], they
have an energy of 100 GeV.

D

focusing magnets

C

bending magnets

electron beam

Molecular Mechanics

How machines are pieced together from single atoms

In 1959 the eminent physicist Richard Feynman set the world the challenge of producing a working electric motor that would fit into a cube with sides less than four-tenths of a millimeter long. It took less than a year for the $1000 prize he offered to be claimed by a machine producing one-millionth of a horsepower. Even this feat of miniaturization seems insignificant compared with modern motors that have dimensions a thousand times smaller. At an even smaller level, instruments now exist that can arrange individual atoms into pictures and patterns.

Machines have become increasingly smaller over recent decades. Thanks to the silicon chip, a modern laptop computer has the processing power that once filled a whole room. It has now become possible to shrink machines even further. *Nanotechnology,* engineering on a scale of nanometers (billionths of a meter), can yield machines assembled from single atoms.

There are two approaches to making these tiny mechanisms. One alternative – known as *bottom-up* design – starts at the atomic scale and works upward, whereas the *top-down* approach depends on making ever smaller versions of existing components. An example of the latter approach is the use of photo-etching techniques, developed by silicon chip manufacturers, to build electric motors that are a hundred times smaller than the machine that won Feynman's prize. But this level of miniaturization is not strictly nanotechnology but microtechnology, the machines being micrometers (millionths of a meter) long.

Building with atoms

True bottom-up nanotechnology has advanced hand in hand with the invention of instruments that can reliably provide pictures of individual atoms. These instruments are the *scanning tunneling microscope* and its close cousin the *atomic force microscope* (AFM). In the AFM, a fine metallic probe is drawn back and forth across the surface of a sample. As it bumps over individual atoms on the sample it causes the path of a laser beam to be deflected. This deflection is detected, measured, and converted into a three-dimensional image by a computer.

Significantly, the AFM probe can also be used to move individual atoms and molecules into new configurations. Using the AFM, scientists have been able to write and draw patterns with single atoms, but the same techniques could be put to more useful work, such as building medical probes and ultrafast computers.

Normal computers rely on the properties of many thousands of electrons moving at once to process digital information, whereas in nanocomputers, these signals are transmitted by the movement of atom-thick carbon rods. Because nanoprocessors can be one thousand times smaller than conventional silicon chips, the time taken for signals to reach their destinations is greatly reduced. This allows calculations to be made much more quickly, resulting in a faster, more powerful computer.

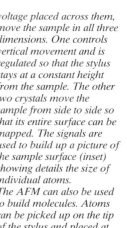

light sensor

laser

cantilever

display

A

sample

computer

stylus

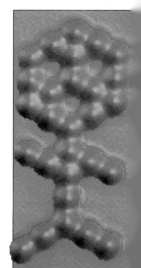

The gentle touch
*Just as the stylus of an LP record player is deflected up and down by the bumps in a disk's groove, the atomic force microscope (AFM) [**A**] tracks the irregularities on a sample's surface with a probe. The stylus of the AFM is much finer – at its tip it narrows to the width of a single atom – and it presses down with a force a million times smaller than that of a record stylus. The AFM stylus, sometimes a tiny shard of diamond, is attached to a silicon dioxide cantilever held rigidly at its other end. The sample moves underneath the stylus [1] so that bumps in its surface make the probe move upward, bending the cantilever [2]. A laser beam is deflected by the moving cantilever onto a light sensor. The fluctuating output of the sensor is fed to a computer in turn linked to the platter on which the sample is mounted. Piezoelectric crystals, which contract in proportion to a*
voltage placed across them, move the sample in all three dimensions. One controls vertical movement and is regulated so that the stylus stays at a constant height from the sample. The other two crystals move the sample from side to side so that its entire surface can be mapped. The signals are used to build up a picture of the sample surface (inset) showing details the size of individual atoms.
The AFM can also be used to build molecules. Atoms can be picked up on the tip of the stylus and placed at precisely determined positions. The stick person (right) was "drawn" using this technique.

Connections: Computers: New Developments 84 Silicon Chip Manufacture 126 Electron Microscopes 136 Microsurgery 158

Micromotor

A silicon chip is built up by depositing layers onto a substrate of silicon and then selectively etching away parts to leave the required shape. Mechanical devices can be made using similar methods. *Surface micromachining* is a method by which a sacrificial layer of silicon dioxide is laid onto the substrate. Another layer of working material is deposited on top and etched to shape. Next, a second etchant completely dissolves the sacrificial layer, leaving the resulting microstructure freely suspended. The technique is used to make the tiny accelerometers that detect car crashes and inflate air bags. The tiny micromotor *(right)* was built using the *LIGA* process, in which X-rays are used to etch away silicon. Its tiny rotor, roughly one-tenth of a millimeter across, is accelerated by electrostatic forces up to speeds of more than 20,000 rpm.

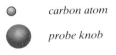

- ○ carbon atom
- ● probe knob
- ●●●● blocking knob
- ◄√√√ spring

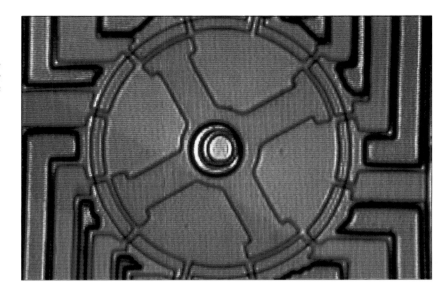

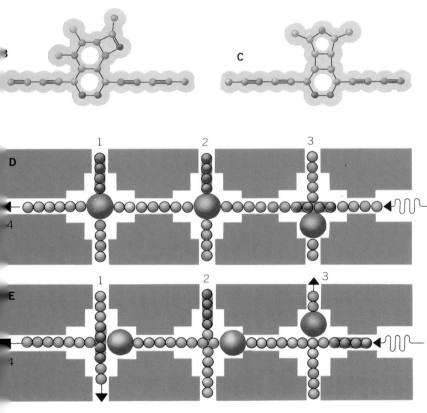

Industries in miniature

Nanotechnology has many other possibile applications, although most of these have not yet progressed beyond the design stage. These include miniature ultralow friction bearings, and electric motors that spin on a cushion of air, thus avoiding frictional losses. The former could be made from two concentric rings of synthetic diamond, coated with layers of fluorine only tens of atoms wide. Balanced *van der Waals* forces, which attract atoms at close distances, would keep the bearing centered.

Pollution-absorbing nanomachines in the atmosphere could clean up the environment, perhaps even reversing the greenhouse effect. In the distant future, the same technology could be used to alter the atmospheres of other planets to suit terrestrial life – a process known as *terraforming*.

The most significant advance of all may be in information technology. The ability to "write" in single atoms will radically effect data storage. The presence or absence of single atoms could be used to encode binary information. In such a form, all the libraries of the world could be stored on a few CD-sized disks. The technology for encoding and reading back information at nanometer scale at high speeds does not yet exist, but is under development.

Micromedicine

Another possible role for miniaturization is in medical treatment. Technologists have already created sensors that can measure blood pressure within an artery. Existing *keyhole surgery* techniques and an armory of miniature scalpels and laser beams enable a surgeon to perform operations through an incision only 2mm across. But in the future it may be possible to build tiny robots, the size of bacteria, that could be injected into the body. There they would directly destroy unwelcome organisms and fat deposits lining arteries.

Nano-nand gate

A nand gate is a logical device that compares two inputs, which can be either "on" or "off." It only gives an "on" output if both inputs are "off." One speculative design would use atom-thick carbon rods. Because of its molecular size, the gate would work at high speeds as part of an ultrafast computer. Each rod can slide into one of two positions for "on" and "off," and is fitted with "knobs." These come in two varieties – probes [B] and blockers [C] – each formed from different compounds. The gate is made of four rods. [1] and [2] are the inputs, [3] the output, and [4] a "test" rod. When the blockers on [1] and [2] are set to "off" (the nand condition), rod [4] can slide to the left, blocking [3] so that it indicates "on" [D]. Sliding [1] down, blocks the movement of [4], so that [3] can move upward, to represent "off" [E].

Keeping Track

How time is measured and defined

Time was traditionally governed by the rhythms inherent in the orbits of the earth and the moon. A day is the time taken by the earth to spin once on its axis, a lunar month the period between consecutive full moons. A year is the time it takes our home planet to complete one orbit of the sun. Yet these fundamental beats waver. Modern atomic clocks are so accurate – to within one second in a million years – that the basic unit of time (the second) is now defined as 9,192,631,770 cycles of one type of radiation released by cesium-133 atoms.

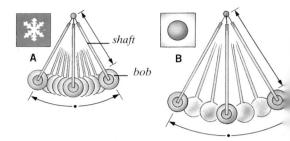

The ancient Egyptian day contained six hours that varied in length according to the season. Today, systems of timekeeping are based on the time taken for the earth to spin once on its axis relative to the steady background of the stars – so-called *sidereal* time. One sidereal day lasts 23 hours, 56 minutes, and 4 seconds, though tiny irregularities in the earth's rotation produce minute variations in this figure.

In the past, different towns in the same country would set their clocks independently. Standardization came with the advent of rail transportation. And as communications became faster, there was a need to standardize time worldwide. The Prime Meridian (0° longitude) was established at Greenwich, England, and the world was divided into 24 time zones a set number of hours ahead of or behind Greenwich time, or *Universal Time* (UT). A refinement of UT, known as *UT1*, corrects for irregularities in the earth's rotation. In practice world timekeeping is today based on *Coordinated Universal Time* (UTC), measured by accurate atomic clocks, calibrated to keep very close to UT1.

Times past

Early timepieces relied on falling water, burning candles, or following the shadows cast by the sun. By medieval times, the first wholly mechanical clocks were being built. These relied on the seesawing of a crossbar, or *foliot*, weighted at both ends, and attached to a spindle on which teeth locked into cogs of an *escapement* wheel. The release of each tooth from a cog turned the spindle and so caused the next swing of the crossbar – which then released the next tooth. Encumbered by friction and calibrated only in hours, these clocks allowed no scope for fine adjustment.

In the seventeenth century, Galileo Galilei discovered the *isochronism* of the pendulum: one of fixed-length swings with a constant period of oscillation, making it the ideal mechanism to drive a timekeeping device. At sea level, a pendulum 9.8in long has a period of exactly one second, but the metal of its rod can expand in hot weather. A *gridiron* pendulum uses the different rates of expansion of steel and brass to avoid this variation.

In 1657, the Dutch scientist Christiaan Huygens improved Galileo's design by using an escapement both to check the speed of the hands, and also to give a tiny push each swing to keep the pendulum going.

Time and length
A pendulum keeps the same period – the time to go through one swinging motion – even if the distance of the swings varies. This makes it an ideal way to govern the speed of a clock. But one factor that does change the period is the pendulum's own length – as the shaft gets smaller so does the period.

*If the shaft is made of a metal it will expand with heat. Thus on frosty days [**A**] the pendulum shortens and swings more quickly than usual, making the clock run fast. On a hot day [**B**] the shaft lengthens and swings with a longer period, so that the clock runs slow.*

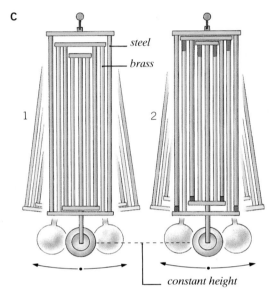

Constant beat
*The problem of thermal expansion is elegantly solved by a grid-iron pendulum. This is made from steel rods, fixed at their upper ends so they expand downward with increasing temperature, and brass rods, which grow upward at the same time [**C**].*

The two metals expand at different rates, so that the lengthening effect of the steel rods is exactly counteracted by the opposite movement of the brass components. This means that the pendulum has a constant length – and hence period – whether it is cold [1] or hot [2].

Time traveling

Accurate timekeeping has always been essential for navigation. To know a ship's longitude accurately, a navigator needs to keep a record of the exact time at the port of departure. By comparing this with the local noon, the time difference, and therefore the longitude, can be found. Even in the era of the satellite Global Positioning System, an accurate time signal is vital. Early portable clocks could not rely on pendulums, but instead used springs. The unwinding of a coiled spring within the clock is governed by an escapement, itself driven back and forth by a tiny hairspring. As the main coil unwinds, it drives a gear train that moves the hands.

Quartz crystals are *piezoelectric* – they vibrate at a precise frequency when excited by an electric current. This makes them a highly accurate way to govern a clock. Even more exact are atomic clocks, accurate to one second in ten thousand billion.

Atomic timing
The outermost electron in an atom of cesium-133 can orbit the nucleus in one of two directions – parallel or antiparallel to the overall spin of the atom. The antiparallel state has a slightly higher energy than the parallel one, and the atom can be made to move from the parallel to the antiparallel level if it is struck by microwave radiation of precisely the right frequency. This property is the basis of how an atomic clock works.

*Inside the clock [**E**], a sample of cesium-133 is heated in an oven. Individual cesium atoms "boil off" the sample and are accelerated into a fast-moving beam. Atoms in the beam are in both parallel (spin left) and antiparallel (spin right) energy states. The beam passes through a specially shaped magnet which separates spin left from spin right atoms. The spin left atoms are then steered into a cavity where they are hit by a beam of*

Connections: Navigation Aids 40 Radio 50 Spectrographs 140 Particle Accelerators 144 Principles 212 226 238 242 246 248 254 256 268

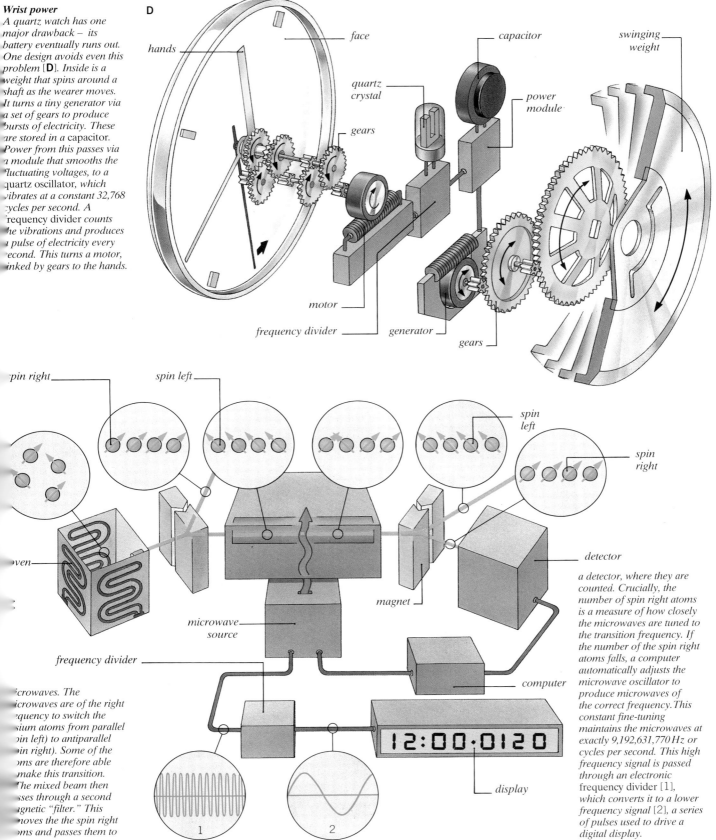

Wrist power

A quartz watch has one major drawback – its battery eventually runs out. One design avoids even this problem [**D**]. Inside is a weight that spins around a shaft as the wearer moves. It turns a tiny generator via a set of gears to produce bursts of electricity. These are stored in a capacitor. Power from this passes via a module that smooths the fluctuating voltages, to a quartz oscillator, which vibrates at a constant 32,768 cycles per second. A frequency divider counts the vibrations and produces a pulse of electricity every second. This turns a motor, linked by gears to the hands.

D

hands
face
quartz crystal
gears
capacitor
power module
swinging weight
motor
frequency divider
generator
gears

spin right
spin left
spin left
spin right

oven
microwave source
frequency divider
magnet
detector
computer
display

microwaves. The microwaves are of the right frequency to switch the caesium atoms from parallel (spin left) to antiparallel (spin right). Some of the atoms are therefore able to make this transition. The mixed beam then passes through a second magnetic "filter." This removes the the spin right atoms and passes them to

a detector, where they are counted. Crucially, the number of spin right atoms is a measure of how closely the microwaves are tuned to the transition frequency. If the number of the spin right atoms falls, a computer automatically adjusts the microwave oscillator to produce microwaves of the correct frequency. This constant fine-tuning maintains the microwaves at exactly 9,192,631,770 Hz or cycles per second. This high frequency signal is passed through an electronic frequency divider [1], which converts it to a lower frequency signal [2], a series of pulses used to drive a digital display.

12:00:0120

1
2

Diagnostic Images
How radiation is used to see inside the body

X-rays – electromagnetic waves with a remarkable ability to penetrate matter – were discovered more than 100 years ago by the German physicist Wilhelm Röntgen. Within a matter of weeks they were being used by physicians to photograph bones and to locate bullets in wounds. Today they remain a vital diagnostic tool, providing detailed images of bone, soft tissues, blood flow, and brain activity. And with the advent of digital detectors and 3D scans, it may not be long before doctors are able to examine holograms of a patient's internal organs.

X-rays are high-energy waves, falling between ultraviolet and gamma radiation in the electromagnetic spectrum. They are produced in a vacuum-filled tube, in which a high voltage (between 40 and 150 kV) accelerates electrons to around half the speed of light. The high-speed electrons are steered into a metal target, and in the resulting collision their energy is converted into high-frequency radiation, or X-rays.

During an examination, a burst of X-radiation is fired at the patient's body. The waves penetrate to a depth that depends on the *atomic number* and *density* of the tissue: some pass right through the body and fall on to a detector – usually a sheet of film – forming an image, or *radiograph*, which is a shadow of the dense parts of the body. Solid structures, such as bones, show up well in this simple type of examination, but detail in soft tissues can be *resolved* only by more sensitive techniques. One such technique – *xeroradiography* – is based on the same principle as the photocopier, and is often used for breast examinations. The detector is not film but an aluminum plate coated with an thin layer of the semiconductor selenium. The layer is given a positive electric charge before being exposed to the X-rays passing through the tissue. Wherever the X-rays strike the plate, they cancel small areas of its positive charge, thereby forming an electrostatic *latent image*. The plate is sprayed with negatively charged toner particles, which stick to the residual areas of positive charge, producing a visible image that can be transferred to a sheet of paper.

Moving pictures
The above techniques can produce only static snapshots of the body. But when coupled with *image intensifiers*, X-rays provide *real-time* images that can be viewed continuously during surgery. In an image intensifier, the X-ray detector is a phosphor screen. When hit by X-radiation, the screen gives out photons of light. This faint light is immediately converted by an adjacent photosensitive layer into electrons, which are in turn accelerated by a high voltage and focused onto a second, smaller phosphor screen, producing a bright, clear image that can be viewed by a TV camera, or digitized for further analysis.

Computer manipulation of digitized images allows X-rays to be used in novel ways. In *digital subtraction angiography* a radiograph is taken of a major blood vessel, then repeated after a dye, opaque to X-rays, has been injected into the bloodstream. The first image is digitally subtracted from the first, leaving behind a clear outline of the blood flow to, for example, the brain or kidneys.

Simple radiographs are sometimes difficult to interpret because the image formed is actually a picture of a number of internal structures superimposed, one on top of another. To separate these structures, a technique known as *computed tomography* is employed. An X-ray beam is passed through a thin slice of the body and is detected by a bank of detectors as it emerges. The beam is then rotated around the subject, and another exposure made. This is repeated until the same slice has been surveyed from all angles. Using a mathematical method known as *back projection*, a computer reconstructs an image of the slice. Many slices can be "stacked" on screen to form a three-dimensional view of a patient's internal organs.

Health hazard
X-rays are a form of ionizing radiation and so are harmful, particularly to dividing cells. Care is therefore taken to minimize a patient's exposure to X-rays during examination.

Recording a radiograph
An X-ray tube [A] is an evacuated glass envelope within which a coiled tungsten filament (the cathode) [1] acts as a source of electrons. A low voltage [2] heats the filament, "boiling" electrons off its surface. These are focused and accelerated to high speed by a large voltage [3] between the target (anode) [4] and the cathode. The electrons collide with the tungsten anode, producing X-rays

and heat as a by-product. To prevent the heat from building up, a motor [5] spins the target to 3000 rpm: in addition, a layer of oil around the tube helps dissipate the heat.

X-rays emerge through an opening in the housing. Before reaching the patient, they pass through a number

of adjustable apertures [6], which limit the size of the X-ray field according to the size of the film. A lamp [7] and mirror [8] provide a beam of light that coincides exactly with the path of the invisible X-rays: this is used to aim the radiation field.

X-ray film [9] is basically the same as photographic film, but has greater sensitivity. It is coated on both sides (rather than just one side) with emulsion [10] that detects X-rays and light alike. The film is sandwiched between two fluorescent screens [11], which emit light when they are struck by X-rays. The light emitted exposes the film, thereby forming a more intense image than X-rays alone. Above the film is a grid of many tiny holes [12]. This allows through those X-rays that have passed straight through the body [13], but not those that have been scattered – deflected by structures within the patient [14]. These scattered rays would otherwise blur the image.

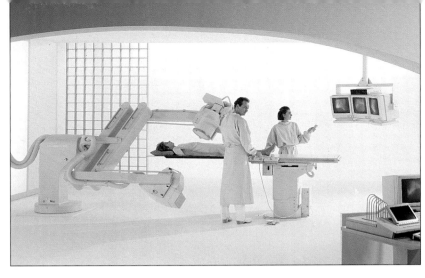

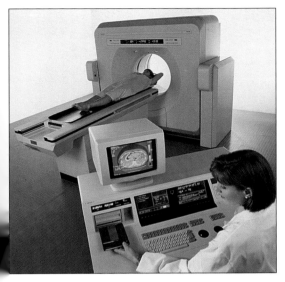

151

CT scan (left)
Computed tomography (CT) is a highly sensitive technique that allows a radiographer to take pictures of "slices" through the body. Using CT, it is possible to see very small differences in tissue density – for example the difference between white and gray matter in the brain. During examination, the patient lies within a doughnut-shaped scanning array. An X-ray source moves 360° around the body, and detectors around the patient take over a million readings of X-ray intensity. A powerful computer translates these readings into an image.

Hi-tech medicine (above)
Electronic image intensifiers allow X-rays to be used in extremely low doses. This means that a patient can safely be exposed to continuous radiation during surgery, affording doctors a real-time internal image that is displayed on screens above the operating table.

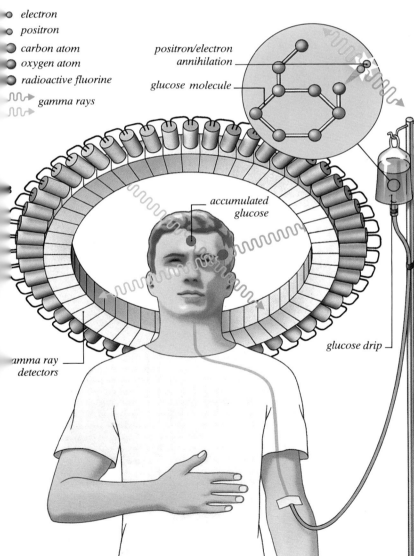

- ○ electron
- ○ positron
- ◐ carbon atom
- ◑ oxygen atom
- ◑ radioactive fluorine
- ∿► gamma rays

positron/electron annihilation

glucose molecule

accumulated glucose

gamma ray detectors

glucose drip

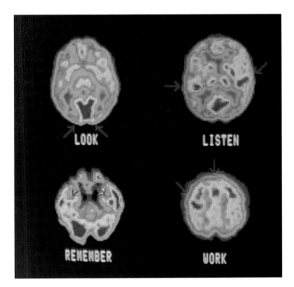

LOOK   LISTEN
REMEMBER   WORK

PET scans
Positron emission tomography (PET) [**B**] is a form of nuclear medicine in which small amounts of radioactive material are introduced into the body as a diagnostic tool. PET scans are often used to examine the brain because they can map its function rather than just exposing its structure.

Active areas of the brain are supplied with "fuel" in the form of glucose. In PET scans, glucose molecules are "labeled" with atoms of radioactive fluorine. As fluorine decays, it emits a subatomic particle called a positron. Almost immediately, this encounters its antiparticle, an electron.

When the two meet, they are annihilated, giving two gamma rays that move in exactly opposite directions. This double emission is the key to the PET scan.

A drip delivers labeled glucose into the patient's bloodstream: this collects in active parts of the brain. A ring of detectors around the head detects the gamma rays leaving the patient, and a computer "searches" for gamma rays that arrive simultaneously at two opposite detectors. These data are synthesized into images that show up those parts of the brain using most fuel – the centers of activity associated with thought processes (above).

Magnetic Medicine
How body scanners work

Using a sophisticated technique known as Magnetic Resonance Imaging (MRI), doctors are able to see deep inside the human body without risking exploratory surgery. MRI can see through bone to form startlingly clear images of the soft tissue beneath, which makes it particularly useful in the diagnosis of brain disease. During an MRI scan, the patient's body is literally magnetized and scanned by radio waves to build up a computerized map of hydrogen nuclei within the body. The technique is safe and painless, with no known side effects.

Forceful fields
*A magnetic resonance scanning department [**A**] is dominated by a huge cylindrical magnet (gray) that sets up a field 70,000 times as strong as that of the the earth. Three lower-powered electromagnets also encircle the patient (mauve, blue, and green).*

The magnetic fields produced by these shim coils are added to the main field to produce gradients in magnetic field strength across the patient's body.
Care is taken to exclude loose metallic objects from the room, because they are turned into projectiles by the intense magnetic field.

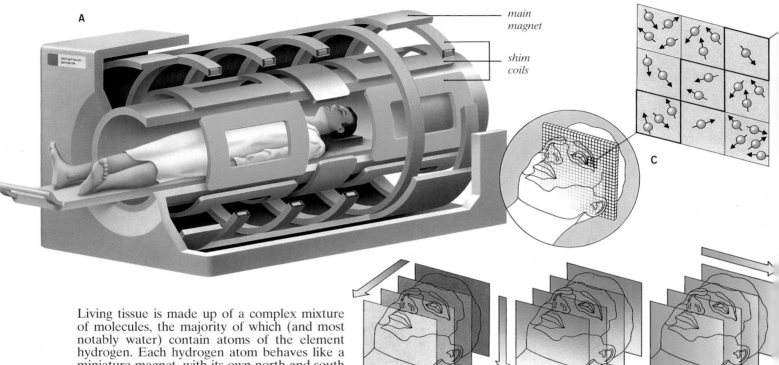

A main magnet / shim coils

C

B   Z field   Y field   X field

Living tissue is made up of a complex mixture of molecules, the majority of which (and most notably water) contain atoms of the element hydrogen. Each hydrogen atom behaves like a miniature magnet, with its own north and south pole. At body temperature, the atoms are constantly buffeted by neighboring molecules, causing their magnetic fields to point in different directions and cancel one another out. Under normal conditions, therefore, the body has no overall magnetic field.

Internal images
To produce an image by MRI, the patient lies on a couch surrounded by the coils of a giant electromagnet. This magnet creates an intense field, which causes the hydrogen atoms in the patient's body to line up in one direction – like iron filings drawn to a hand-held magnet – and temporarily magnetizes the patient.

Crucially, the hydrogen atoms in the body do not line up *exactly* with this powerful field, but wobble or *precess* around it, in the same way that spinning tops can be seen to wobble around their axes. The frequency of the precession (the rate at which the atoms wobble) is known as the *Larmor frequency* and depends on the strength of the magnetic field applied. In the next stage of imaging, a pulse of radio

waves "tuned in" to the Larmor frequency is fired at the tissue. Because the two frequencies match exactly, the precessing atom *resonates* and is momentarily thrown out of alignment. As it falls back into line with the original magnetic field, its proton reemits a radio signal of the same frequency as that which threw it out of alignment in the first place.

However, before the atoms have time to realign, the magnetic field is changed from a constant-strength field to a gradient. Atoms at different points along this magnetic gradient therefore realign to fields of slightly different strength and correspondingly give out radio signals of different frequency.

Each frequency is characteristic of a particular field strength, which in turn can be localized to a particular part of the patient's body. By monitoring the reemitted radio frequencies, a two or three dimensional picture of hydrogen atom density can be built up. In practice, this is

Mapping the body
*Each of the three shim coils sets up a magnetic gradient in a different direction [**B**] across the patient's body. The Z coil sets up a gradient in which the magnetic field diminishes from head to toe (mauve); the Y coil from top to bottom (blue); and the X coil from left to right (green). Every part of the human body can therefore be located by its X, Y, and Z magnetic coordinates.*
*To make an image of a section through the human head, the tissue is divided up into a grid of tiny boxes, or voxels, by the instrument's computer [**C**]. Each voxel has its own magnetic "address," and each one*

Connections: Radio 50 Spectrographs 140 Medical Radiation 150 Medical Ultrasound 154 Principles 212 214 248 262

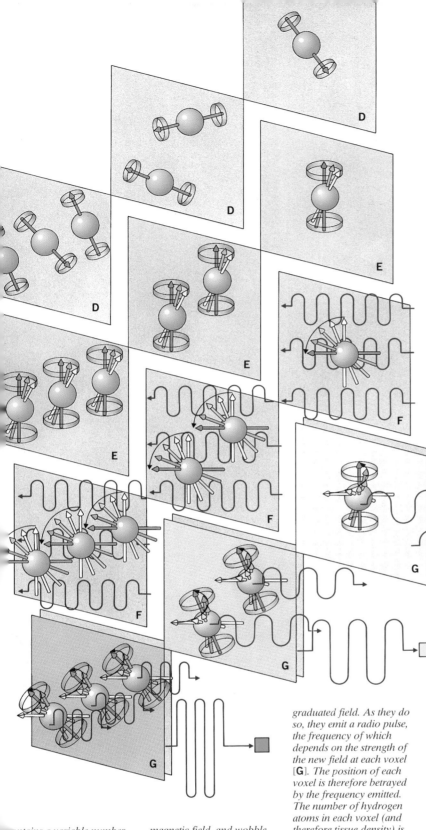

done by a powerful computer capable of storing and synthesizing all the positional information. The final image is displayed on a screen: colors added by computer allow areas of high hydrogen atom density to be distinguished clearly from less dense regions. Density is highest in body fluids, followed by soft tissues, then cartilage and membrane. Atoms immobilized in bone produce no detectable signal in MRI: so bone itself does not show up on the scan, although its outline is delineated by surrounding tissues.

During MRI, the patient is not exposed to any potentially harmful radiation; moreover, MRI produces far clearer images than X-ray or gamma-ray techniques. It allows doctors to visualize minute tumors, and to distinguish between the brain's white and gray matter. However, the technique does have some drawbacks: the patient must remain still for the full duration of the scan – up to 30 minutes – and some, especially small children, may require sedation or anesthesia.

Just as with plain X-rays, MRI images can be enhanced using contrast agents. The element gadolinium is sometimes injected to alter the body's local magnetic environment. This has the effect of changing the frequencies emitted by the protons, thus enhancing the image obtained.

ntains a variable number f hydrogen atoms.

Before the main magnet is vitched on, the hydrogen oms in all voxels are inning, but are aligned at ndom [D]. The main agnet is switched on [E] d the Z coil activated to t up a field gradient which cks out the plane of erest. The hydrogen oms line up with the

magnetic field, and wobble about their axis of spin. A pulse of radio waves, tuned into the frequency of wobble, is fired at the atoms, "knocking" them over [F]. Just before the atoms have a chance to realign themselves to the field, the X coil is switched on. Only then do the atoms "flop" back, aligning themselves to the new,

graduated field. As they do so, they emit a radio pulse, the frequency of which depends on the strength of the new field at each voxel [G]. The position of each voxel is therefore betrayed by the frequency emitted. The number of hydrogen atoms in each voxel (and therefore tissue density) is betrayed by the amplitude of the radio wave: high amplitudes mean high densities and vice versa. By measuring frequencies and amplitudes, a map of tissue density across the X axis is built up. A similar process is repeated with the Y coils switched on to get a two-dimensional section through the patient's head.

Acoustic Images

How ultrasound is used in medicine

The first picture of a child to appear in a family album is often a "sonic snapshot" taken months before birth. Formed by the echoes of inaudible sound waves beamed into the body, this *ultrasound* image provides doctors with vital diagnostic information about the developing fetus. Though most familiar from obstetric scans, ultrasound is also used to image the adult body, even providing moving pictures of blood flow through the beating heart. Comfortable, safe, and inexpensive, ultrasound has replaced conventional X-rays in many clinical applications.

The term "ultrasound" is applied to any frequency of sound that lies beyond the upper limit of human hearing (around 20kHz). In a medical scanner, brief "chirps" of ultrasound of between 1 and 15MHz are transmitted into the body, and the returning echoes are detected. Both of these operations are carried out by a *transducer,* the most important component of which is a crystal made of the synthetic *piezoelectric* material lead zirconate titanate (PZT).

A piezoelectric crystal has remarkable properties. When a voltage is applied, it changes shape; as soon as the voltage is switched off, the crystal snaps back to its original conformation. An oscillating voltage therefore makes the crystal vibrate, producing high-frequency sound as it does so. In the scanner, the transducer generates an ultrasound chirp lasting less than one microsecond, which is directed into the patient. A fraction of a second later, returning echoes of the pulse strike the transducer crystal, which then acts in reverse, converting the sound into electrical signals. These can be displayed on an oscilloscope screen for interpretation by a physician. This type of scan is called an A-scan: it gives simple information about depth and density of internal organs.

Sound scans

Making a recognizable image using ultrasound is less straightforward, because the ultrasound beam must be scanned, using many readings taken over a section of tissue. In early instruments, this was done by moving a single transducer by hand. Today the sound beam is focused and scanned by using an electronically controlled multielement transducer. The intensity of the echo at each point of the scan is converted electronically into a shade of gray, which is displayed on a screen. Relatively large echoes are sent back from organ boundaries, while small structures within the tissue give "grainy" low-level echoes. Fluid-filled cysts have a dark echo-free appearance allowing them to be distinguished from healthy tissue: similarly, tumors produce a characteristic echo.

The higher the frequency of ultrasound used, the shorter its wavelength, and the smaller the details that can be resolved. However, high frequencies are rapidly absorbed by tissue, and soon disappear beneath the electronic "noise" created by the scanner. The frequency used for scans is 3MHz – a compromise that yields resolution of around 1mm to reasonable depth.

Depth and density scan
*The probe used in the simplest type of ultrasound scan [**A**] consists of a piezoelectric transducer, and an acoustic lens, which crudely focuses the sound emitted. Pulses of ultrasound are emitted about one-thousandth of a second apart. In the intervening periods the transducer "listens" for returning echoes from the boundaries of different tissue types. Echo intensity and return time are detected and displayed on a screen. The position of the pulses reveals the depth of organs, and the height of the peaks shows the tissue density.*

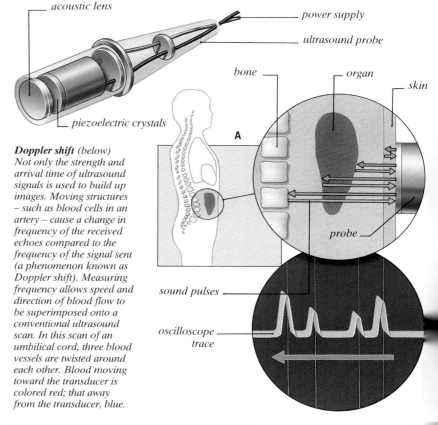

Doppler shift *(below)*
Not only the strength and arrival time of ultrasound signals is used to build up images. Moving structures – such as blood cells in an artery – cause a change in frequency of the received echoes compared to the frequency of the signal sent (a phenomenon known as Doppler shift). Measuring frequency allows speed and direction of blood flow to be superimposed onto a conventional ultrasound scan. In this scan of an umbilical cord, three blood vessels are twisted around each other. Blood moving toward the transducer is colored red; that away from the transducer, blue.

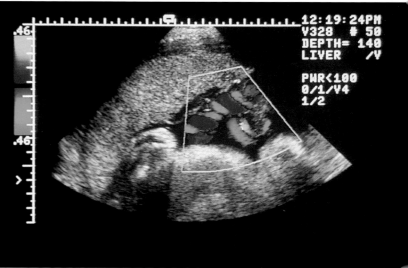

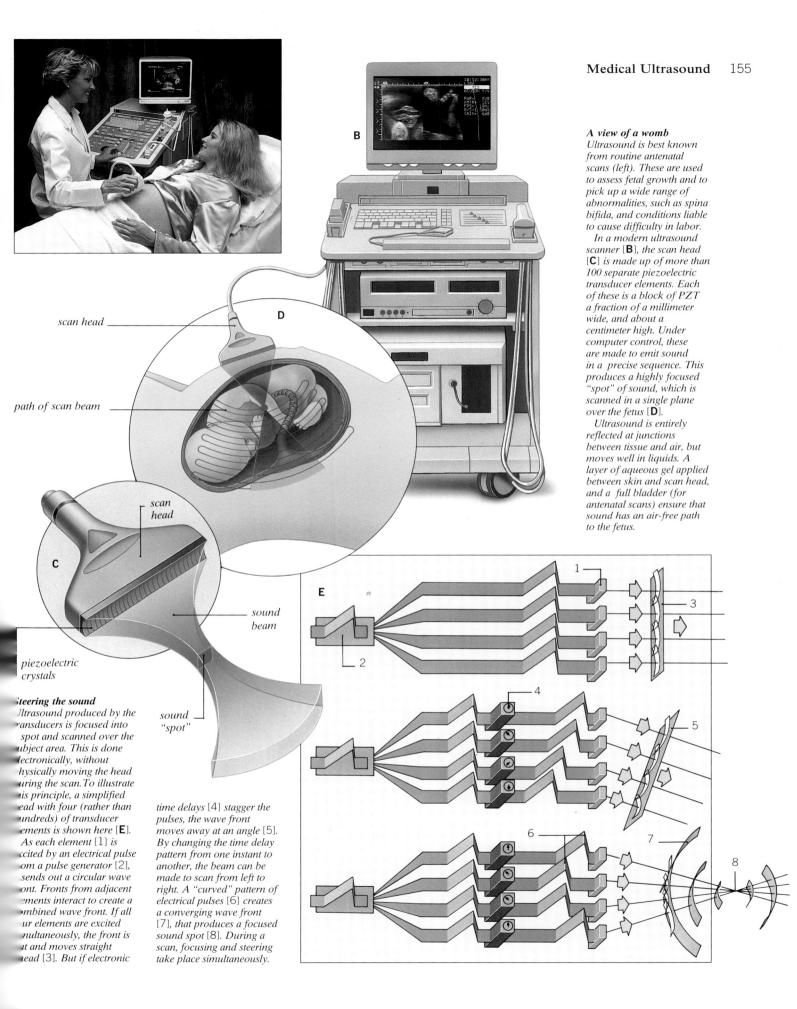

B

A view of a womb
Ultrasound is best known from routine antenatal scans (left). These are used to assess fetal growth and to pick up a wide range of abnormalities, such as spina bifida, and conditions liable to cause difficulty in labor.

In a modern ultrasound scanner [**B**], the scan head [**C**] is made up of more than 100 separate piezoelectric transducer elements. Each of these is a block of PZT a fraction of a millimeter wide, and about a centimeter high. Under computer control, these are made to emit sound in a precise sequence. This produces a highly focused "spot" of sound, which is scanned in a single plane over the fetus [**D**].

Ultrasound is entirely reflected at junctions between tissue and air, but moves well in liquids. A layer of aqueous gel applied between skin and scan head, and a full bladder (for antenatal scans) ensure that sound has an air-free path to the fetus.

scan head

path of scan beam

D

scan head

C

sound beam

piezoelectric crystals

sound "spot"

E

Steering the sound
Ultrasound produced by the transducers is focused into a spot and scanned over the subject area. This is done electronically, without physically moving the head during the scan. To illustrate this principle, a simplified head with four (rather than hundreds) of transducer elements is shown here [**E**]. As each element [1] is excited by an electrical pulse from a pulse generator [2], it sends out a circular wave front. Fronts from adjacent elements interact to create a combined wave front. If all four elements are excited simultaneously, the front is flat and moves straight ahead [3]. But if electronic

time delays [4] stagger the pulses, the wave front moves away at an angle [5]. By changing the time delay pattern from one instant to another, the beam can be made to scan from left to right. A "curved" pattern of electrical pulses [6] creates a converging wave front [7], that produces a focused sound spot [8]. During a scan, focusing and steering take place simultaneously.

Traces of Life

How the body's vital signs are monitored

The staff of a hospital's intensive care unit are aided in their work by ever-vigilant eyes – electronic sensors that continually monitor a patient's heart rate, blood pressure, oxygen level, and body temperature, giving instant warnings of developing problems. Their output can be recorded on a computer, building up a complete history of the patient's condition. This can even be transmitted down a telephone line, allowing a specialist in another hospital, possibly in another part of the world, to diagnose a patient's condition.

The human body is alive with electrical activity. Nerve cells (*neurons*) carry information around the brain and body in the form of *action potentials* – 100mV electrical pulses that last around one millisecond and move along nerve fibers at speeds of 220mph. Similarly, when a muscle contracts, each of its component fibers produces an action potential. In both cases, the action potentials produce measurable electrical signals that can be detected at the surface of the body by sensitive electronic circuits called *biopotential amplifiers*. Their main clinical application is in electrocardiography, which reads the electrical signals produced by nerves and muscles in the beating heart, and records them as an electrocardiogram (ECG).

The simplest type of ECG monitor is the *cardioscope,* which is attached to the patient via three electrodes. This is used to monitor heart rate and function following a heart attack, accident or serious illness, and during surgery. The more sophisticated *diagnostic* electrocardiograph uses a dozen or more electrodes placed around the chest to provide a three-dimensional "map" of the heart's electrical activity. A similar technique called *electroencephalography* measures and maps faint neural activity in the brain and records this as an electroencephalogram (EEG).

In the blood

Oxygen is carried around the body by the blood. In the lungs, atmospheric oxygen binds to molecules of hemoglobin, which are packed inside red blood cells (*erythrocytes*). The heart then pumps this *oxygenated* blood to respiring tissues. Whereas ECG can alert a physician to irregularities in heart function, other techniques are needed to gauge whether sufficient oxygen is being delivered to tissues around the body.

One such technique is *oximetry*. This makes use of the fact that oxygenated blood is brilliant crimson in color, while blood cells carrying less oxygen are a duller red/brown. Small samples of blood removed from a patient have their color "read" in a *spectrometer*, which reveals the proportion of oxygenated cells. The same measurements can be made by fiberoptic cables inserted directly into a vein or artery, or entirely noninvasively by *pulse oximetry*. Here, light is beamed through relatively transparent tissues, such as that in a finger, the nose, or an earlobe, and the emerging wavelengths are detected and converted into oxygen-content readings.

million times before display. Although EEG traces lack the regular appearance of the ECG, their component waves can be recognized by an experienced operator. Low-frequency rhythms called alpha waves are present in a relaxed state. During drowsiness or sleep, they are replaced by lower-frequency theta and delta waves. Higher-frequency beta waves are associated with activity in frontal areas of the brain.

The EEG is used to map regions of abnormal brain activity associated with tumors and epilepsy, and also to identify the incidence of "brain death."

Brain waves
To record an EEG, about 20 equally spaced electrodes are stuck to the patient's scalp (right). The patterns of electrical activity are traced out by a pen recorder or displayed on a screen (below). A typical recording takes between 30 and 60 minutes. The electrical signals on the surface of the brain result from the synchronized activity of millions of neurons. They are detected on the scalp and amplified about one

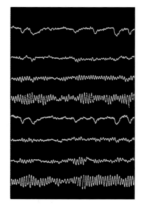

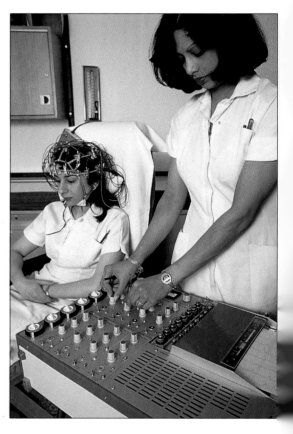

Vital signs

With each "beat," the heart goes through a pumping phase as it pushes blood at high pressure around the body (*systole*), and a resting phase when it refills with blood and pressure falls (*diastole*). By measuring blood pressure a physician can therefore determine if the heart is efficiently delivering blood to body tissues.

Blood pressure is measured with a *sphygmomanometer*. A cuff, wrapped around the arm is inflated to a known pressure, blocking off the flow of arterial blood. The cuff pressure is then slowly reduced. When it falls below the systolic pressure, restricted flow in the artery resumes. This makes a faint "rushing" sound that can be detected by a microphone placed on the limb. As the cuff pressure falls below the diastolic pressure, flow becomes completely silent. Blood pressure is expressed as two values, which represent the systolic over the diastolic pressure measured in millimeters of mercury (mmHg).

The color of blood
*The pulse-oximeter [**F**] measures oxygen levels in the blood. It consists of two light-emitting diodes producing red [1] and infrared [2] light, and two photodiodes sensitive to red [3] and infrared [4] wavelengths. Red light is absorbed by deoxygenated blood cells (brown) [5] but allowed to pass to the photodiode by oxygenated cells (red) [6]. Conversely, infrared is absorbed by oxygenated cells [7] but let through by deoxygenated [8] cells. From a comparison of the strengths of the signals from the two photodiodes, a measure of blood oxygen is calculated.*

Connections: Spectrographs 140 Chemical Analysis 142 Medical Radiation 150 Microsurgery 158 Principles 222 232 246 250 266

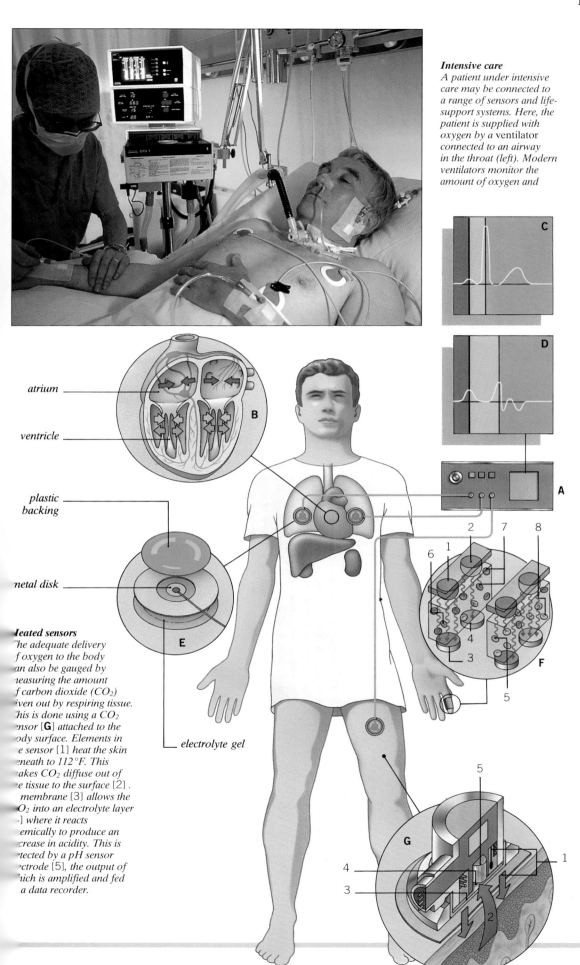

Intensive care
A patient under intensive care may be connected to a range of sensors and life-support systems. Here, the patient is supplied with oxygen by a ventilator connected to an airway in the throat (left). Modern ventilators monitor the amount of oxygen and carbon dioxide in the exhaled air, sounding an alarm if normal respiration is interrupted.

Numerous sensors keep track of the patient's vital signs. The cardioscope [**A**], for example, monitors electrical activity of the heart by detecting tiny electrical signals from the chest.

Three electrodes connect the patient to the cardioscope. Two are on the chest above the heart and the third is a "reference" connection on the leg.

The signal collected is only a few millivolts at the skin surface. This is amplified before being displayed on an oscilloscope screen. The trace produced not only reveals the rate at which the heart beats, but can also help diagnose heart disorders. In a normal heart [**B**], blood enters chambers called atria at the top. The atria contract (red arrows), forcing blood into the larger ventricles below. The ventricles then contract (green arrows), pumping blood around the body and then relax (blue arrows). Each part of this cycle produces its own characteristic "spike" on the cardioscope display (red, green, and blue areas respectively) [**C**]. In some types of heart disease, the distance between the first spike (corresponding to atrial contraction) and the second (ventricular contraction) is noticeably greater than normal [**D**]. This observation suggests that the nerve bundles that coordinate the contraction of the atria and ventricles are disrupted by inflammation or infection.

The disposable electrodes that pick up signals from the heart [**E**] consist of a metallic disk that makes contact with the skin via a conductive electrolyte gel. A plastic backing insulates the disk from external contacts.

For safety's sake, the cardioscope, which uses high voltage electricity, is isolated from the patient. This is achieved by optical breaks, which use light, rather than wires, to carry signals from the patient to the cardioscope.

atrium

ventricle

plastic backing

metal disk

Heated sensors
The adequate delivery of oxygen to the body can also be gauged by measuring the amount of carbon dioxide (CO_2) given out by respiring tissue. This is done using a CO_2 sensor [**G**] attached to the body surface. Elements in the sensor [1] heat the skin beneath to 112 °F. This makes CO_2 diffuse out of the tissue to the surface [2]. A membrane [3] allows the CO_2 into an electrolyte layer [4] where it reacts chemically to produce an increase in acidity. This is detected by a pH sensor electrode [5], the output of which is amplified and fed to a data recorder.

electrolyte gel

Microsurgery
How new technology is revolutionizing medical practice

The laser is the ultimate cutting edge of modern surgery. First developed for use in eye surgery in the 1960s, the medical laser now has numerous applications: laser light is used to vaporize brain tumors without harming surrounding tissue, and to "weld" together nerves and blood vessels in transplant surgery. New techniques in laser and other forms of surgery can deliver precise, delicate treatments without the hazards of a conventional operation. Even some heart surgery can now be performed while the patient is still conscious, and with little or no blood loss.

The use of the laser as a surgical instrument depends on its ability to deliver a narrow, highly concentrated beam of light. Although a medical laser may have a power output of only 60W – comparable to a domestic lightbulb – all the energy it produces is focused into a spot with a diameter as small as one thousandth of a millimeter (1 µm). The laser beam is said to have a high *power density* (measured in watts per square centimeter, W/cm^2).

Healing light
When laser light strikes biological tissue, it is absorbed by water or pigments in the tissue and converted into heat. If the power density of the laser is high (around $1kW/cm^2$), the heating effect is sufficiently intense to vaporize the target cells instantly, and cut through the tissue. At low power densities (less than $500mW/cm^2$), the temperature rises enough to kill and coagulate, but not vaporize, the target cells. The surgeon can change the effect of the laser beam on the target tissue by altering the power density of the laser light. This is most conveniently achieved by bringing the laser into and out of focus on the target: a *defocused* beam forms a wide spot with a low power density, suitable for coagulation, whereas a focused beam forms a narrow spot with high power density, and is used for cutting.

Laser light can be delivered to surface tissues directly, and to internal organs through an endoscope inserted through a natural opening.

The bloodless scalpel
The effect of laser light on tissue also depends on its wavelength. This is determined by the type of laser used. For example, light from the carbon dioxide (CO_2) laser has a wavelength of 10.6 µm. At this wavelength, light is absorbed by water in the target cells and its energy is quickly converted into heat at the surface, so the effect of the laser light is highly localized. The CO_2 laser is therefore useful where precision cutting is called for. Neurosurgeons use it to remove otherwise inoperable brain tumors, where the slightest damage to adjacent nerve tissue would prove catastrophic.

Light from the Nd–YAG laser (in which the lasing medium is based on the element neodymium, Nd) has a shorter wavelength, which is less readily absorbed by water. It passes through surface cells, penetrating more deeply into underlying tissue. The Nd–YAG laser

Keyhole surgery
*Using an endoscope [**A**], a surgeon is able to carry out minor operations, without cutting through overlying tissues. The endoscope consists of a long, flexible tube attached to a handset. The tube is inserted through a body opening and its tip is "steered" to its destination by turning wheels on the handset, which pull on wires running the length of the tube [1]. Bundles of optical fibers [2] inside the endoscope transmit light to the tip. An image is formed on a CCD (an array of light-sensitive cells [3] at the tip), sent up the tube along an electrical cable [4], and fed into a video monitor. A channel [5] carries air and water, making it possible to wash and dry the surgical site. Miniature surgical instruments, such as forceps, are controlled by a cable running through a parallel channel [6]. A wide variety of instruments can be fitted to the endoscope: toothed biopsy forceps [7] allow samples of tissue to be removed for analysis; metal "snares" [8] carry high-frequency electric current that can coagulate tissue.*

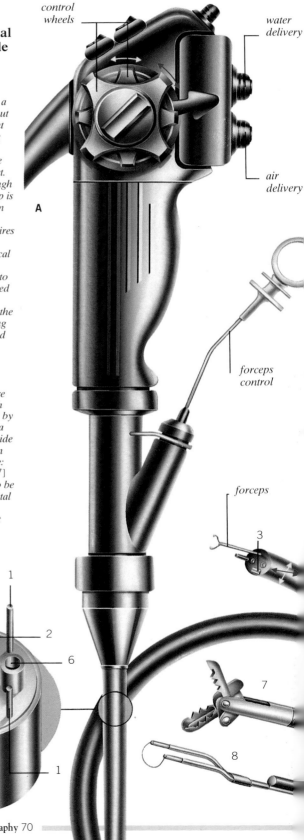

control wheels

water delivery

air delivery

forceps control

forceps

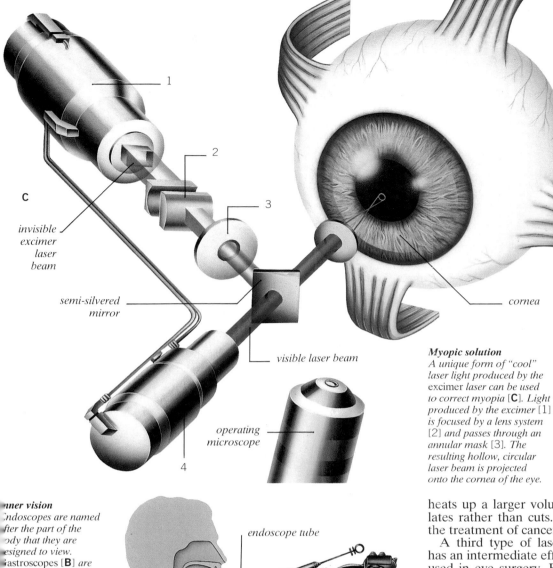

C

invisible excimer laser beam

1

2

3

semi-silvered mirror

visible laser beam

operating microscope

4

cornea

Myopic solution
A unique form of "cool" laser light produced by the excimer laser can be used to correct myopia [C]. Light produced by the excimer [1] is focused by a lens system [2] and passes through an annular mask [3]. The resulting hollow, circular laser beam is projected onto the cornea of the eye.

In a myopic eye, the cornea is too thick, so that light comes to a focus in front of the retina, resulting in a blurred image. The excimer sculpts away layers of the cornea to restore perfect vision. The light produced by the excimer is invisible, so the laser is physically linked to a helium-neon laser [4], which produces colored light. The surgeon uses its visible light as an aiming guide.

Inner vision
Endoscopes are named after the part of the body that they are designed to view. Gastroscopes [B] are used to examine the stomach. Here the image reveals an ulcer in the stomach wall.

B

endoscope tube

stomach

water supply

heats up a larger volume of tissue, and coagulates rather than cuts. It is most often used in the treatment of cancerous growths.

A third type of laser, the visible argon-ion, has an intermediate effect on tissue, and is often used in eye surgery. For example, a complication of diabetes is that blood vessels do not supply the retina with sufficient oxygen. To compensate, abnormal vessels grow forward into the eye chamber, impeding vision. The laser is used to coagulate the edges of the retina, preventing new blood vessel formation, and thus reducing the demand for oxygen. The procedure limits the field of vision slightly but prevents blindness.

Squeeze and freeze surgery
Balloon surgery is a technique that can clear arteries blocked with deposits of fat. Previously, this was done by highly invasive bypass surgery.

A tube is passed through the artery until its tip comes into contact with the fatty deposits. A balloon in the tip of the tube is then inflated, compressing the fat against the wall of the artery, thereby clearing a channel for the blood.

Cryosurgery uses freezing temperatures to destroy tissue. Liquid nitrogen, which has a boiling point of –319°F, is sprayed onto the tissue, either directly (as in the treatment of warts) or via a hollow probe that is inserted into the tissue (cancerous tumors can be destroyed this way).

This artificially colored image taken by the earth-observing Landsat satellite shows the San Francisco bay area.

6
Space

Liftoff

How rockets escape the pull of gravity

Poised on its pad, just north of the equator in French Guiana, a 187-foot-high Ariane 4 rocket is prepared for launch. Its liquid-fueled main engines along with its four strap-on boosters produce enough thrust to lift a payload of over four tons into a 250-mile orbit. But this European launcher is dwarfed by the massive Russian Energiya rocket, which can carry 100-ton satellites into low-earth orbit. Despite a cost of US$100 million per launch, most of a rocket will be used only once, abandoned to burn up as it falls back through the atmosphere.

A rocket is a simple demonstration of Newton's third law of motion: for every action (or force) there is an equal and opposite reaction. Inside the rocket *fuel* reacts violently with an *oxidizer* to produce very hot waste gases. These are forced backward at high speed out of the nozzle, producing a reaction force that pushes the launcher upward.

Escaping the earth's atmosphere takes a huge amount of energy. Consequently, most of the weight of a rocket launch vehicle at liftoff is fuel – the Russian *Energiya* rocket weighs more than 2200 tons at launch, but carries a maximum payload of only 100 tons.

Engines and motors

The fuel burned comes in two forms. Liquid-fueled rockets need complicated pipework and turbopumps to force huge volumes of liquid fuel (often hydrogen) and oxidizer (pure oxygen) together in a combustion chamber. Here they burn explosively, and with such heat that the nozzle has to be cooled by the liquid fuel. Hydrogen and oxygen are known as *cryogenic fuels* because they have to be kept at very low temperatures (–423°F for hydrogen). Fuels easier to store can be used, such as *hydrazine*, but do not give as much thrust per pound.

Solid-fuel rocket motors are more simple. Much like a firework rocket, in these the fuel and the oxidizer are mixed with a rubbery binder to form a cylindrical charge, or *grain*. A spark ignites one end and the explosive reaction drives the gas products out of the nozzle. The amount of thrust produced depends on how much surface area of the grain is exposed at any one time. Most grains have a hole running through them from end to end. If this is circular, a progressively greater area burns, so that the thrust increases during the flight. If the hole is star-shaped, the thrust remains constant. Complicated combinations of sections can be designed that vary the boost a rocket motor gives throughout a mission.

These motors are very powerful for their weight but suffer from the major disadvantage that once they have been lit they cannot be extinguished until all the fuel has been used up. In contrast, liquid burners can be throttled and turned on and off to suit different needs during the various stages of a mission. It is also much easier to steer these more complex machines by pivoting the relatively small engine around a central bearing.

Rocket-powered delivery
*Ariane 5 [**A**] is the European Space Agency's latest launch vehicle. Its main liftoff power comes from two solid-fuel rocket booster motors, each producing 600 tons of thrust. Each is a 100ft-long "firework", filled with a cylinder of solid fuel. An igniter lights the fuel at liftoff, and the boosters burn until their fuel is exhausted.*

At the top of each is a parachute cannister, which enables the booster to be recovered after a mission. In the middle is the central rocket engine, which burns oxygen and hydrogen, and controls thrust at critical stages of flight. Once these stages burn out, the upper stage takes over, carrying payloads of around 7 tons into high, geostationary orbit, or 22 tons into a low earth orbit.

The split fairing can be of different sizes to take varying payloads – one or two different satellites or even manned missions.

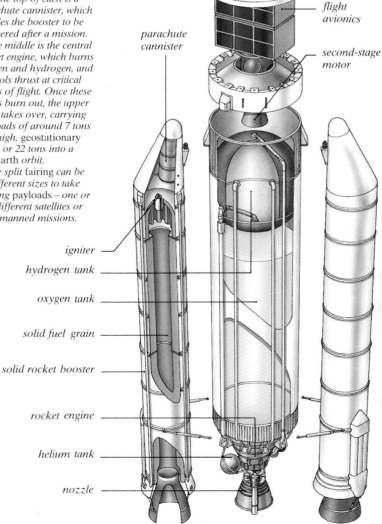

payload

A

fairing

flight avionics

parachute cannister

second-stage motor

igniter

hydrogen tank

oxygen tank

solid fuel grain

solid rocket booster

rocket engine

helium tank

nozzle

Connections: Space Shuttle 164 166 Space Flight: New Developments 180 Advanced Materials 194 Principles 220 224 234 238 240 242

Liquid engine

The schematic [D] shows a typical liquid-fueled engine, burning liquid hydrogen (fuel) and oxygen (oxidizer). The two liquids are pumped from their tanks by a pair of turbopumps before meeting in the combustion chamber. An explosive reaction takes place here, the products of which shoot backward out of the nozzle. This is kept cool by the liquid hydrogen pipes that wrap around it along their route. Small amounts of fuel and oxidizer are tapped off to power the turbopumps. They burn in a gas generator, the exhaust products driving a turbine connected by a shaft to the turbopumps.

A liquid-fueled rocket engine has the great advantage over a solid-fueled rocket motor that it can be throttled – the thrust it produces can be varied. Valves in the pipelines are used to alter the amounts of fuel and oxidizer burnt.

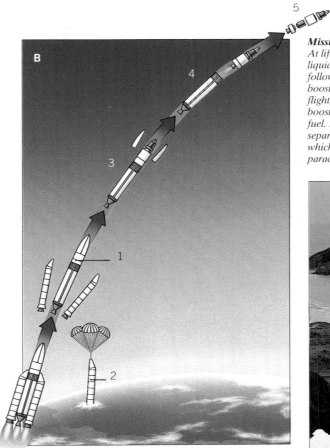

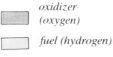

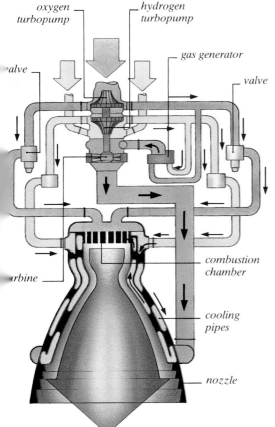

oxygen turbopump

hydrogen turbopump

valve

gas generator

valve

turbine

combustion chamber

cooling pipes

nozzle

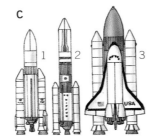

C

Rockets of the world

Ariane 5 [1] and the Japanese H-II [2] can launch satellites for a fraction of the cost of the manned American Space Shuttle [3] [C]. H-I (above right), another Japanese launcher, blasts off from the Japanese island of Tanega.

☐ oxidizer (oxygen)

☐ fuel (hydrogen)

☐ turbine gas

☐ combustion products

Mission profile

At liftoff, Ariane 5's main liquid engine lights first [B], followed by the solid rocket boosters. After two minutes' flight, 37 miles up [1], the boosters have used up their fuel. Explosive bolts fire, separating the spent rockets, which fall gently to earth on parachutes [2].

At 68 miles the fairing splits and falls away [3]. The main engine burns for 615 s, taking the rocket to 90 miles. Here it separates from the upper stage [4] and burns up in the atmosphere. The upper stage continues into space, powered by its own rocket engine, to place the payload into orbit [5].

Stages to orbit

A typical launch vehicle makes use of several different rocket motors. Typically these are made of a liquid-fueled main stage fired at the same time as a number of solid-fueled boosters, which detach from the main vehicle when their fuel runs out and return to earth on parachutes. The main engines continue burning for several minutes until a stable low earth orbit is achieved. A second stage then fires to insert the payload into its correct orbit even farther from the surface of the earth.

The location of the launch base itself can have a critical effect on the amount of fuel that a rocket needs to carry. The European Space Agency launches from Kourou in French Guiana, which is almost on the equator. This means that vehicles gain the maximum help from the spin of the earth, and need less fuel to reach high orbits compared with vehicles launched from centers farther north or south.

The Spaceplane
How NASA's Space Shuttle is launched into orbit

The maiden flight of NASA's Space Shuttle *Columbia* on April 9, 1981 marked the start of a new era in the space age. About the size of a small airliner, the Shuttle carries thirteen times the payload of a conventional Delta rocket into orbit, at only one and a half times the cost. The spaceplane is propelled into orbit by the combined thrust of liquid- and solid-fueled rockets, and maneuvered into position by two additional rocket systems. After completing its mission, it glides back to a runway landing and can be made ready for another launch in 100 working days.

Throughout its development in the late 1960s and 1970s, the Space Shuttle was seen as a vital component of the ambitious US space program. Regular shuttle flights were planned, with the spaceplane ferrying personnel and materials to permanent space stations – themselves staging posts for the exploration and exploitation of the Solar System. Although budget cutbacks and technical problems clipped the wings of this ambitious venture, the shuttle survived, albeit in a more modest form. Today it is used mainly to launch nationally or scientifically important payloads. Six orbiters have now been built: the prototype *Enterprise, Columbia, Discovery, Atlantis,* the ill-fated *Challenger,* and its replacement, *Endeavor.* A total of six to eight flights a year is planned for the foreseeable future.

Launch-pad line up
The shuttle can be launched from either of two sites depending on the type of mission being flown. Launches from Cape Canaveral, Florida, head eastward over the Atlantic, picking up extra energy from the earth's rotation, to enter an orbit above the equator. This trajectory is used to launch communications satellites and space probes. Alternatively, the shuttle can launch southward from Vandenberg Airforce Base in California, entering a polar orbit where it carries out environmental monitoring missions or launches remote sensing satellites.

Countdown to orbit
The Space Shuttle Main Engines (SSMEs) – three advanced liquid-fuel rockets situated at the rear of the orbiter – are fired 3.8 seconds before launch. The Solid Rocket Boosters (SRBs) fire almost simultaneously, and the SSMEs are throttled back to keep acceleration below a comfortable 3g. Fifty seconds after launch, the shuttle breaks the sound barrier. Within two minutes it reaches a speed of Mach 4.5 and an altitude of 28 miles. At this point, the SRBs cut out, their fuel exhausted, and fall back into the sea. After splashdown, the engine casings can be recovered and reused.

The SSMEs throttle up, taking the shuttle to Mach 15 and 80 miles, before it enters a six-mile dive at the end of which the main engines cut off, and the external tank is jettisoned. The Orbital Maneuvering System (OMS) – a rocket assembly fueled by propellant carried in the orbiter itself – then fires, placing the shuttle into orbit.

labels: orbital maneuvering system (OMS); fuel tanks; oxidizer tanks; helium tank; OMS/RCS pod; RCS thrusters; rudder; three main engines (SSMEs); elevon; solid rocket booster (SRB); United States; A; B

Modular spacecraft
*Prepared for launch, the Space Shuttle [**A**] consists of four units. The orbiter – a 125 ft spaceplane – sits atop a giant liquid-fuel tank flanked by two Solid Rocket Boosters (SRBs). The whole assembly is 184 ft high, and weighs more than 2200 tons, 82 tons of which is the orbiter itself.*

The SRBs are the largest solid-propellant motors ever flown, and the first designed to be reused. Each is packed with aluminum powder mixed with a powerful oxidizing agent and an iron oxide catalyst. When this mixture is ignited, the SRB behaves like a giant firework, burning down from top to bottom at a predetermined rate. The thrust it produces cannot be controlled after launch.

In contrast, the shuttle's three main engines (SSMEs) are liquid-fueled.

Because the flow of fuel can be regulated, the SSMEs can be throttled to vary their thrust. In each of the main engines, liquid oxygen and hydrogen, drawn from the vast External Tank (ET), are mixed and ignited.

The ET holds 142,560 gallons of liquid oxygen in its forward tank, and 400,000 gallons of liquid hydrogen in its aft tank. The two fuel tanks are separated by an intertank containing most of the electrical

components for directing fuel flow. The ET is heavily insulated to withstand the extremes of launch, but burns up in the atmosphere after it is jettisoned.

The wings and tail fin of the spaceplane have an aluminum framework of ribs and spars. Elevons on each wing and a rudder on the fin control the orbiter as it glides in for landing.

Connections: Aircraft: Aerodynamics 26 Rockets 162 Space Shuttle: In Operation 166 EVA: The Spacewalk 168 Living in Space 170

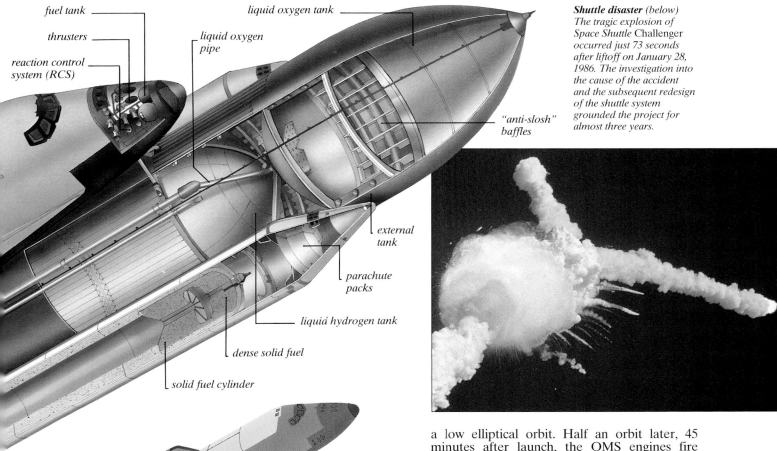

fuel tank

thrusters

reaction control
system (RCS)

liquid oxygen tank

liquid oxygen
pipe

"anti-slosh"
baffles

external
tank

parachute
packs

liquid hydrogen tank

dense solid fuel

solid fuel cylinder

C

carbon compound

high-temperature tiles

low-temperature tiles

felt insulator

Shuttle disaster (below)
*The tragic explosion of
Space Shuttle* Challenger
*occurred just 73 seconds
after liftoff on January 28,
1986. The investigation into
the cause of the accident
and the subsequent redesign
of the shuttle system
grounded the project for
almost three years.*

a low elliptical orbit. Half an orbit later, 45
minutes after launch, the OMS engines fire
again to lift the shuttle into a standard circular
orbit 250 miles up, where it circles the earth
once every 92 minutes.

Controlled descent
Shuttle flights usually last seven days, after
which the orbiter enters the descent phase. The
44 small rockets of the reaction control system
(RCS) fire to rotate the orbiter 180° in space, so
it is traveling tail-first at around 16,780mph.
The OMS engines are then fired to brake the
shuttle, and it begins to descend. The shuttle is
then turned again, and its nose raised to expose
the heavily insulated tiles on its underside.

At an altitude of around 75 miles friction
against the earth's upper atmosphere begins to
generate large amounts of heat, and temperatures on the shuttle exterior can reach over
2700°F. The air around the spacecraft is
stripped of electrons, creating an area of ionization which blacks out radio communication for
twelve minutes. As the atmosphere becomes
thicker, the shuttle's behavior changes from a
spacecraft's to a plane's, and its airfoil surfaces
are used to control descent.

Approach and landing are similar to a conventional aircraft's. The shuttle glides down to a
landing at 210mph, and its braking system can
be assisted by deploying an airbraking parachute from the tail section.

Keeping cool
*The shuttle is covered with
thousands of heat-insulation
tiles to protect it from the
fierce heat of reentry [**C**]. Its
most vulnerable parts – the
leading edges of the wings
and the nose cone – are
sheathed in a carbon
compound that withstands
temperatures up to 3000°F.
The craft's underside, where
temperatures reach 2325°F,
is covered in silica-based
tiles, while the upper and
mid-surfaces are shielded by
lighter tiles that withstand
temperatures up to 1200°F.
Part of the upper side,
which does not exceed
700°F, is covered in heat-
resistant felt.*

ace maneuvers
*hrust for orbital insertion
d de-orbit is provided by
o engines of the Orbital
aneuvering System
MS), housed in "pods"
the rear of the orbiter [**B**].
lium gas forces fuel and
idizer from their tanks
o the OMS engines where
y mix and ignite.*

*Once in orbit, delicate
adjustments to the shuttle's
path are carried out by the
Reaction Control System
(RCS) – 44 small rocket
engines, grouped in three
clusters; one on each of the
OMS pods and one on the
orbiter's nose. The RCS
engines use the same
propellants as the OMS.*

Orbital Operations
How the Space Shuttle is put to work

On April 10, 1984, Space Shuttle *Challenger* drew alongside the crippled Solar Max satellite in orbit 288 miles above the earth. The shuttle's 50-foot robot arm extended, grabbed the satellite, and pulled it into the spaceplane's giant cargo bay. After repair by space-suited astronauts, Solar Max was returned to its orbit. Performing *in-situ* repairs to costly satellites is just one of the shuttle's many capabilities. It is also used as a launch vehicle and to carry scientific experiments; in the late 1990s it will play a central role in the building of the International Space Station.

The key to the Space Shuttle's success is its great flexibility. The spaceplane's cargo bay is 60ft long and 15ft across – large enough to swallow a school bus – allowing it to carry various commercial, scientific, and military cargos, or *payloads*. The shuttle also carries up to seven astronauts – four crew and three scientists – who can perform many different tasks in a single mission, thereby making full use of precious time in orbit.

A ferry to space
The primary goal of a shuttle mission is often to deploy a space probe or satellite. During shuttle launch, the valuable payload (a communications satellite may be worth US$ 500 million, and the Hubble Space Telescope cost more than US$1.5 billion) is held securely in support cradles. Once in orbit, the shuttle's cargo bay doors are opened, and "clamshell" heat shields are extended over the payload to protect it from solar radiation. The shuttle then maneuvers into position using the rocket engines of its Reaction Control System, and the payload is either ejected from the cargo bay by springs, or lifted out by a robot arm.

Because the shuttle was designed only to reach low earth orbit (300 miles high), payloads destined for geostationary orbit (22,250 miles) or above carry additional rockets, which fire after the satellite is clear of the shuttle.

Flight deck (far right)
The crew and passengers of the shuttle occupy a cabin at the front of the orbiter. The cabin has two levels: a flight deck, from where the orbiter and its payload are controlled; and a lower mid deck level where the crew live and relax.

The forward control panels of the flight deck contain the orbiter flight systems, operated by the commander and pilot. The rest of the flight deck houses systems for in-orbit operations. Payload-handling controls are at the rear of the deck, as are the systems that maneuver the orbiter into position for satellite retrieval.

When Spacelab is carried, its subsystems are controlled from an area to the left of the aft flight deck, while science experiments are carried out on the right.

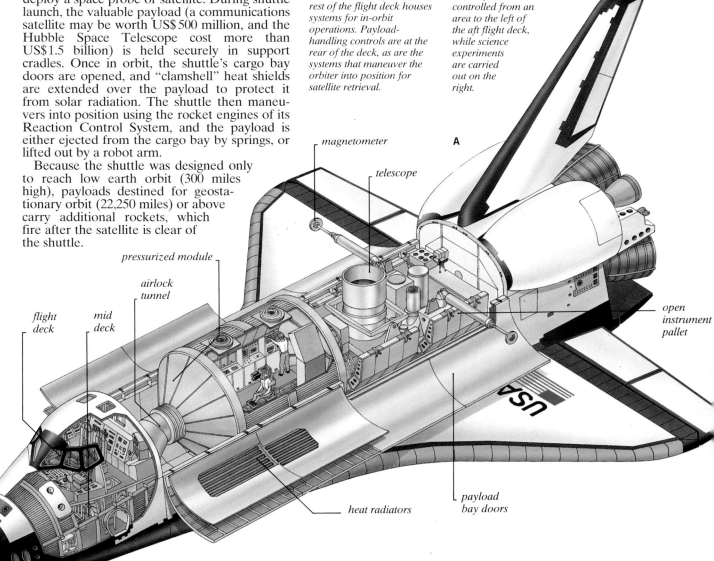

magnetometer

telescope

A

pressurized module

airlock tunnel

flight deck

mid deck

open instrument pallet

heat radiators

payload bay doors

Satellite launch

The shuttle is incapable of lifting satellites directly into geostationary orbit. To reach this higher orbit, the payload itself is fitted with a propulsion system, such as the Inertial Upper Stage (IUS) [**B**]. This comprises two solid rocket motors separated by an interstage. On reaching low earth orbit the IUS/satellite system is raised at an angle of 58° out of the cargo bay. The whole assembly is ejected by springs at approximately 4in/s. An hour later, after the shuttle has moved a safe distance away, the first-stage motor is ignited and the satellite heads away from the earth. Six hours later,

when it has reached geostationary distance, the second-stage motor fires to circularize the satellite's orbit.

Robot repairs

The shuttle can been used to service satellites in orbit or even return them to earth for repair.

Operations outside the shuttle may be carried out by suited astronauts, or by remote control, using a 50ft long articulated arm [**C**]. This Remote Manipulator System (RMS) is used to pull satellites in toward the shuttle. A standardized docking target, or grapple, on all shuttle-launched payloads is grasped by a

capture mechanism inside the "hand" of the arm. Once the grapple is inside the hand, a triangle of three wires closes, trapping the payload grapple.

The RMS is also used to lift very large satellites, such as the Hubble Space Telescope (left), out of the payload bay at launch.

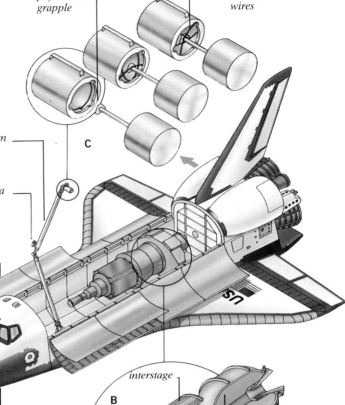

payload grapple

closing wires

remote manipulator arm

television camera

C

interstage

satellite interface

B

solid rocket motors

Shuttle and Spacelab

Spacelab is a modular laboratory which remains in the shuttle's cargo bay for the duration of a mission. It has no power source of its own, and relies on the spaceplane for heat and light. Experiments can be carried out in the vacuum of space, on pallets, or in

pressurized modules, which provide a comfortable working environment for up to three scientists. Different combinations of modules and pallets allow a variety of scientific experiments to be carried out.

The cylindrical pressurized module consists of a "core" segment, which holds life-support and data-processing equipment and some working space. A second pressurized segment can be added if more room is required for experiments. Scientists wearing normal clothing transfer from the spaceplane's mid deck to the pressurized module through an airlock tunnel. Experiments in the physical

sciences carried out in the modules include growing crystals in "microgravity" conditions, under which they remain flawless. Spacelab has also been used for studying the effects of weightlessness on living creatures. Often the astronauts themselves are the experimental subjects, though animals and plants are also taken into orbit.

The pallets are U-shaped to support and protect equipment, typically astronomical instruments and/or earth remote sensing monitors. An instrument pointing system mounted on the pallets enables accurate orientation of instruments relative to the shuttle.

Temporary space station

In the absence of a permanent US space station, the shuttle has taken on the role of orbiting laboratory. Small-scale space science can be performed in the orbiter's two-tier cabin. Larger experiments are carried out aboard Spacelab, a purpose-built European laboratory that sits in the shuttle's cargo bay. The shuttle can also launch a free-flying experimental pallet, called the Long Duration Exposure Facility. This can carry numerous experiments in orbit for more than a year, before it is retrieved by another shuttle mission.

During operations in orbit, the shuttle stays in contact with the ground via a constellation of five communications satellites, known as the Tracking and Data Relay Satellite system. From the vantage point of geostationary orbit, these satellites relay command signals and data between the shuttle and the White Sands ground terminal in New Mexico.

Connections: Rockets 162 Space Shuttle: In Flight 164 EVA: The Spacewalk 168 Living in Space 170 Satellites 172 174 176 Principles 236 242 262

A Walk Above the Earth

How a spacesuit gives vital protection to an astronaut

Among the most enduring images of space exploration's early years is that of man walking on the moon. As the exploitation of space has increased, so the importance of the spacewalk – or extravehicular activity (EVA) – has grown. Astronauts often have to leave their spacecraft to carry out repairs and perform experiments, or to develop new techniques – such as those needed to build the International Space Station in the late 1990s. Removed from the safety of their craft, astronauts depend on spacesuits to shield them from the hostile environment of space.

Beyond our planet's 60-mile-thick atmosphere, there is no air to breathe, and without the pressure exerted by the atmosphere, blood boils in human veins and arteries. The atmosphere filters out most of the harmful radiation emitted by the sun and diffuses its heat. In space, the human body receives the full intensity of the whole spectrum of the sun's radiation. This contains high levels of ultraviolet rays, X-rays, and gamma rays, which can burn the skin and cause skin cancer and blindness. The side of a body exposed to the sun is heated to over 212°F, while the shaded side receives no heat at all and, without insulation, radiates its heat into space and freezes.

Micrometeoroid impact is another hazard for astronauts. Many millions of tiny particles – most smaller than a grain of sand – swarm around the planet at speeds of up to 11,200 mph. Some are leftovers from the formation of the Solar System 4.5 billion years ago, but many are human-made – relics of satellite or rocket explosions. Spacesuits used in EVA must protect against micrometeoroids and radiation, keep temperature within tolerable limits, and provide a breathable atmosphere, while allowing an astronaut the mobility to perform delicate operations.

Early spacesuits were not designed for EVA, but were intended as backups, to be used in the event of cabin depressurization. When inflated, they became rigid, like a balloon: any movement by the astronaut reduced the suit's volume, increasing pressure, which made further movement even more difficult. The shuttle spacesuit, or Extravehicular Mobility Unit (EMU) is the first spacesuit designed specifically for EVA.

Solo flight
*A one-person spacecraft [**B**] called the Manned Maneuvering Unit (MMU) enables an astronaut to fly free in space without any link to the shuttle. The MMU is propelled by 24 nitrogen thrusters arranged in clusters of three on each of the eight corners of the unit. The nitrogen gas is kept under high pressure in two heavily sheathed aluminum tanks. When the astronaut operates the controls at the ends of the two side-arms, gas is fed to one or more of the thrusters. Gas rushing out in one direction propels the MMU slowly in the opposite*

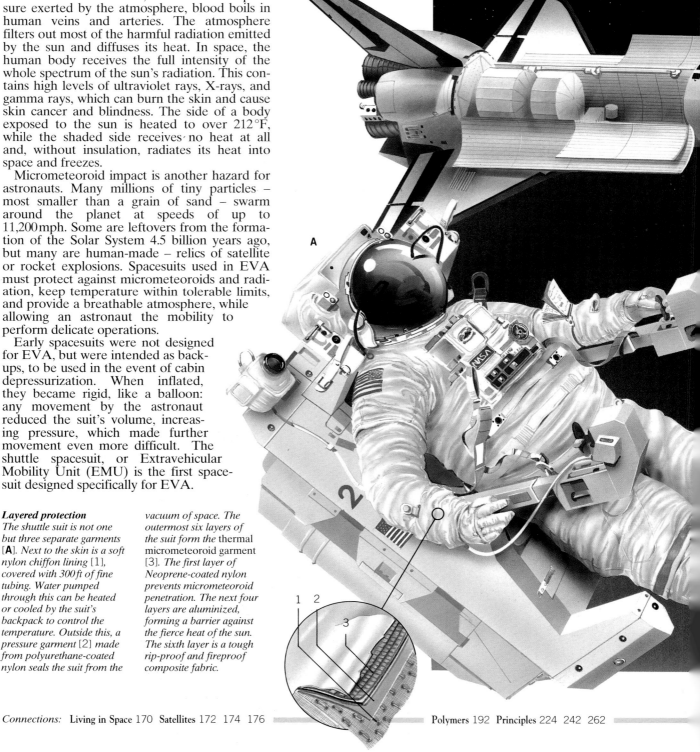

Layered protection
*The shuttle suit is not one but three separate garments [**A**]. Next to the skin is a soft nylon chiffon lining [1], covered with 300 ft of fine tubing. Water pumped through this can be heated or cooled by the suit's backpack to control the temperature. Outside this, a pressure garment [2] made from polyurethane-coated nylon seals the suit from the vacuum of space. The outermost six layers of the suit form the thermal micrometeoroid garment [3]. The first layer of Neoprene-coated nylon prevents micrometeoroid penetration. The next four layers are aluminized, forming a barrier against the fierce heat of the sun. The sixth layer is a tough rip-proof and fireproof composite fabric.*

Connections: Living in Space 170 Satellites 172 174 176 Polymers 192 Principles 224 242 262

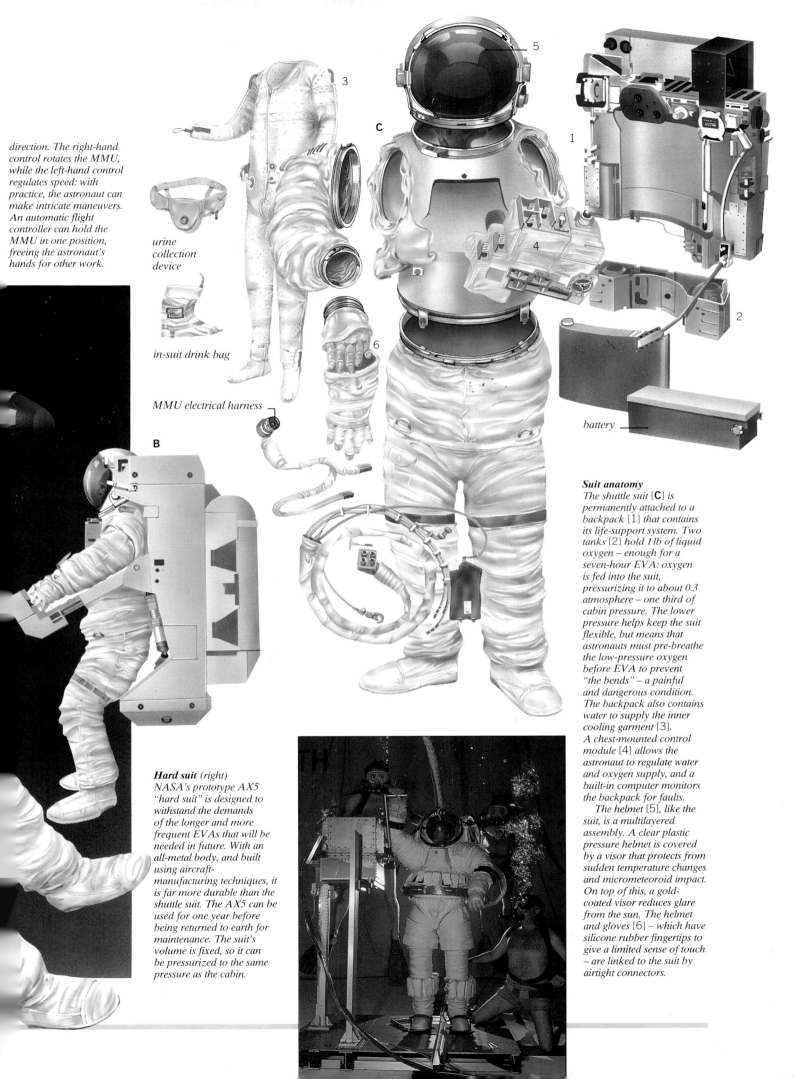

direction. The right-hand control rotates the MMU, while the left-hand control regulates speed: with practice, the astronaut can make intricate maneuvers. An automatic flight controller can hold the MMU in one position, freeing the astronaut's hands for other work.

urine collection device

in-suit drink bag

MMU electrical harness

B

C

Hard suit (right)
NASA's prototype AX5 "hard suit" is designed to withstand the demands of the longer and more frequent EVAs that will be needed in future. With an all-metal body, and built using aircraft-manufacturing techniques, it is far more durable than the shuttle suit. The AX5 can be used for one year before being returned to earth for maintenance. The suit's volume is fixed, so it can be pressurized to the same pressure as the cabin.

battery

Suit anatomy
The shuttle suit [C] is permanently attached to a backpack [1] that contains its life-support system. Two tanks [2] hold 1 lb of liquid oxygen – enough for a seven-hour EVA: oxygen is fed into the suit, pressurizing it to about 0.3 atmosphere – one third of cabin pressure. The lower pressure helps keep the suit flexible, but means that astronauts must pre-breathe the low-pressure oxygen before EVA to prevent "the bends" – a painful and dangerous condition. The backpack also contains water to supply the inner cooling garment [3]. A chest-mounted control module [4] allows the astronaut to regulate water and oxygen supply, and a built-in computer monitors the backpack for faults.

The helmet [5], like the suit, is a multilayered assembly. A clear plastic pressure helmet is covered by a visor that protects from sudden temperature changes and micrometeoroid impact. On top of this, a gold-coated visor reduces glare from the sun. The helmet and gloves [6] – which have silicone rubber fingertips to give a limited sense of touch – are linked to the suit by airtight connectors.

Space for Living
How space stations provide a platform for science and exploration

In 1988, Musa Manarov and Vladimir Titov became the first cosmonauts to spend an uninterrupted year away from the earth. Their long residence onboard the Mir station was more than a demonstration of the Soviet mastery of space. Orbiting stations provide a unique environment for pure and applied science, and allow space agencies to develop the techniques needed for future lengthy missions to the moon and planets. Early in the next millennium, Russian expertise and American technology will be combined in a new international space station.

After the early successes of the Soviet and American space programs, the establishment of permanently manned orbiting stations was seen as a logical continuation of humankind's push into space. The first space station, the Soviet Salyut 1, was launched in 1971 and was followed by six more Salyuts, the last of which was "mothballed" in orbit in 1986. Salyut 6 was occupied for two years, helping space scientists to develop the technology needed to sustain crews in orbit for long periods.

In 1973, NASA launched Skylab, a space station made out of a converted hydrogen tank from a Saturn V moon rocket. In all, four Skylab missions were flown, the longest stay by one crew being 84 days. Both Salyut and Skylab were of fairly simple design, with combined living quarters and experimental areas.

In contrast, the Soviet Mir space station, launched in 1986, was far more sophisticated and spacious, incorporating individual pressurized modules devoted to different purposes. The station is modular in design, with six docking ports to accommodate a resupply vessel and up to five laboratory modules. The first science module to be linked with Mir was the Kvant astrophysics lab, which contained ultraviolet, X-ray, and gamma-ray instruments exposed to the vacuum of space.

Space station Freedom
On January 5, 1984, President Ronald Reagan announced the beginning of America's most ambitious venture into space since the moon landings. At a cost of more than US$8 billion, a giant space station called *Freedom* was to be assembled in orbit by the mid 1990s. Since Reagan's announcement, plans for the station have been redrafted many times, reflecting successive cuts in the project's budget. In its new form, the space station will be an international venture, combining the original American design with elements of the Russian Mir space station and laboratory modules built by the European and Japanese space agencies. The station will orbit at an inclination of 52° to the equator, making it well placed for looking down on the earth, as well as for performing astronomical observations. Its laboratory modules will be fitted with 33 standard "user racks," so that a number of experiments can be carried out simultaneously.

The low-gravity (or *microgravity*) conditions in earth orbit open up new avenues of scientific research. Fundamental processes can be

Onboard exercise *(right)*
The rowing machine on board the Shuttle Atlantis *is used by astronauts for taking routine exercise as well as for evaluating the effects of weightlessness. Experiments have been carried out to assess how the cardiovascular system, bone and muscle adapt to weightless conditions. One of the main medical problems that must be addressed if long stays in space are to become routine is* bone demineralization. *This involves the loss of calcium and phosphate from weight-bearing bones under weightless conditions, and causes a dangerous weakening of the skeleton.*

Joint venture *(below right)*
The cost of the largest space projects has become so great that even the wealthiest countries cannot afford to finance them alone. Development of the Russian successor to the Mir space station has run into financial problems, as has NASA's ambitious Freedom station.

The space station of the future may therefore take the form of a joint venture that fuses the best elements of the two projects. In this artist's impression the basic American design is teamed with Russian supply vehicles, guidance and docking systems, and elements of the Mir station.

observed without the interference of gravity: the laws of motion, for example, may be examined without frictional effects. Mixtures of two compounds can be made in which the denser component does not settle to the bottom: this has commercial implications because it allows higher quality semicondutors to made, to create ever more efficient microprocessors. Single crystals can be grown at accelerated speeds to larger sizes than possible on the earth, which has implications for pharmaceutical research.

Astronauts onboard space stations often use their own bodies as experimental subjects. Experiments involve physiological monitoring to assess the detrimental effects of long stays in space, such as bone demineralization and the progressive rundown of the immune system. These effects need to be understood and controlled if the longer-term goals of human space exploration – colonization of the moon, and a mission to Mars – are ever to be accomplished.

Connections: Silicon Chip Manufacture 126 Space Shuttle 164 166 Space Flight: New Developments 180 Shaping Metals 190

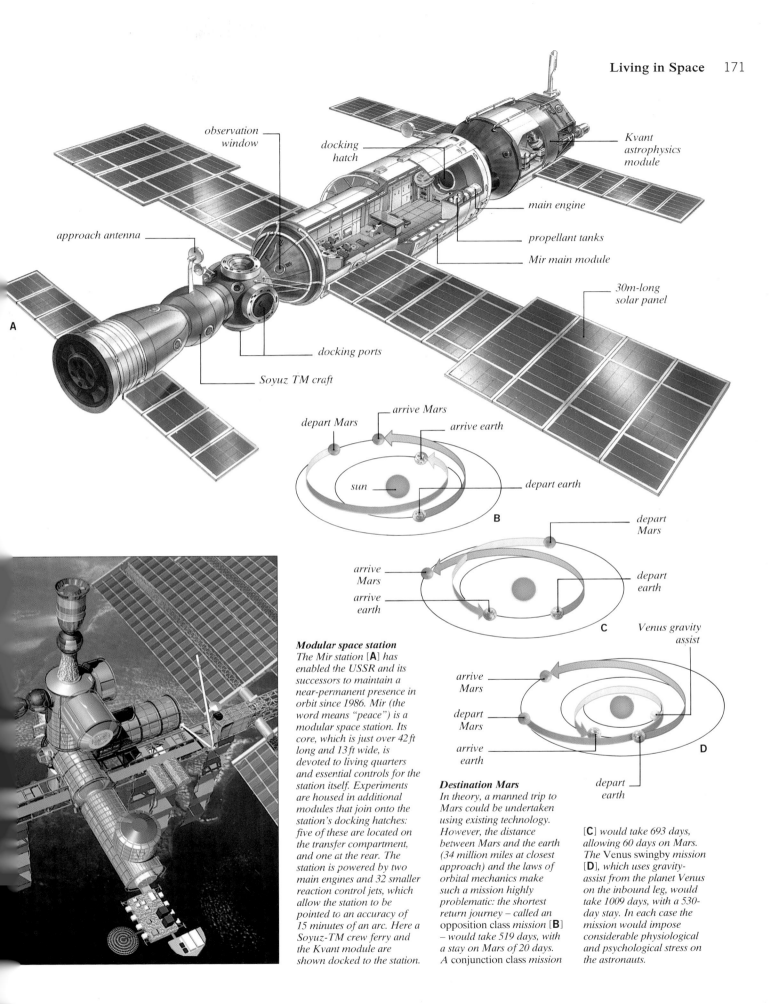

observation window

docking hatch

Kvant astrophysics module

main engine

propellant tanks

Mir main module

approach antenna

30m-long solar panel

docking ports

Soyuz TM craft

arrive Mars

depart Mars

arrive earth

depart earth

sun

B

depart Mars

arrive Mars

arrive earth

depart earth

C

Venus gravity assist

arrive Mars

depart Mars

arrive earth

depart earth

D

Modular space station
The Mir station [**A**] has enabled the USSR and its successors to maintain a near-permanent presence in orbit since 1986. Mir (the word means "peace") is a modular space station. Its core, which is just over 42 ft long and 13 ft wide, is devoted to living quarters and essential controls for the station itself. Experiments are housed in additional modules that join onto the station's docking hatches: five of these are located on the transfer compartment, and one at the rear. The station is powered by two main engines and 32 smaller reaction control jets, which allow the station to be pointed to an accuracy of 15 minutes of an arc. Here a Soyuz-TM crew ferry and the Kvant module are shown docked to the station.

Destination Mars
In theory, a manned trip to Mars could be undertaken using existing technology. However, the distance between Mars and the earth (34 million miles at closest approach) and the laws of orbital mechanics make such a mission highly problematic: the shortest return journey – called an opposition class mission [**B**] – would take 519 days, with a stay on Mars of 20 days. A conjunction class mission

[**C**] would take 693 days, allowing 60 days on Mars. The Venus swingby mission [**D**], which uses gravity-assist from the planet Venus on the inbound leg, would take 1009 days, with a 530-day stay. In each case the mission would impose considerable physiological and psychological stress on the astronauts.

Circling the Earth
How satellites stay in orbit

Since the voyage of Sputnik 1 in 1957, more than 3000 satellites have been launched into earth orbit. Today, our planet is surrounded by a halo of space hardware. Night and day, hundreds of artificial satellites traveling at speeds of up to 5 miles per second circle the earth, relaying telephone and television signals among the continents, and scrutinizing our world and the universe beyond. The orbit of each satellite is carefully tailored to its function, and continually adjusted by onboard propulsion motors.

A communications satellite races around the earth at almost 2 miles per second – ten times as fast as a jet aircraft. Like any moving object, its tendency is to move in a straight line, in a path that would take it farther and farther away from the earth. This tendency is countered by gravitational forces, which constantly accelerate the satellite back toward the ground. The satellite's path is therefore continually "bent" by gravity into a circle (or ellipse) that rings the earth.

Time for revolution
The earth's gravitational field is strongest at the planet's surface, but diminishes rapidly with altitude, dwindling to near zero at a distance of 620,000 miles. The stronger the gravitational force acting on a satellite, the faster it must fly to avoid spiraling down to the ground. Satellites in low orbits (around 250 miles up) move at around five miles per second and complete one rotation of the earth approximately every two hours. Those in higher orbits move at slower speeds and have a longer orbital *period*. Few satellites are placed into orbits lower than 185 miles because at these altitudes they rub against the upper atmosphere. Slowed down by frictional forces, they gradually lose height, and therefore have a short operational life.

Options for orbit
A satellite placed into orbit 22,300 miles above the equator takes exactly one day to complete an orbit. In this time, the earth beneath it turns exactly once on its axis: the satellite therefore rotates in perfect time with the earth, always remaining in the sky above exactly the same point on the planet's surface. The science-fiction writer Arthur C. Clarke was the first to suggest a practical use for this *geostationary* orbit: three satellites placed in a ring at this height could act as radio relays between any two points on the earth's surface. Today, geostationary satellites broadcast everything from phone calls to navigation signals: in fact, the orbit is becoming so overcrowded that its use has to be governed by international regulations.

Large, costly rockets are needed to launch satellites into high geostationary orbits, but less powerful launch vehicles (such as NASA's space shuttle) can place satellites into lower orbits far more economically. The "cheapest" orbit to achieve is *low earth orbit*. Here, the satellite circles the earth above the equator at a height of around 250 miles. The cost is further reduced by firing the satellite in an easterly direction so it "steals'" some of the earth's rotational energy at launch. Low earth orbit is used to "park" satellites before they are boosted into higher, more useful orbits; and some earth-monitoring and astronomical satellites also occupy this lower orbit.

Pole to pole
A *polar orbit* carries a satellite at high speed and relatively low altitude in a circle over the poles. Typically, the satellite completes 14 revolutions of the planet in a single day. As the satellite traverses the globe from north to south (more or less parallel to lines of longitude), the earth rotates beneath it from west to east. This means that the satellite "sees" every part of the earth's surface at close range, though it passes over the same point relatively infrequently. Polar orbits are particularly suited to *remote*

Many moons
Of all satellite orbits [A], the geostationary one – 22,300 miles above the equator – is perhaps the most useful. It is occupied by weather satellites, such as Meteosat [1], which have a constant view of an entire hemisphere of the planet, and by communications satellites, such as Intelsat [2]. A message can be sent from earth's surface to one of the Intelsat constellation, then amplified and transmitted via a second or even third Intelsat [3] to any part of the globe. TDRS [4] is a specialized communications satellite that relays messages from various spacecraft to ground control. Launching a heavy satellite like TDRS into high geostationary orbit is accomplished step-by-step. It is first launched into a low equatorial "parking" orbit, 250 miles up [5]. Booster motors then fire to take it into an elliptical transfer orbit [6]. At geostationary height, the motors fire once again to circularize the elliptical orbit [7].

Although geostationary communications satellites cover most of the globe, they transmit poorly to high latitudes. Areas such as Siberia get 24-hour coverage from Molniya satellites [8] in angled, elliptical orbits.
Remote sensing satellites circle the globe from pole to pole as the Earth rotates underneath. Landsat 3 [9], 560 miles up, orbits the earth 14 times a day, and "sees" the entire globe once every 18 days.
Astronomical satellites, such as the Gamma Ray Observatory [10], usually occupy low, "cheap" orbits that elevate them above the earth's atmosphere.

A

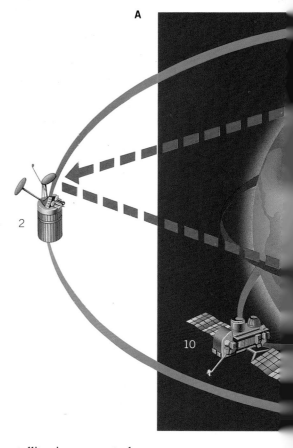

***Pass the message** (right) Satellites carry advanced communications systems. The TDRS satellite, which relays signals from the earth to numerous spacecraft, has seven antenna systems. The two large umbrella-like dishes relay data at a high rate (300 megabits per second) to and from one spacecraft at a time. The central multiple-access antenna can send messages to 20 satellites at a time, though data are sent at a slower rate (50 kilobits per second). Commercial communication satellites ca simultaneously transmit 1200 telephone calls or one television signal.*

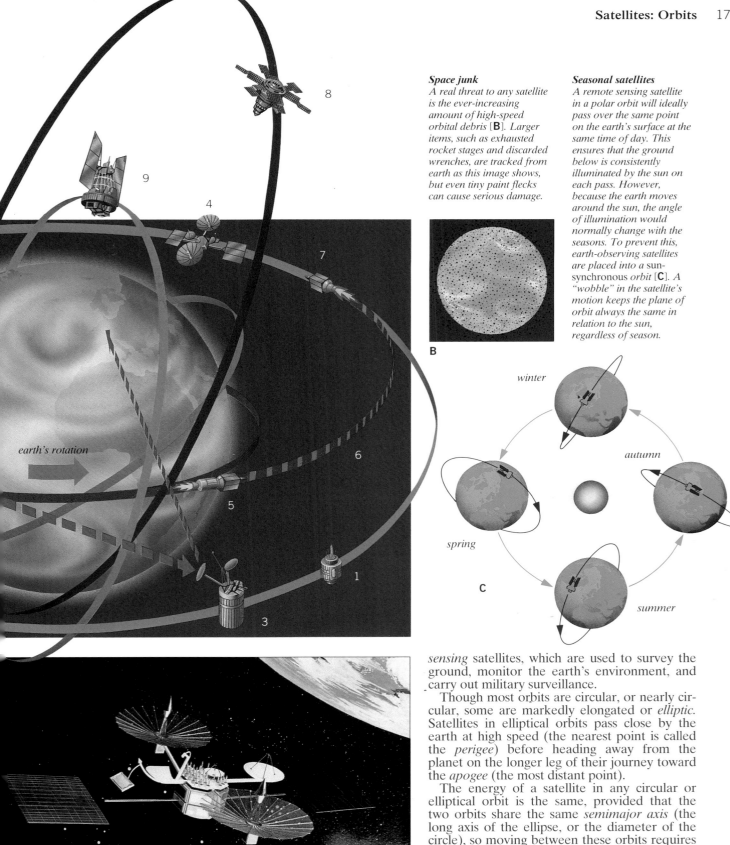

earth's rotation

8

9

4

7

6

5

1

3

Space junk

A real threat to any satellite is the ever-increasing amount of high-speed orbital debris [B]. Larger items, such as exhausted rocket stages and discarded wrenches, are tracked from earth as this image shows, but even tiny paint flecks can cause serious damage.

B

Seasonal satellites

A remote sensing satellite in a polar orbit will ideally pass over the same point on the earth's surface at the same time of day. This ensures that the ground below is consistently illuminated by the sun on each pass. However, because the earth moves around the sun, the angle of illumination would normally change with the seasons. To prevent this, earth-observing satellites are placed into a sun-synchronous orbit [C]. A "wobble" in the satellite's motion keeps the plane of orbit always the same in relation to the sun, regardless of season.

winter

autumn

spring

C

summer

sensing satellites, which are used to survey the ground, monitor the earth's environment, and carry out military surveillance.

Though most orbits are circular, or nearly circular, some are markedly elongated or *elliptic*. Satellites in elliptical orbits pass close by the earth at high speed (the nearest point is called the *perigee*) before heading away from the planet on the longer leg of their journey toward the *apogee* (the most distant point).

The energy of a satellite in any circular or elliptical orbit is the same, provided that the two orbits share the same *semimajor axis* (the long axis of the ellipse, or the diameter of the circle), so moving between these orbits requires only a short engine burn at the right point in the orbit. Elliptical orbits are therefore used as a "cheap" way of taking satellites far away from the earth, for example, to study the magnetic fields and radiation belts around our planet.

Eyes in the Sky

How the earth and atmosphere are monitored from space

From their vantage point hundreds of miles above the earth, satellites are ideally placed to observe our environment. Bristling with advanced sensors, they provide a continually improving picture of our home planet and what we are doing to it. Satellite-borne instruments can distinguish among different types of vegetation, track weather systems, and generate detailed maps of uncharted areas. In recent years their ever-vigilant eyes have identified the Antarctic ozone hole, and tracked forest fires and the spread of environmental pollution.

The value of earth observation (or *remote sensing*) from space was first recognized in 1960 – just 30 months after the pioneering voyage of Sputnik – when the United States launched its Tiros 1 weather satellite. However, the full potential of remote sensing was only realized several years later, when the first astronauts reported seeing roads, individual buildings, and smoking chimneys from orbit. Given this impetus, the space agencies rapidly developed the techniques and equipment needed to survey the planet from space. The following years saw the launch of satellites gathering information of scientific, commercial, and military value.

Seeing in a different light

Remote sensing instruments carried by satellites may be *passive* or *active*. Passive sensors record radiation reflected or emitted by the earth; in contrast, active instruments probe the planet with pulses of radio waves. Passive sensors include TV and photographic cameras that record visible light, and infrared scanners that detect heat emitted from the earth's surface.

Arrays of passive detectors may be combined in a single scanner, such as the Thematic Mapper carried on board the American Landsat satellites. The Thematic Mapper works by focusing radiation onto seven sensors, each sensitive to a different wavelength band, from the visible to the far infrared. Each band carries specific information about the terrain below. For example, Band 1 (visible light) can distinguish bare soil from vegetation, and so is used for forestry mapping. The amount of green light reflected by vegetation is measured on Band 3, allowing diseased crops to be identified. Longer infrared wavelengths (Bands 6 and 7) can determine the moisture of soil and plants, warning of impending drought damage. They also differentiate between different types of rock, and so are used in geological surveys.

The main drawback of passive monitoring is that target areas are frequently obscured by clouds. In addition, passive instruments depend on the illumination of the ground by sunlight, which varies in direction and intensity from one place to another. This means that measurements from different areas cannot be compared directly. Active instruments are not subject to the same limitations: they send out radio pulses of consistent intensity at frequencies that penetrate cloud cover. Directly comparable images can therefore be made in all weathers.

Sea view (below)
Passive sensing in the visible and infrared parts of the spectrum is often carried out using a scanning radiometer. *Radiation reflected from, or emitted by, the earth is focused on to a detector by an oscillating mirror that scans the surface below the* satellite. *Images are constructed from a series of adjacent scan lines. The image below was taken by the Coastal Zone Color Scanner on board NASA's Nimbus 7 satellite. Added color indicates plankton distribution: orange shows a high plankton abundance; purple low abundance.*

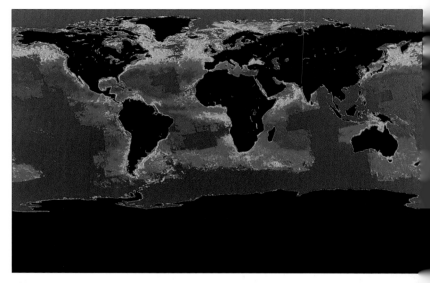

European eye
In 1991, the first European Remote Sensing Satellite (ERS-1) [**A**] was launched into a polar orbit. Equipped with an array of active radar sensors, it has returned much valuable data about land and seas.

The radar signals sent toward the earth by ERS-1 from its orbit 485 miles up have to be strong to produce detectable echoes, so large antennae are needed. As a result, the satellite is heavy, weighing 2.6 tons. It draws 1 kW of electrical power, supplied by a 40 ft-long solar panel.

The satellite's exact position above the earth must be known for its data to have value. Two systems are used to track the satellite; a microwave ranging system (PRARE), and a passive reflector, which reflects pulses of laser light directed at the satellite from the ground.

The satellite carries an infrared radiometer (the ATSR) that allows precise measurement of sea temperatures and of the water vapor content of the atmosphere. But the largest sensor on ERS-1 is the Active Microwave Instrument (AMI). This comprises two radar systems: the wind scatterometer, with which wind speed and direction can be measured; and the Synthetic Aperture Radar (SAR). This produces highly detailed radar maps of 60-mile-wide strips of the earth's surface.

The large amounts of scientific data from these instruments are transmitted via the the Instrument Data Handling and Transmission (IDHT) system.

Far-sighted satellites

Earth-observing satellites are normally placed into polar orbit between 155 and 620 miles up. As they move from pole to pole, the earth rotates beneath their path, so that they pass over the same point on the planet's surface only once every 16 days. However, their low altitude means they can produce high-resolution images (down to 100 ft) of this point. In contrast, weather satellites are usually placed into much higher geostationary orbits, where they have a continuous but low-resolution (around half a mile) view of the same broad area. This type of view is more useful for tracking weather fronts, which change direction from hour to hour.

Remote sensing data are digitized and stored magnetically on board the satellite. When the craft passes over a ground station, the information is sent down as a radio signal. Alternatively, data can be relayed in "real time" via geostationary communications satellites.

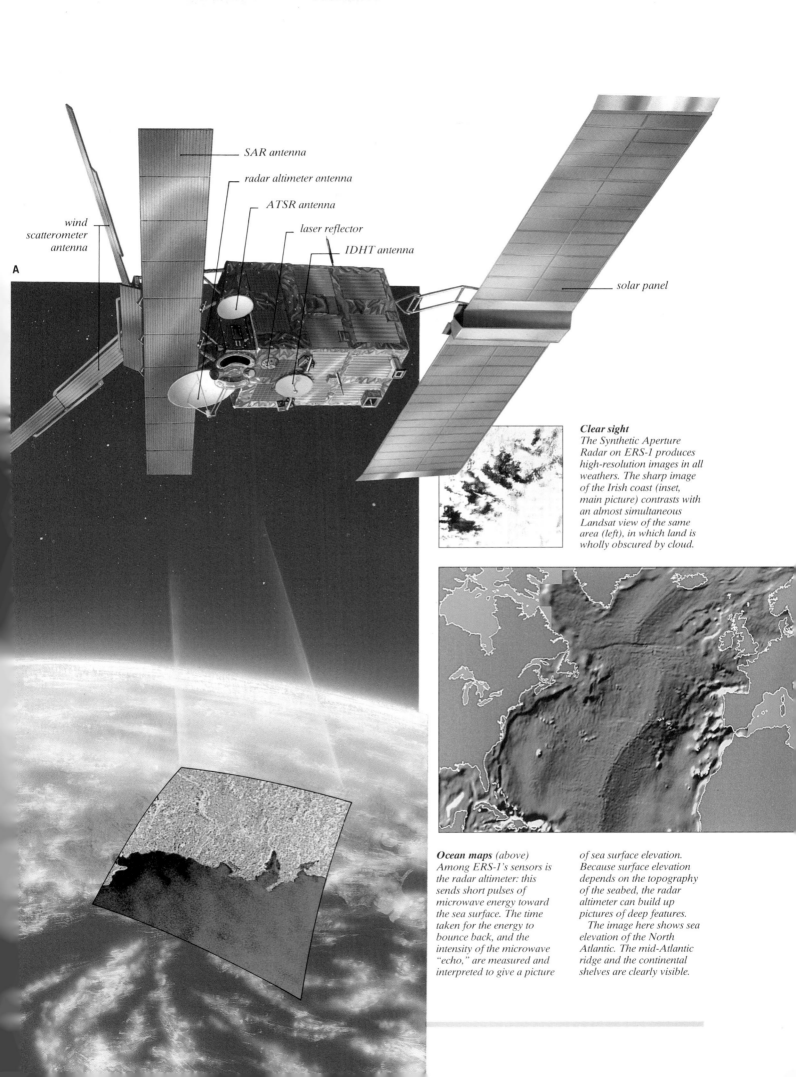

SAR antenna

radar altimeter antenna

ATSR antenna

laser reflector

IDHT antenna

wind
scatterometer
antenna

A

solar panel

Clear sight
The Synthetic Aperture
Radar on ERS-1 produces
high-resolution images in all
weathers. The sharp image
of the Irish coast (inset,
main picture) contrasts with
an almost simultaneous
Landsat view of the same
area (left), in which land is
wholly obscured by cloud.

Ocean maps (above)
Among ERS-1's sensors is
the radar altimeter: this
sends short pulses of
microwave energy toward
the sea surface. The time
taken for the energy to
bounce back, and the
intensity of the microwave
"echo," are measured and
interpreted to give a picture
of sea surface elevation.
Because surface elevation
depends on the topography
of the seabed, the radar
altimeter can build up
pictures of deep features.
   The image here shows sea
elevation of the North
Atlantic. The mid-Atlantic
ridge and the continental
shelves are clearly visible.

Overhead Observatories

How scientific satellites improve our knowledge of the universe

Earth-based astronomy has been compared to staring at the sky from the bottom of a murky pond. Our planet's atmosphere is a formidable barrier to observation, obscuring and distorting radiation from distant stars. But observatories in orbit outside this turbulent layer are now giving astronomers an unhindered view of the universe. The Hubble Space Telescope, for example, improves on the resolution of the best terrestrial telescopes by the same amount that Galileo's telescope bettered the resolution of the unaided eye.

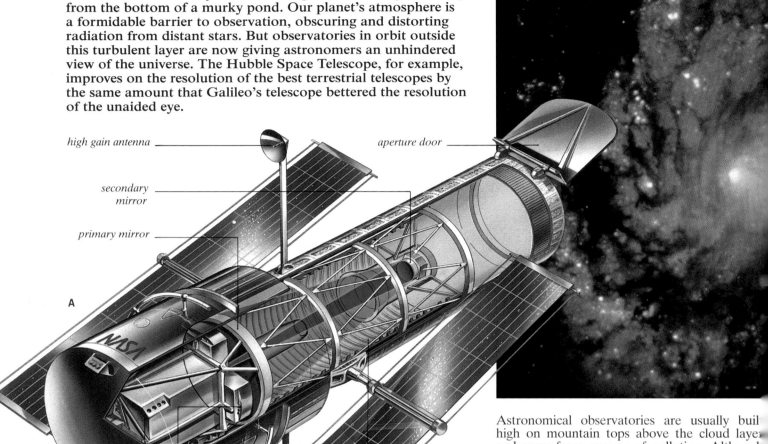

high gain antenna

aperture door

secondary mirror

primary mirror

A

NASA

light baffles

scientific instruments

fine guidance sensors

solar array

Astronomical observatories are usually buil high on mountain tops above the cloud layei and away from sources of pollution. Althougl atmospheric effects can be minimized by carefu siting, they cannot be eliminated. Pockets of ai of different density act as lenses, warping ligh before it reaches the telescope; and particles ir the atmosphere scatter light rays, making fain objects difficult to observe, even with the larges ground-based instruments. Furthermore, the atmosphere acts as a giant filter, admitting some frequencies while blocking out others. The mos energetic radiation (X-rays, gamma rays, ane ultraviolet light) is absorbed, as is much infrare light. Objects that emit radiation mainly a these frequencies are invisible from the earth.

Clarity of vision

Astronomers have overcome the limitations c observing from earth by putting telescopes int orbit. The most ambitious astronomy satellite t date is the Hubble Space Telescope, launche by NASA in 1990 at a cost of more than US$ billion. Hubble's early performance was marre by a fault in its 7.8 ft-diameter primary mirro and only after the installation of correctiv optics in 1993 was its full potential realized.

Hubble is able to see objects 100 times faint than those detectable by the largest telescope on earth. It can focus light onto any of fiv detectors, giving astronomers information abou the structure and composition of the universe.

The space telescope
The 43 ft-long Hubble Space Telescope [**A**] detects visible and ultraviolet radiation. Light falls onto the 7.8 ft concave primary mirror and is reflected back up to the secondary mirror. From here, the light is directed through a hole in the center of the primary mirror and is

brought to a focus in the scientific instrument section. Here the light is analyzed by a number of detectors. Two spectrographs split light into its component wavelengths to provide information about the composition of celestial objects. A faint-object camera gives high-resolution images of very

dim objects, while a wide-field camera gives a broader view. Collected data is converted into digital form and transmitted back to the ground station.

Angled baffles inside the "barrel" of the telescope prevent stray light entering its field of view, while an aperture door closes to protect the space telescope's optics and instruments when it is pointing at objects near the sun.

The image (above right) taken by the Hubble's wide-field camera after the telescope's repair in 1993 shows the spiral galaxy M100, which is thought to be about 50 million light-years away from the earth.

Connections: Telescopes 138 Spectrographs 140 Space Shuttle 164 166 Living in Space 170 Principles 224 262 264 266

Seeing the invisible

Hot regions of space – such as atmospheres of massive stars and the clouds in which stars form – give out energy as ultraviolet light. Upon reaching the earth's atmosphere, much of this radiation is absorbed by the ozone layer and cannot be detected by an observer on the ground. This part of the electromagnetic spectrum has been explored by a succession of satellites from the 1960s onward – most successfully by the International Ultraviolet Explorer (IUE), launched in 1978.

Whereas the IUE uses a conventional 18 in parabolic mirror to form an image, telescopes observing at higher frequencies have a fundamentally different design. X-rays and gamma rays are so energetic that they would pass straight through a conventional reflecting mirror. Instead, they are made to graze off the surface of a gold-coated mirror, positioned almost parallel to the incoming beam, before reaching sets of detectors. Such telescopes are used to examine high-energy objects such as white dwarfs, black holes, and neutron stars. NASA's Compton Gamma Ray Observatory, launched in 1991, detects gamma rays, which are the most energetic form of electromagnetic radiation. Gamma rays are emitted during some of the most violent events in the cosmos, such as supernova explosions.

Space telescopes can also be tuned into infrared radiation: this is less energetic than visible light, and has a slightly longer wavelength. Infrared radiation is absorbed by water vapor in the earth's atmosphere, but can be detected by orbiting observatories, allowing astronomers to see small stars and other cooler objects. The Infrared Astronomical Satellite (IRAS), which mapped the sky in 1983, discovered the previously unknown class of cool brown dwarf stars, and detected dust rings around nearby stars – possibly the first stages in the formation of new planetary systems.

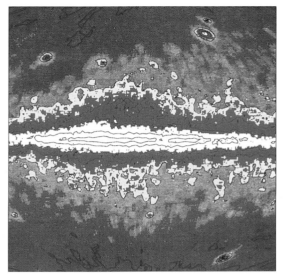

Violent universe *(above) Instruments onboard the Compton Gamma Ray Observatory produced this gamma-ray map of the whole sky. White regions correspond to the strongest emissions; blue to the weakest. The white band down the center shows the gamma-ray emissions of our galaxy, the Milky Way. Other bright sources are pulsars – spinning neutron stars – and distant quasars. Compton's detectors incorporate a material that produces a flash of light when struck by a gamma ray. The light is then recorded electronically to produce a digital signal.*

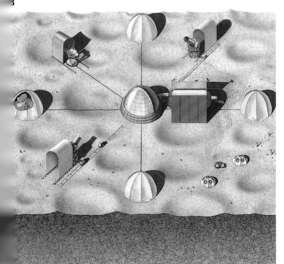

A return to the moon *Although orbiting observatories have greatly improved our view of the universe, they are still hindered by dust and gas in the upper atmosphere, slowed down by atmospheric drag, and threatened by high-speed debris circling our planet. Moreover, their observing instruments contract and expand as the satellite passes into and out of the earth's shadow, complicating delicate measurements.*

For these reasons, the first lunar bases, which may be in place early in the next millennium, may well be observatories. Situated on the far side of the moon,

*they would have an unimpeded view of the universe. Furthermore, in the low gravity of the lunar surface, the size and weight restrictions on optical telescopes would be reduced. This artist's impression [**B**] shows one possible configuration of a future lunar observatory.*

Hidden skies *Some wavelengths are completely absorbed by the earth's atmosphere. This makes whole classes of objects and activities, such as radio universes and gamma-ray bursts (which radiate solely in these wavelengths), invisible to astronomers on the the earth's surface. The chart [**C**] plots the full range of electromagnetic wavelengths against atmospheric transparency – that is, the percentage of radiation that passes through the atmosphere. Note the "windows" of observation – the areas of high transmission in the visible and microwave regions.*

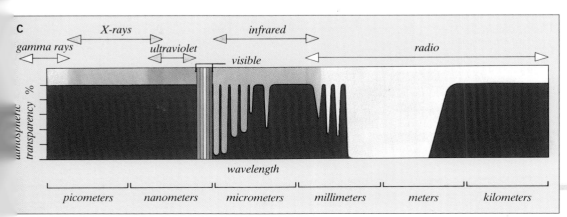

C
gamma rays   X-rays   ultraviolet   visible   infrared   radio

atmospheric transparency %

wavelength

picometers   nanometers   micrometers   millimeters   meters   kilometers

Mission to the Planets
How spacecraft probe the Solar System

Moving at a speed of more than 18 miles per second, the *Ulysses* spacecraft is the fastest object ever built. Ulysses is an unmanned probe, and its mission is to observe the sun's north and south poles, which are normally invisible from our planet. Since the 1960s, space probes like Ulysses have explored the Solar System, visiting every planet except for Pluto. The data they collect continues to improve our understanding of how the Solar System formed, and gives valuable information about the possible fate of our own planet.

Exploration of the Solar System began in 1959, with a flypast of the moon by the unmanned Soviet Luna spacecraft. Luna was followed to the moon by a succession of probes, which tested the technology needed for the manned moonshots of the late 1960s and 1970s. These early missions collected valuable data about the *solar wind* – the stream of charged particles that flows from the sun – and about conditions on the lunar surface. The surface of the moon, which has remained basically unchanged for billions of years, holds a record of past solar activity, so these studies provided information with many implications for earth history.

Next-door neighbors
The first successful planetary probe, Mariner 2, was launched by the United States in 1962. Passing within 21,750 miles of Venus, its instruments were able to measure the planet's surface temperature (800°F) and pressure (90 earth atmospheres), and detect the absence of a magnetic field. More than twenty probes have since visited Venus, some flying by the planet, and others penetrating its dense carbon-dioxide atmosphere. Data gathered by these missions showed that the fierce Venusian climate was the result of a runaway *greenhouse effect*, similar in principle to the phenomenon that threatens to alter the earth's climate.

The earth's other "neighbor", Mars, was also visited by a series of Mariner spacecraft. These returned spectacular images of the red planet that showed a cratered landscape, carved up by giant valleys and peppered with extinct volcanoes. By measuring the strength of radio signals received from the probes as they passed behind Mars, scientists were able to measure the scattering effect of its atmosphere, and thus gain valuable information about its density.

Our knowledge of Mars was greatly expanded by the Viking missions of 1976. Two spacecraft were dispatched to the planet, each consisting of an orbiter, which mapped the planet from space, and a lander, which parachuted down to take measurements from the surface. The landers relayed data about the atmospheric composition (primarily carbon dioxide, but with traces of nitrogen, oxygen, and inert gases) and humidity. Each lander was equipped with a mechanical arm to scoop up soil samples. These were analyzed by sophisticated laboratories inside the lander to reveal the chemistry of the Martian soil, and to search (unsuccessfully, in

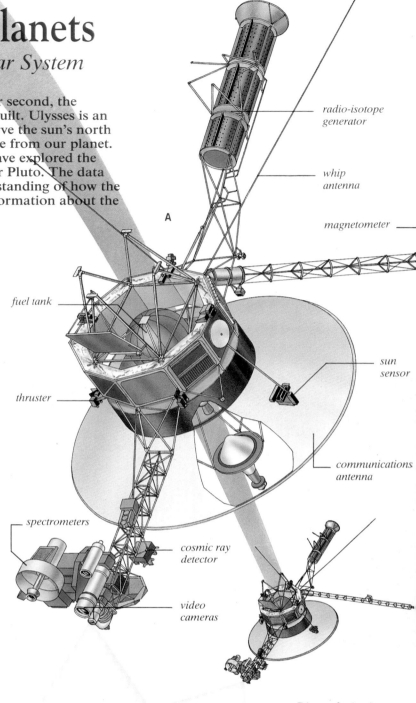

radio-isotope generator

whip antenna

magnetometer

A

fuel tank

sun sensor

thruster

communications antenna

spectrometers

cosmic ray detector

video cameras

Distant destinations
*The outer planets were surveyed by two Voyager probes [**A**], launched in 1977. Each was fitted with an array of sensors: two television cameras recorded visual images; spectrometers determined composition, temperature, and pressure of planetary atmospheres; whip antenna picked up radio emissions from the planets and elsewhere in space; a 43 ft- long boom held a magnetometer to probe magnetic fields around the planets. Power came from a nuclear generator, becau*

Connections: Rockets 162 Living in Space 170 Satellites: Orbits 172 Space Flight: New Developments 180 Principles 238 240 242

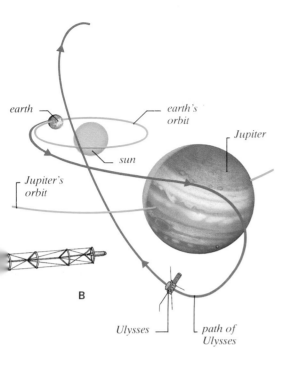

earth

earth's orbit

Jupiter

sun

Jupiter's orbit

B

Ulysses

path of Ulysses

Cosmic catapults

*The Ulysses craft, launched in 1990, used gravity-assist from the giant planet Jupiter to lift itself over the sun's poles [**B**]. The probe was launched away from the sun, toward Jupiter, which deflected it out of the ecliptic (the plane in which the earth and the other planets orbit the sun). The gravitational "slingshot" also increased the speed of the probe to 18 miles per second.*

Ulysses passed over first the south, then the north pole of the sun in 1994/5, making a detailed study of the solar wind and providing the first-ever view of the sun's extremities.

this case) for traces of life on the planet. Recent Mars exploration has been ill-fated, probes such as NASA's Mars Observer, malfunctioning before arrival. However, a series of further missions planned for the 1990s is set to continue study of the planet. The Martian surface will be sampled by robotic roving vehicles, capable of traversing over half a mile per day; by *penetrators* (probes fired into the surface); and by sensors carried by balloons. It is envisioned that each balloon will inflate in the heat of the day, carrying its probe through the atmosphere, and cool at night, descending to the surface to place the sensors in contact with the soil.

Voyage to the giants

Space probes targeted at Venus take around 145 days to reach their destination, Mars missions around 260 days. Visits to the outermost planets would, however, take many decades using even the most powerful rockets. This time is considerably reduced using a technique known as *gravitational slingshot*. The spacecraft flies close to one planet, and uses its gravitational field to accelerate it towards the next. Use of this technique depends on the planets' being in the right positions at the right time, so launch windows to the distant planets are very infrequent. For example, the Voyager 2 mission to Saturn, Uranus, and Neptune depended on a planetary configuration that had last occurred 180 years before launch.

Saturn flyby *(left)*

This computer-enhanced image of Saturn was taken by Voyager 2 on July 12, 1981 – four years after the spacecraft's launch. The television picture, made up of 800 lines (and 800 pixels per line), clearly shows variation in the composition of Saturn's seven rings.

Parachuting probes

*The Cassini craft [**C**], due for launch in 1997, will explore the Saturnian system during its four-year mission. While the main body of the spacecraft [1] orbits the planet itself, the Huygens probe [2] will land on Titan, Saturn's largest moon [**D**].*

As Huygens approaches Titan, its speed of 4.5 miles/s will be reduced to 890 ft/s as its 10 ft-diameter decelerator dish rubs against gases in the moon's atmosphere. At an altitude of around 110 miles, the decelerator will be jettisoned and its parachutes deployed. This will expose the main sensors of the probe which will map the surface of the moon and analyze its dense nitrogen atmosphere during the two- to three-hour descent. Huygens is not expected to survive impact with the surface, but mission scientists hope that it will "live" long enough to collect data and even analyze soil before contact with the orbiter fails.

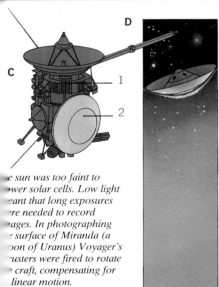

D

C

1

2

...e sun was too faint to ...wer solar cells. Low light ...eant that long exposures ...re needed to record ...ages. In photographing ...e surface of Miranda (a ...oon of Uranus) Voyager's ...usters were fired to rotate ...e craft, compensating for ...linear motion.

Thrust for the Future

How space vehicles will be powered in the 21st century

Each flight of NASA's Space Shuttle costs in excess of half a billion dollars and demands the combined skills of 40,000 scientists and engineers. The commercial exploitation of space – which may multiply dramatically in the 21st century – cannot be sustained by the present generation of expensive, temperamental rockets. Future demand will be met by new vehicles, many of which are already on the drawing board. Utilizing new materials and technologies, these craft will be capable of reaching orbit with the frequency, reliability, and economy of modern airliners.

Building rockets is a hugely expensive business. It is not surprising, therefore, that rocket scientists tend to be conservative, drawing upon proven technology when designing new launch vehicles. The result is that today's rockets are essentially high-performance descendants of the ballistic missiles first built in the 1960s. Their complex design requires precision engineering and expensive materials to reduce weight and increase performance. These engines have very narrow operational margins, meaning that even a small engineering hitch can lead to the postponement or cancellation of a launch.

Cheaper, simpler, and safer

In the near future, rocket design will be led by a new philosophy, in which low weight and high performance will be sacrificed for reliability, simplicity, and low cost. NASA, for example, is planning its National Launch System – a family of unmanned vehicles that would operate from pared-down launch facilities. These vehicles would be powered by new Space Transportation Main Engines (STMEs), with far fewer components than present liquid-fueled rockets.

These engines still follow the standard pre-space-age design, but other propulsion systems that depart radically from this model are also being considered. *Hybrid* rockets, in which liquid oxygen oxidizes a charge of solid fuel, could offer many advantages over traditional designs. They are safer than conventional solid-

A breath of fresh air
*Hypersonic space planes, such as the British design HOTOL (HOrizontal TakeOff and Landing vehicle) [**A**], use radical engine designs to accelerate into orbit from a conventional runway takeoff. Around the size of a Boeing 747, HOTOL is propelled by four hydrogen-burning engines. While in the atmosphere, air is taken in and used to oxidize the fuel: only at high altitude (around 18 miles) does the engine switch to liquid oxygen carried in a tank on board. The weight saved by carrying only a little liquid oxygen greatly reduces launch costs.*

*Air is drawn in through a duct located under the fuselage, cooled (by passing through heat exchangers), and compressed before being fed into the four engines. The air intake has variable geometry to cope with the widely differing flight speeds. At supersonic speeds [**B**], too much air is scooped up: a fraction (green) is directed to the engines, while the excess (blue) is "spilt" out of the rear of the engine. At hypersonic speeds, all the air is directed to the engines [**C**]. During reentry and descent through the atmosphere [**D**], the intake doors are closed as the craft glides to a runway landing.*

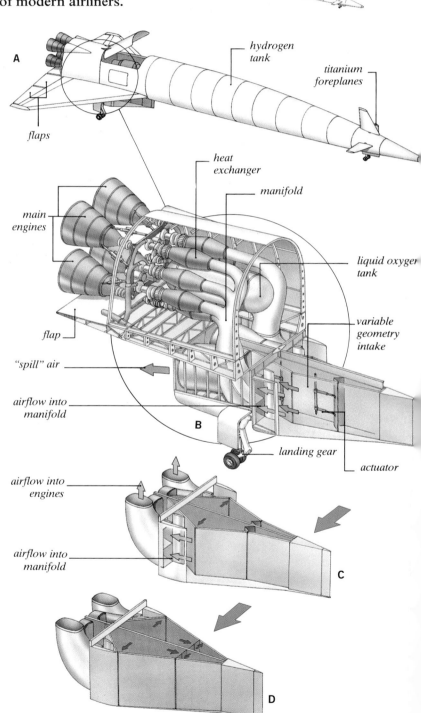

A

flaps

hydrogen tank

titanium foreplanes

heat exchanger

manifold

main engines

flap

"spill" air

airflow into manifold

B

liquid oxygen tank

variable geometry intake

landing gear

actuator

airflow into engines

airflow into manifold

C

D

Connections: Jet Engines 24 Aircraft 26 32 Rockets 162 Space Shuttle: In Flight 164 Principles 220 230 236 238 240

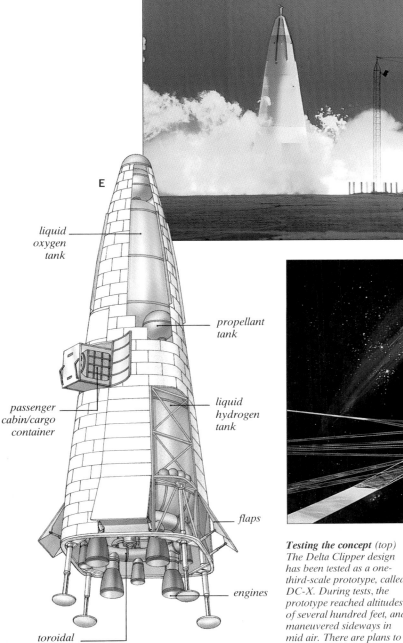

E

liquid
oxygen
tank

propellant
tank

passenger
cabin/cargo
container

liquid
hydrogen
tank

flaps

engines

toroidal
combustion
chamber

Sun sails (below)
Ultimately, the power of the sun itself may be harnessed for propulsion. This artist's impression shows a spacecraft propelled by a solar sail. The sails, made of lightweight polymer, capture the momentum of photons given out by the sun. Although the sail accelerates very slowly (just 2mm per second per second), this acceleration is continuous as long as the sun is sufficiently close. It is estimated that a round trip to Mars would take 300 days. Solar sails can be steered (even toward the sun) by tacking into the solar photon stream, just as mariners tack into the wind.

Testing the concept (top)
The Delta Clipper design has been tested as a one-third-scale prototype, called DC-X. During tests, the prototype reached altitudes of several hundred feet, and maneuvered sideways in mid air. There are plans to build a full-size model by the end of the century.

Accessible orbits
The Delta Clipper [**E**], currently being developed by McDonnell Douglas Aerospace, may become the world's first fully reusable single-stage rocket in the next millennium. Designed to carry both manned and automated missions, the 82 ft-tall craft could be turned around between missions very quickly. The rocket will be powered by eight engines around a doughnut-shaped (toroidal) combustion chamber: this means that failure of a single engine could be tolerated. The craft will take off and land vertically, and maneuver in mid air using flaps around its base.

fuel motors because thrust can be readily controlled by manipulating the flow of oxygen; and they are cheaper to manufacture and operate than liquid-fueled engines because the fuel injection and feeding system is much simplified.

Single stage to orbit

In the longer term, the commercialization of space depends on fully reusable craft, lifted into orbit by single-stage rockets or even exotic variants of the jet engine. Taking off and landing at existing airports, these craft could be turned around between missions in the course of a single day. Significantly, they would have a lifespan of at least 20 years, and need a major overhaul after every 200 flights – figures comparable to a modern airliner's. Such durable, economical "space buses" are currently in development and a one-third-scale prototype – called the Delta Clipper X (for experimental) – has already been successfully flown.

The full-size Delta Clipper will carry 11-ton payloads into orbit. It will be powered by an advanced liquid hydrogen/oxygen engine known as an *aerospike*, and so will still suffer from one of the main disadvantages of rocket engines – a substantial percentage of weight at launch is made up of oxidizer, instead of useful payload. This drawback may be overcome with the advent of the *hypersonic* vehicle. Taking off horizontally from a runway, such a craft could accelerate to Mach 25 and an altitude of 165,000ft propelled by *scramjets*, engines that "breathe" air from the atmosphere. It would then fire a set of small rocket engines to take it out of the atmosphere and into orbit.

Nuclear travels

Further in the future, nuclear-powered rockets may provide the prolonged thrust needed for trips to Mars and beyond. The *Nuclear Thermal Rocket*, for example, is launched conventionally, before starting up its nuclear generator at a safe height. The generator is used to heat liquid hydrogen fuel, which is ejected, creating thrust.

More radical nuclear designs use *Nuclear Electric Propulsion* (NEP). Here, a small reactor generates electricity, which is used to produce a stream of charged ions from an inert "fuel." Exhaust speeds of up to 45 miles/second can be generated, but acceleration to high speeds takes a long time, making NEP most suitable for long-distance missions.

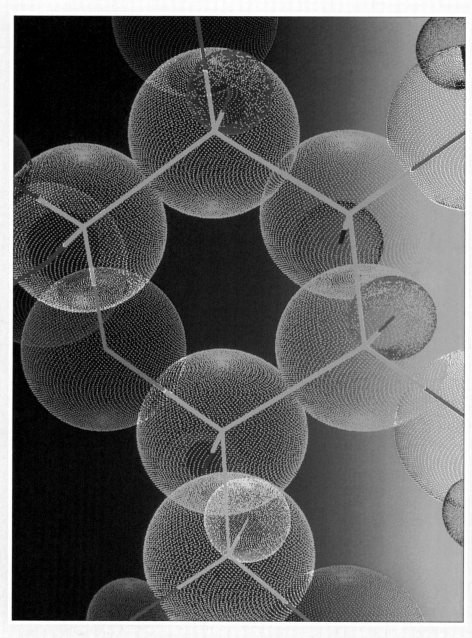

*Computer generated models of biological compounds, such as this insulin molecule,
are used to help develop new pharmaceuticals.*

7
Molecules and Materials

Chemical Engineering

Oil Refining

Steel Manufacture

Shaping Metals

Polymers

Advanced Materials

Biotechnology

Brewing

Pharmaceuticals

Genetic Engineering:

Lower Organisms

Genetic Engineering:

Higher Organisms

Genetic Engineering: Humans

Industrial Reactions

How chemicals are made on a large scale

The Haber-Bosch process for the synthesis of ammonia is among the most significant innovations of the 20th century. Developed in 1913, it allowed chemists to make ammonia cheaply, on a huge scale, from the elements hydrogen and nitrogen. As a raw material for fertilizer and explosives, ammonia was instrumental in shaping human history, helping to feed a growing population in peacetime, and fueling the destruction of war. Although its products may not be glamorous, chemical (or process) engineering influences all aspects of modern life.

Not only fertilizers and explosives, but plastics, fibers, refrigerants, and many other commodities use ammonia in their manufacture. Ammonia is one of around 50 chemical raw materials (others include sulfuric, nitric, and phosphoric acids, sodium hydroxide, chlorine, oxygen, and nitrogen) that form the foundations of modern industry. All these substances are manufactured in huge quantities (140 million tons of sulfuric acid are produced worldwide each year) and all are widely used in diverse applications. Because of their great economic importance, the processes used to make these chemicals are constantly being refined to make them more productive, cheaper, and safer.

Men and molecules

Chemistry on the industrial scale is a science of compromise, the aim of which is to maximize production while keeping costs down. Chemical engineers use the agents of heat, pressure, and catalysis, and the ingenious design of equipment to make a reaction proceed faster, more productively, and therefore economically. Some of the challenges facing the industrial chemist are exemplified in the *contact process* – the main means of making sulfuric acid.

In the contact process, sulfur dioxide (SO_2) reacts with oxygen to produce sulfur trioxide (SO_3). This is then combined with water to give sulfuric acid (H_2SO_4). However, the reaction between SO_2 and oxygen to make SO_3 is not a one-way street. As SO_3 is formed, some of it is converted back into the reactants. Eventually an *equilibrium* is reached, at which the proportion of reactants to products remains constant. This equilibrium can be shifted in favor of the desired product (SO_3) by keeping reaction temperature low. But at low temperatures the reaction proceeds very slowly and becomes uneconomic. It is speeded up by using a metallic *catalyst*: however, the catalyst only becomes active at temperatures above 842°F. The best yield of SO_3 is obtained by compromise – keeping the reaction temperature at just above 842°F.

Many other techniques are used to prepare the chemicals needed by industry, and not all involve chemical reactions. For example, oxygen and nitrogen (which are present in vast quantities in the earth's atmosphere) are purified by liquefying and then distilling air. The process by which this is achieved is essentially the same as the operation of a refrigerator. Air

Chlorine cells
Electrolysis – the use of electricity to split a compound into simpler substances – is the basis of chlorine manufacture [A]. A current is passed through molten sodium chloride [1] (or a salt solution), in which sodium and chlorine ions are free to move.

Negative chlorine ions are attracted to the positively charged anode [2], where chlorine gas [3] is given off. The negative cathode [4], which is made of liquid mercury, attracts positively charged sodium ions. The resulting mixture of mercury and sodium is pumped into a second vessel [5]. Water is added [6], forming sodium hydroxide (caustic soda) [7] and hydrogen gas [8] – both valuable by-products of chlorine manufacture, with many uses in the chemical industry. Chlorine is produced commercially from brine in electrolysis cells (right). It is used as a bleaching agent, and to make many organic compounds, such as PVC.

is first compressed, and then cooled by passing it through a heat exchanger – a maze of thin pipes that increase the surface area over which the air can lose heat to the surroundings. The gas is then made to expand rapidly, which produces a strong cooling effect (due to Charles' Law). Some of the gas liquefies, and the remainder is passed back to the compressor to repeat the process. The liquids are left to warm: nitrogen, with a boiling point of –320°F evaporates before oxygen (–297°F) and the two substances can be separated. Nitrogen has many uses. This inert gas is a fire suppressant, and can be used to preserve highly reactive compounds in a safe environment. Liquid nitrogen is also used to fast-freeze food. Many industrial reactions use atmospheric oxygen: in some cases, use of pure oxygen increases their speed and intensity, improving the yield for the time the equipment is used. These savings may outweigh the extra cost involved in buying pure oxygen.

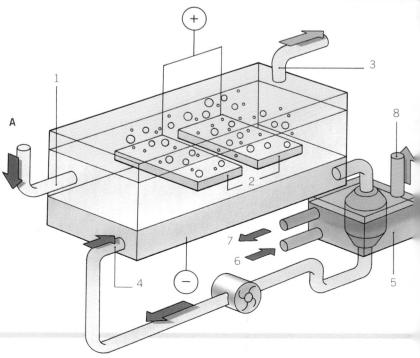

Connections: Chemical Analysis 142 Oil Refining 186

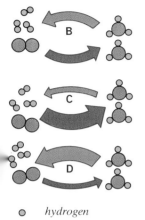

hydrogen

nitrogen

ammonia

Setting an equilibrium

In the Haber-Bosch process, nitrogen (N_2) is directly combined with hydrogen (H_2) to produce ammonia, NH_3. This reaction is reversible – just as hydrogen and nitrogen combine into ammonia, so ammonia dissociates into nitrogen and hydrogen [B]. In any particular set of conditions (temperature, pressure, concentration of reactants and products), the forward and backward reactions reach a state of equilibrium. At equilibrium, the proportions of reactant to product remain constant.

Chemical engineers try to maximize the proportion of product. They are guided by

Le Châtelier's Principle, which dictates that chemical equilibrium shifts in a direction that opposes any changes made to the system. In practice, this means that if ammonia is removed from the system, more is synthesized to take its place. Similarly, if heat (also considered to be a product of the reaction) is removed, then more heat (and incidentally more ammonia) is made to compensate.

Under conditions of low temperature and high pressure, equilibrium

"shifts" toward the products [C]. Conversely, with high temperature and low pressure, the reactants are favored [D].

However, the point of equilibrium is not the sole consideration for a chemical engineer. Also important is how quickly equilibrium is reached: at room

temperature, the reaction between N_2 and H_2 is so slow as to be imperceptible. Raising the temperature speeds up the rate of the reaction, but according to Le Châtelier's Principle, it also lowers the percentage of product. It is the chemical engineer's job to find a favorable compromise.

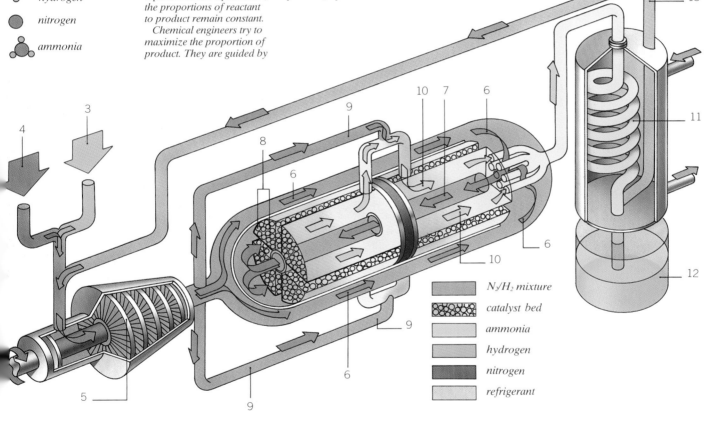

N_2/H_2 mixture

catalyst bed

ammonia

hydrogen

nitrogen

refrigerant

Making ammonia

Ammonia is still made in a process similar to that devised by the German chemists Haber and Bosch. The reaction takes place at a temperature of around 0 °F and a pressure of 0 atmospheres. This gives 0 percent conversion to ammonia, while still proceeding at an economic rate. The reaction is speeded further by a catalyst in the form of iron pellets. The reactants (H_2 and N_2) bind the surface of the pellets

[E]. The bonds that hold the molecules together break apart, and the atoms recombine as ammonia [2].

The reaction takes place in a steel vessel [F]. H_2 [3] and N_2 [4] gases are drawn in, mixed, and pressurized to 400 atmospheres by a compressor [5]. The gases pass around the reaction vessel [6], picking up heat. They are carried by a duct through the middle of the catalyst beds [7], picking up more heat. The gases then pass over the catalyst in the

first catalyst bed [8]. As ammonia is formed, the reaction produces heat and the mixture becomes so hot as to make the reaction less productive. To restore productivity, cool reactants are added to the mixture [9] before it passes into the second catalyst bed [10]. Here, more ammonia is produced. The mixed gases are refrigerated [11], liquefying the ammonia, which is collected [12]. The unreacted H_2 and N_2 are recycled [13].

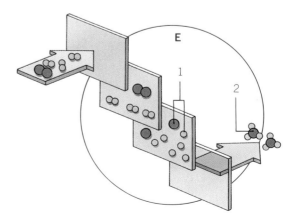

A Refined Industry

How crude oil is split into a range of valuable products

Oil is central to almost every industrial process. It is a lubricant, a raw material and a fuel, which today accounts for **40 percent** of the world's energy demand. But when extracted from the ground in its crude state, oil is of little use. It consists of a variable mixture of liquids and dissolved gases, which must be physically separated and chemically treated to gain commercial value. These processes take place in refineries, large complexes that may be the size of a small town. More than **900 refineries** worldwide produce the huge array of oil products demanded by modern societies.

Crude oil is not a single substance but a variable cocktail of hydrocarbons – molecules composed primarily of carbon and hydrogen atoms (sometimes incorporating sulfur, nitrogen, and oxygen). These molecules vary greatly in size: some are simple and unbranched, composed of less than ten atoms; whereas others are branched or ring-shaped, and comprise more than 100 atoms. In general, the small molecules are more *volatile* than the larger, meaning that they boil at lower temperatures. The basic process used in an oil refinery – *fractional distillation* – makes use of these differences in boiling point to separate the different components or *fractions* of the crude oil.

At the refinery, distillation takes place in a steel fractionating tower up to 263 ft in height. Horizontal perforated trays are fitted inside the tower at various heights, and the temperature within the tower is controlled so that at the bottom it is about 750°F and at the top 86°F.

Crude oil is heated in a furnace to a temperature of 750°F, and the resulting mixture of hot liquids and gases is pumped into the base of the tower. The liquids sink to the bottom, but the vapors rise, cooling as they go. When the temperature falls below their boiling points, they condense and settle on the trays. When the trays fill, the oils overflow on to the next tray down where they vaporize once again. A continuous process is set up in which hot crude oil is fed into the fractioning tower, vapors rise,

Distilled products

*The heart of an oil refinery [**A**] is a fractionating tower [1]. Crude oil, heated in a furnace to around 750°F – a temperature at which a portion of it becomes vapor – is pumped into the base of the tower [2]. The oil vapors rise through the tower, cooling as they go, and condense at levels determined by their boiling points. The liquids settle on trays [3] arranged at intervals of about 1.5 ft up the tower. Bubble caps [4] on each tray force the rising gases to bubble through liquids that have already condensed. The hot, rising vapor heats the liquid in the trays, causing the more volatile components to evaporate and join the remaining gases rising up the tower. This simple arrangement splits the crude oil into a number of fractions, each of which is a mixture of hydrocarbons with very similar boiling points. The lightest fraction – made up of small, simple molecules – leaves the tower*

as a gas [5]. Heavier fractions are tapped off the column at different levels [6]; some are used directly, others are further purified or chemically altered before use. The heaviest fraction made up of liquids that did not vaporize when the crude oil was initially heated, is sent to a vacuum pipestill [7]. Here, a partial vacuum lowers the boiling point of the component compounds, allowing them to be distilled at a far lower temperature than would be needed for distillation at standard atmospheric pressure.

The heavy fractions obtained by vacuum distillation are sometimes chemically treated to produce smaller, more useful molecules. Among the most significant of these processes is catalytic cracking in which the heavy fractions are broken into smaller units in the presence of a powdered catalyst.

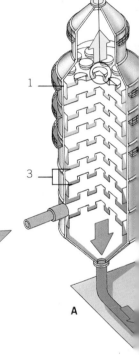

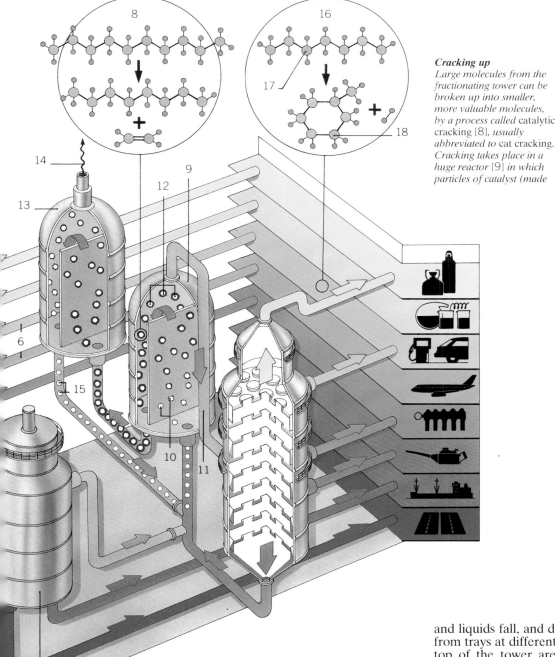

Cracking up
Large molecules from the fractionating tower can be broken up into smaller, more valuable molecules, by a process called catalytic cracking [8], *usually abbreviated to cat cracking. Cracking takes place in a huge reactor [9] in which particles of catalyst (made of powdered minerals such as silica, alumina, and zeolites) [10] are mixed with feed from the vacuum pipestill at a temperature of around 930°F. The cracked vapors comprising the smaller molecules leave through the top of the vessel and pass to a second fractionating tower to be separated [11].*

During cracking, catalyst particles become choked with a layer of carbon [12], which prevents them from taking any further part in the reaction. The choked catalyst passes to a second reactor [13], in which the carbon is burned off at a temperature of 1337°F. Waste carbon dioxide gas leaves the vessel [14], and the regenerated catalyst is returned to the reactor [15] to be used again.

Many processes other than cat cracking are used to alter the size, shape, purity, and therefore the properties, of molecules produced in a refinery. For example, the quality of gasoline produced is enhanced by a process known as reforming [16]. *During reforming, linear hydrocarbons such as heptane [17] are converted to ring-shaped (aromatic) molecules [18], which burn more smoothly, causing less "knocking" in the engine cylinders. A gasoline with a high content of such aromatic molecules is said to be* high octane.

and liquids fall, and different oils can be tapped from trays at different levels. Fractions from the top of the tower are called "light" and those from the bottom "heavy."

Purer products
Distillation is a physical process that separates the fractions without changing their chemical identity. After distillation each fraction is purified or altered chemically to further enhance its commercial value. For example, large molecules in the heavier fractions may be *cracked* into smaller, valuable products like gasoline. Conversely, lighter gaseous molecules may be *polymerized* – linked together – to make larger ones. Fuels such as diesel oil and kerosene are chemically cleaned to reduce their sulfur content: this lowers harmful sulfur emissions when they are burned. And some fuels are *isomerized* – the shape of the molecules is changed – to improve combustion properties.

Production plants (left)
The largest refineries cover areas of 2500 acres or more and have as many as 15 fractionating towers and 0 storage tanks. Most are situated at deepwater ports that allow access to supertankers carrying more than 440,000 tons of crude oil.

Oil applications
Refineries around the world process around 3 billion tons of crude oil a year into an enormous range of products.
The most vaporous fraction that rises highest in the fractionating column is called refinery gas. This is burned at the refinery to provide power. The next lightest fraction, which contains up to 4 carbon atoms per molecule, is liquefied petroleum gas (LPG). This is bottled for distribution and sale. Other light fractions are gasoline (5–8 carbon atoms) and naphtha (6–10 carbon atoms). The latter is an important raw material for the chemicals industry. The heavier fractions include kerosene (jet fuel) and gas oil (10–20 carbon atoms), which is used for heating and as a fuel for diesel engines. Heavier still are lubricating and fuel oils, and the bitumens, with more than 20 carbon atoms per molecule, which are used to surface roads.

Inner Strength

How steel is manufactured and shaped

Steel gives shape to the inventions of the modern world. Its great durability, tensile strength, and low cost make it the most widely used of all metals, and the basis of countless industries, from shipbuilding to watchmaking. Steel is made from iron hardened by the addition of small amounts of carbon and other elements. Careful adjustment of these trace elements gives rise to a vast number of steel alloys, each with distinctive properties and applications. And because it is made from iron, steel is easily separated from other waste by magnets, to be efficiently recycled.

After aluminum, iron (Fe) is the most abundant metal in the earth's crust. It occurs naturally in many different minerals, but its most important *ores* are magnetite (Fe_3O_4) and hematite (Fe_2O_3), in which iron is chemically bonded to oxygen. Almost one billion tons of iron ore is mined each year, and much of this production goes to supply the steel industry.

Steel manufacture is a two-step process. Iron is first extracted from ore in a blast furnace, and then refined and chemically modified into steel.

In the blast furnace, hot, oxygen-enriched air is passed over a mixture of iron ore, coke, and limestone. Oxygen reacts with coke to form carbon monoxide (CO) gas and more heat, which raises the temperature in the furnace to 3450°F – hot enough to melt the ore. The carbon monoxide then reacts with the molten ore, stripping away its bound oxygen in a chemical process called *reduction*. This gives carbon dioxide gas (CO_2) and molten iron. Limestone is added to the furnace because it binds with impurities, concentrating them in a liquid *slag* that floats on top of the molten metal, and which can be readily removed. The resulting iron (*pig iron*) is about 95 percent pure, the remainder being mainly carbon with traces of sulfur, phosphorus, manganese, and silicon.

Refined materials

Steel is made by reducing the carbon content of pig iron to between 0.03 and 1.7 percent. In practice, this is done by removing virtually all the carbon present in the iron, and readding controlled quantities of the element.

The *basic oxygen process* is the most common way of making bulk steel, used in consumer goods, and the engineering and construction industries. Specialist steels and alloys are more often made in the *electric arc* furnace, which allows more precise control over steel composition. The techniques differ in that the basic oxygen process uses molten pig iron mixed with scrap steel as raw materials, whereas the arc furnace uses only cold scrap metal, melted by a powerful electric current. In both cases, carbon is first "burnt off" by a stream of oxygen, and reintroduced, together with additives, just before the molten steel is poured off, or *tapped*.

When mixed with molten steel, carbon bonds with iron to form an extremely tough compound called *cementite*. So, in general, the higher the carbon content, the harder the metal. Steel alloys with less than 0.15 percent

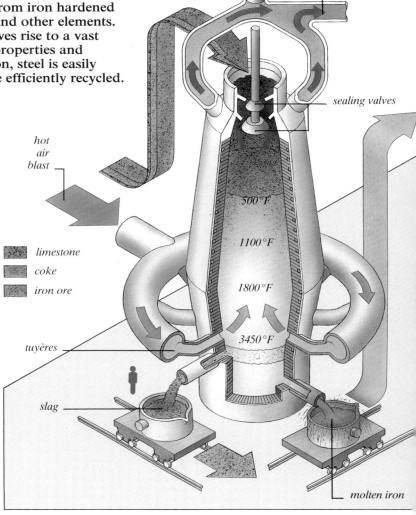

A

waste gas to cleaning plant

sealing valves

hot air blast

500°F

1100°F

1800°F

3450°F

limestone

coke

iron ore

tuyères

slag

molten iron

carbon are malleable, ductile, and resistant to impact – useful in automobile bodies. "Mild" steels, used in construction, have more carbon and are stronger but less ductile, while machine tools, such as rock drills, contain up to 1.6 percent carbon and so are very hard-wearing.

By adding other metals, steels with special properties can be produced. Mixtures of chromium and nickel give stainless steel, valued for its attractive, lustrous appearance and high resistance to corrosion; manganese increases the hardness of steel, giving an alloy used to make soldiers' helmets; and by varying the proportions of nickel and cobalt, steel can be made that expands at the same rate as glass – essential in the manufacture of window frames.

The physical properties of steel also change depending on the way it is cooled. Slowly cooled, or *annealed*, steel is malleable and resistant to mechanical shock. In contrast, rapidly cooled steel is hard and brittle.

Making iron
*Iron ore, limestone, and coke, together known as the "charge," are fed in through the top of the blast furnace [**A**]. Hot air is blasted into the furnace via nozzles known as* tuyères. *Iron ore is reduced to iron in the heat of the furnace. The molten iron sinks to the bottom of the furnace and is tapped off. The slag also sinks, but floats above the iron and is removed separately. Waste gases rise to the top of the furnace to be directed to a gas cleaning plant.*

Blast furnaces run nonstop for about ten years, after which the heat-resistant brick lining deteriorates and must be replaced.

Connections: Mining 100 Chemical Engineering 184 Shaping Metals 190 Principles 216 218 220 222 226

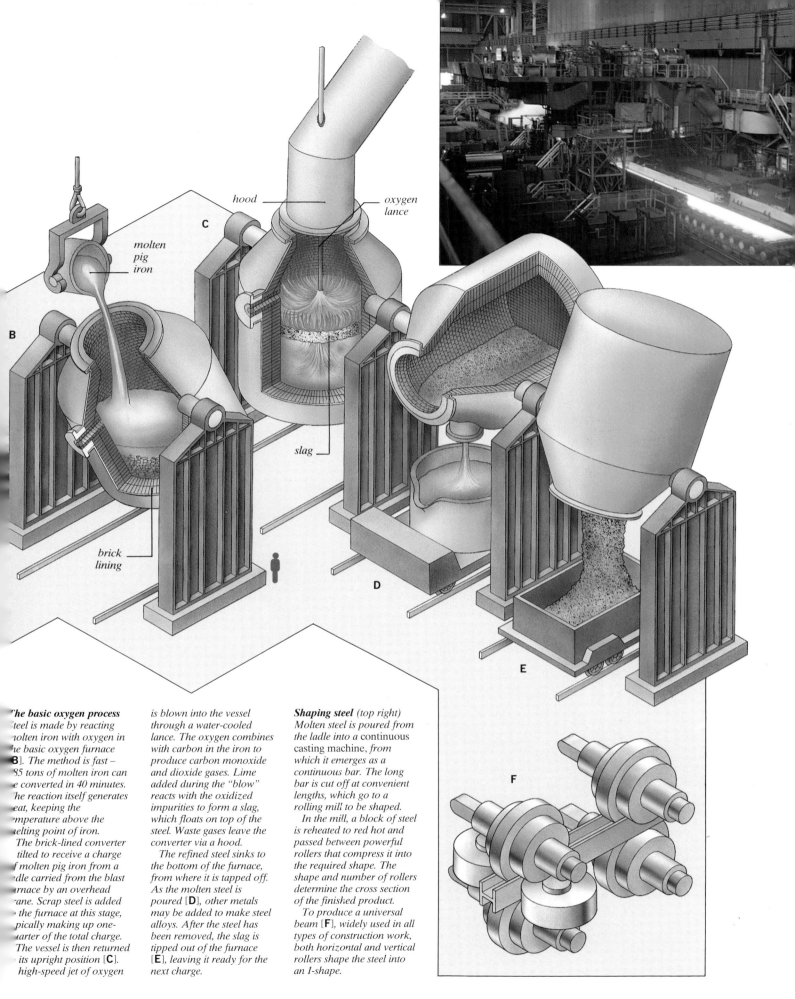

Labels on diagram:

hood

oxygen lance

molten pig iron

C

B

slag

brick lining

D

E

F

The basic oxygen process
Steel is made by reacting molten iron with oxygen in the basic oxygen furnace [**B**]. The method is fast – 85 tons of molten iron can be converted in 40 minutes. The reaction itself generates heat, keeping the temperature above the melting point of iron.

The brick-lined converter tilted to receive a charge of molten pig iron from a ladle carried from the blast furnace by an overhead crane. Scrap steel is added to the furnace at this stage, typically making up one-quarter of the total charge. The vessel is then returned to its upright position [**C**]. A high-speed jet of oxygen

is blown into the vessel through a water-cooled lance. The oxygen combines with carbon in the iron to produce carbon monoxide and dioxide gases. Lime added during the "blow" reacts with the oxidized impurities to form a slag, which floats on top of the steel. Waste gases leave the converter via a hood.

The refined steel sinks to the bottom of the furnace, from where it is tapped off. As the molten steel is poured [**D**], other metals may be added to make steel alloys. After the steel has been removed, the slag is tipped out of the furnace [**E**], leaving it ready for the next charge.

Shaping steel (top right)
Molten steel is poured from the ladle into a continuous casting machine, *from which it emerges as a continuous bar. The long bar is cut off at convenient lengths, which go to a rolling mill to be shaped.*

In the mill, a block of steel is reheated to red hot and passed between powerful rollers that compress it into the required shape. The shape and number of rollers determine the cross section of the finished product.

To produce a universal beam [**F**], widely used in all types of construction work, both horizontal and vertical rollers shape the steel into an I-shape.

Malleable Media

How metals are extracted and used

The interior of a jet engine is one of the most ferocious environments on earth. Exhaust gases, heated to 3600°F, rush past turbine blades making them spin many thousands of times per minute. To withstand these conditions, the blades must be extremely resistant to heat and mechanical stress, as well as being light in weight to maximize efficiency. Such exacting requirements are met only by advanced alloys – combinations of aluminum and exotic metals like titanium – and by special casting methods that produce blades made from single metal crystals.

Pure metals, such as 24-carat gold, are rarely used in industry. Metal components, from automobile bodies to electrical wires, are usually made from *alloys* – mixtures of two or more different metals. Depending on the metals mixed and their proportions the alloy can be brittle or ductile, malleable or extremely hard. Alloys can be more chemically stable than pure metals, and retain their properties over a wider range of temperatures. Most of the elements in the Periodic Table are metals, each with its own set of physical and chemical properties. There is thus a huge number of possible alloys.

Making an alloy is a relatively simple process. Molten metals are mixed and allowed to cool and solidify. Such a mixture of metals is a *solid solution*, in which the atoms of one metal (the solute), are incorporated into the crystal lattice of another (the solvent). An alloy is therefore not a compound because the different metals are not joined by chemical bonds.

From ore to high purity
Aluminum is the most abundant element in the earth's crust. Its most important ore is bauxite, a rock composed of aluminum hydroxides. The extraction of aluminum metal from bauxite is a two-step process [A]. Bauxite is first refined to aluminum oxide (alumina) in the Bayer process; and alumina is then split by electricity to give pure aluminum.

Bauxite [1] is ground to a powder in a mill, then passed to a vessel where it is mixed with lime, soda ash, and water, forming a slurry containing alumina [2]. Heating [3] dissolves the alumina, but none of the impurities. The slurry then passes to a settlement tank [4], where most of the solid impurities sink to the bottom of the liquid. They are removed as a red mud [5], a dangerous pollutant that must be disposed of very carefully.

The solution then passes through a filter [6] before being cooled. This makes a crystalline form of alumina (combined with water) precipitate out of solution. This is collected by another settlement tank and filter [7,8]. The crystals pass to a revolving kiln [9], where heat drives off the water, leaving high-purity alumina, a white powder [10].

To produce aluminum metal, the alumina is dissolved in molten cryolite, another aluminum mineral, in an electrolysis bath. This has a carbon lining [11], which acts as a cathode, connected to the negative terminal of a power supply [12]. Carbon anodes connected to the positive terminal dip into the surface [13], and as a current flows between the electrodes, a layer of molten aluminum forms on the bottom of the bath [14]. This is siphoned off and cast into long ingots of pure aluminum [15].

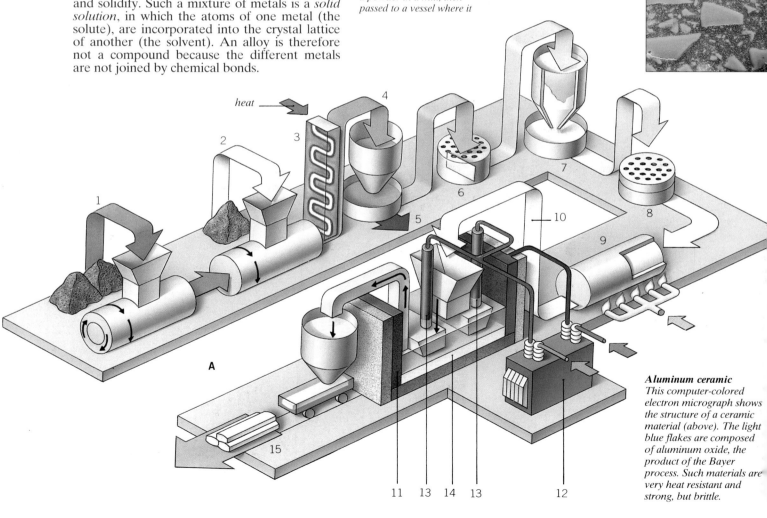

Aluminum ceramic
This computer-colored electron micrograph shows the structure of a ceramic material (above). The light blue flakes are composed of aluminum oxide, the product of the Bayer process. Such materials are very heat resistant and strong, but brittle.

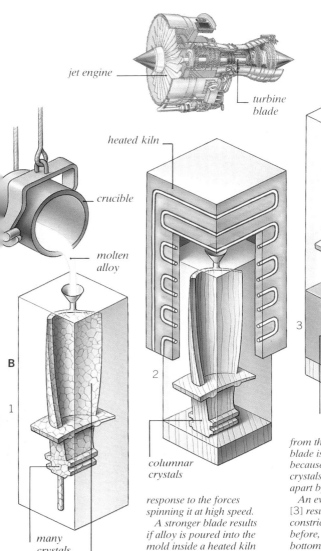

jet engine

turbine blade

heated kiln

crucible

molten alloy

single crystal

B

1

2

3

constriction

columnar crystals

many crystals

turbine blades

response to the forces spinning it at high speed.

A stronger blade results if alloy is poured into the mold inside a heated kiln [2]. Only the bottom surface is cool enough for crystals to form, so as the mold is slowly withdrawn from the kiln, long thin crystals grow

from the base to the tip. The blade is extremely strong because its column-shaped crystals cannot be pulled apart by the spinning forces.

An even stronger blade [3] results if the mold has a constriction near its base. As before, crystals form at the bottom of the mold. But the constriction allows only one crystal to continue growth. The result is a single-crystal blade that has no weakening grain boundaries at all.

Hot metal
The turbine blades of a jet engine are made from "superalloys" based on nickel and aluminum, which are able to withstand extremes of temperature and mechanical stress [B].

When a turbine blade is cast in a conventional manner [1], molten alloy is poured into a cool mold. Small crystals start forming on all the surfaces in contact with the mold, each growing in a slightly different direction. When the blade finally solidifies, it is made up of numerous small, randomly-oriented grains. This structure is prone to creep – a gradual lengthening of the blade in

Alloys, crystals, and mechanical properties

As a mixture of molten metals cools and begins to solidify, small crystals start to form throughout the liquid. With further cooling, more atoms are added to the individual crystals or *grains*, which grow in size until they eventually form boundaries with their neighbors. Metals are stronger within grains than along boundaries. When a metal is beaten, it changes shape. Some of this deformation results from individual crystals changing shape, but much is caused by neighboring crystals sliding past one another into new positions. During beating, the internal structure of the crystals is altered so that they become less able to slide past one another in the future. The result is that the metal becomes harder and more brittle.

The properties of an alloy are also influenced by the way in which the component metal atoms fit together inside each crystal. Brass, an alloy of copper and zinc, exists in two distinct

Strong and stable
The sea is a highly corrosive environment, so a ship's propeller (above) must be cast from a strong, stable metal. Aluminum bronze – an alloy of copper and 10 per cent aluminum – is just such a material. Air and water quickly react with the aluminum to form a protective layer of oxide on the propeller. The alloy also resists cavitation, which occurs as the turning blades create areas of low pressure in the water. Under low pressure, bubbles of vapor form in the water. As these implode, they cause shock waves that can damage softer materials.

crystalline forms – *alpha*, which contain small proportions of zinc, and *beta*, with more than 35 percent zinc. In alpha brasses, atoms of the solute (zinc) incorporate directly into the crystal lattice of the solvent (copper). Alpha brasses are malleable, allowing them to be easily worked when cold by pressing or hammering. They are used to make household fittings and electrical contacts. Beta brasses, in contrast, form less regular crystals. When the alloy is stressed, these crystals move past each other with great difficulty, making beta brasses stronger but more brittle than alpha brasses. These hard, corrosion-resistant alloys are cast into ships' propellers.

Many other familiar metals are alloys of copper. On its own the metal is an excellent conductor of electricity but is relatively weak. For this reason, copper cables contain 0.1 percent silver, which stops the wire from softening as it conducts large electrical currents. And copper alloyed with 4–10 percent tin becomes bronze – a far stronger material that is easier to cast. Modern bronzes contain traces of phosphorus, making them still harder-wearing, and so ideal for use as bearings and gears.

Aluminum is lightweight, malleable, and easily cast. Weight for weight it conducts electricity better than any other metal. It is used to make high-voltage power lines because it allows for wider gaps between pylons. And light aluminum alloys are widely used in aircraft.

Plastic Molecules

How polymers are made and put to use

Polymers are the giants of the chemical world. Made from thousands of smaller molecules linked together, they possess unique properties that have been put to use by humans and nature alike. Naturally occurring polymers include DNA, the information store at the heart of most living cells, and lignin, the main component of wood. Synthetic polymers, more often called plastics, are found everywhere in modern technological societies. They are tailor-made to suit particular needs, ranging from clothing and packaging to automobile and aircraft components.

Polymers get their name from the Greek words *polys* (many) and *meros* (parts). They are giant molecules made up of small subunits, or monomers, repeated anywhere between one hundred and more than a million times. Although most commercially important polymers are organic, that is, compounds of carbon, some are inorganic. For example, many industrial lubricants are made of polymers based on the element silicon.

Some polymers, called *homopolymers,* are composed of only one type of monomer; for example, polythene contains just ethene subunits. In contrast, copolymers, are made of two or more types of monomer: for example, many synthetic rubbers are compounds of butadiene and styrene.

Form and function

The properties of a polymer depend not only on the nature of its constituent monomers but also on the way these are linked together in three-dimensional space. Polymers may be linear and made up of monomers joined together like beads on a string or branched. In some cases, side branches of adjacent molecules link up to form a giant lattice. By controlling conditions during the synthesis of a polymer *(polymerization)*, chemists can determine its size and shape, and thereby "design" molecules with a range of useful properties such as toughness, elasticity, or the ability to form fibers.

Polymer molecules with few side branches to interrupt their regular three-dimensional structure tend to form crystals, interlocking with their neighbors to form a dense, tough material. Molecules with many side branches cannot pack closely together: these *amorphous* polymers are less dense and weaker in tension. Similarly, changing the degree of cross-linking between chains alters a polymer's response to heat or mechanical stress.

The properties of polymers are strongly influenced by additives. For example, PVC (polyvinylchloride), which is used to make gutters, pipes, and panels, can be altered by the addition of a *plasticizer*. The resulting substance is soft enough to be made into clothing.

Unlike natural materials, polymers do not rust or rot. Whereas this is often an advantage because it prolongs the life of plastic products, it poses problems in the disposal of waste. One solution is to incorporate sugar molecules into the polymer chain. Bacteria digest the sugar,

Building a chain
*To make polythene [**A**], ethene monomers are added one by one to the end of a chain. Ethene gas [1] pressurized to 20 atmospheres and heated to 300°F, is bubbled through petroleum spirit [2], which contains granules of titanium catalyst [3]. As an ethene molecule nears the catalyst, one of its bonds is induced to break [4,5]. This bond then pairs with a free bond on the end of the polymer chain [6]. This process is repeated up to 50,000 times to produce a polythene molecule. The chain "grows" only from the surface of the catalyst, giving an unbranched molecule. This "interlocks" with adjacent chains giving high-density polythene.*

Connections: Chemical Engineering 184 Oil Refining 186

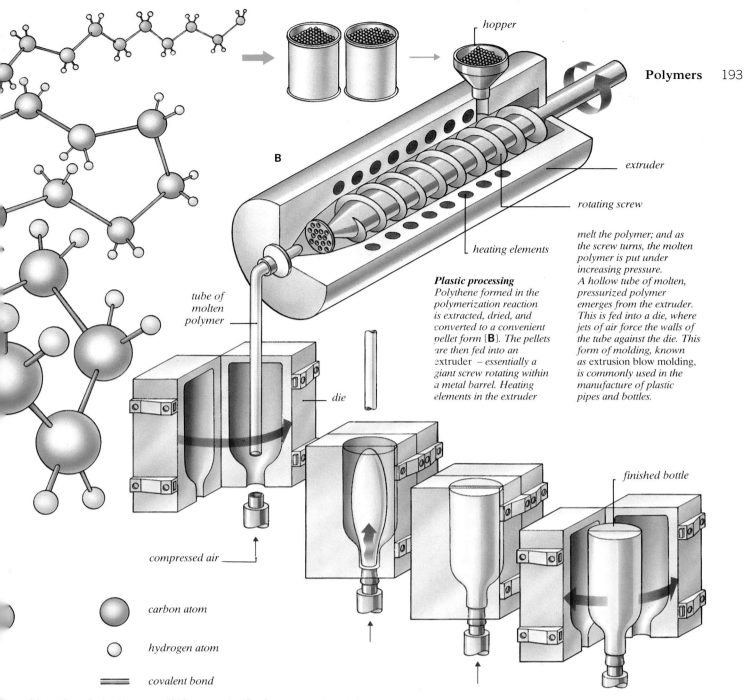

hopper

extruder

rotating screw

heating elements

tube of molten polymer

die

compressed air

finished bottle

Plastic processing
Polythene formed in the polymerization reaction is extracted, dried, and converted to a convenient pellet form [B]. The pellets are then fed into an extruder – essentially a giant screw rotating within a metal barrel. Heating elements in the extruder melt the polymer; and as the screw turns, the molten polymer is put under increasing pressure. A hollow tube of molten, pressurized polymer emerges from the extruder. This is fed into a die, where jets of air force the walls of the tube against the die. This form of molding, known as extrusion blow molding, is commonly used in the manufacture of plastic pipes and bottles.

● carbon atom

○ hydrogen atom

▭ covalent bond

reaking the chain into small fragments. Such iodegradable polymers are also used in medical science. "Stitches" that break down fter they have done their job, and drug cap- ules that dissolve at a controlled rate in the atient's body, are made of similar materials.

lastic computers
he emerging science of *photonics*, in which hotons, rather than electrons, are used to rocess, store, and transmit information, epends on polymer technology. Transparent lastics are used to conduct photons in the same ay that metals conduct electrons in conven- onal circuits. Computers using photonic rcuits can handle data more rapidly than elec- onic devices because photons travel much ster than electrons. Moreover, plastic compo- ents are lighter than metals, can store formation more compactly, and are not bject to magnetic or electrical interference.

Cross-linked chains
Polymers with little cross-linking between chains tend to soften when heated and deform irreversibly when stretched [C]. Such materials are called thermoplastics: they are used in products such as plastic bottles, polyethylene film, and nylon fiber. Elastomers have more cross-linkages and snap back to their original shape after being deformed [D]: they are used for automobile tires. Thermosets harden permanently once molded because of strong links between the chains [E]. These tough compounds are good thermal and electrical insulators, ideal for electrical appliances.

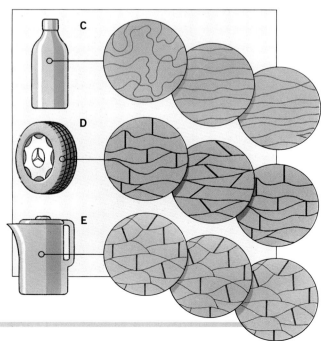

C

D

E

Material Gains

How new composites and ceramics are used

The greatest leaps in technological innovation occur as improved materials become available. Stone, bronze, and iron heralded new ages of invention; and today, rapid advances in materials science continue to transform our world. All manner of products, from aircraft and automobiles to tennis rackets and scissors, make use of "tailormade" ceramics and composites. And so-called "smart" materials, which can actually respond to changes in their environment or "remember" their shape are being applied to the technological challenges of the 21st century.

A *composite* is a material made up of two or more components that confer different types of strength or resilience. It is therefore stronger than the sum of its parts. A familiar example is fiberglass. This is made up of strands of spun glass, which have high tensile strength (they resist stretching), embedded in a solid resin. On their own the fibers are flexible; locked in the resin, however, they resist both tension and compression. And on its own, the resin is brittle and cracks easily: but when part of a composite it becomes tough enough to be made into auto-mobile body panels and boat hulls.

Stronger composites are based on high-tensile carbon fibers, made by charring a *precursor polymer*, usually rayon or polyacrylonitrile (PAN). Once these are fixed in place within a resinous matrix, materials are produced that weight-for-weight have four times the strength of steel, and are extensively used in the manu-facture of aircraft and sports equipment.

Layers of strength
Some of the strongest composite materials contain twisted strands of carbon fiber encased in a matrix of resin. Two main types of carbon fiber composite are manufactured [A].

In the first, many continuous, long strands of fiber [1] are pulled from spools and aligned. After surface treatment, the fibers are drawn through a bath of resin, becoming generously coated. The sticky fibers are then pressed onto a backing tape, and the reinforced sheet is rolled up. The sheet is strong only in the direction that the fibers

run. To make tough materials called laminates, several sheets are layered on top of one another, with the fibers running at 45° angles in successive sheets. Such laminates can be built up into complex shapes, such as that of an IndyCar monocoque [2] (a combined body and chassis). At points on the structure where stresses are greatest, extra layers of laminate are added.

Once the shape has been built up, the monocoque is cured by heat into a completely rigid structure that is stronger and lighter than its metal equivalent.

The other form of carbon fiber composite is strong in all directions within a single sheet. To make this composite, carbon fibers are drawn from spools and cut into short lengths by a set of rotating blades [3] in a chopper. The pieces of fiber fall onto backing tape that has been coated with a layer of resin, and are then covered by a second resin-coated tape. This sandwich of tape and fiber then passes through a series of rollers that consolidate it. Finally it is rolled onto a cylinder to await use. This form of laminate lacks the great strength in one direction of a continuous fiber laminate, but its short lengths of fiber give it moderate strength in all directions.

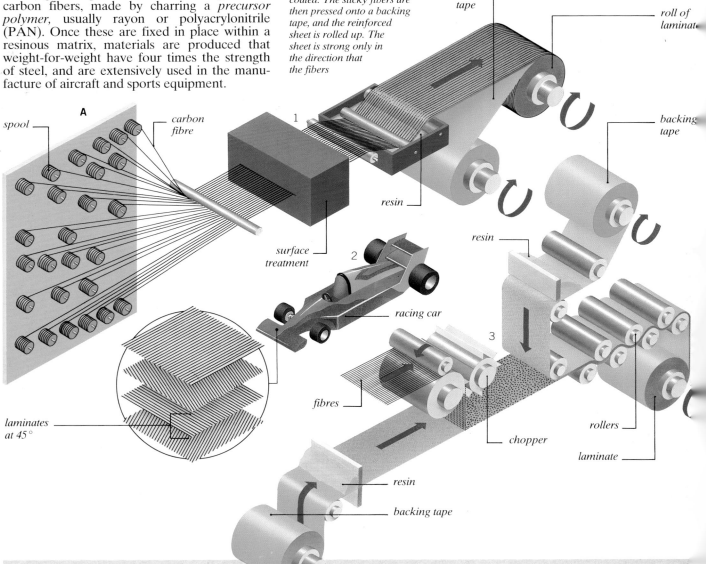

spool — carbon fibre

A

surface treatment

1

resin

2 — racing car

fibres

resin

backing tape

laminates at 45°

backing tape — roll of laminate

backing tape

resin

3

rollers

chopper

laminate

resin

backing tape

Baking strength
Some new composite materials are designed to combine the qualities of ceramics – very high compressive strength and rigidity – with those of metals – high strength in tension and ductility. The picture (right) shows a sample of cermet *being removed from a furnace. This material is made from boron carbide, a particularly strong ceramic material, mixed with aluminum. When the two are heated up to 1500°F, the molten metal infiltrates the ceramic structure, forming a matrix that is lighter than aluminum but stronger than steel.*

Flexible conductor
High-temperature superconductors are being developed that can be made into complex shapes. A flexible sheet of the material (below) is formed in a mold and then fired to remove the plasticizers that give the sheet its flexibility, leaving a rigid product.

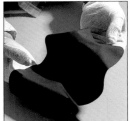

Getting back in shape
Special alloys have been made that have the ability to regain their shape after they have been bent. These "shape memory" alloys are mixtures of nickel, titanium, and copper. The photograph (below) shows three samples that were originally made into S-shapes, then deformed. When each twisted specimen is gently heated it springs back into its original shape – this has already happened to the piece on the right. The central one is in the process of untwisting, while the sample on the left has not yet been heated. Such metals could be used to make self-healing car bodywork.

B

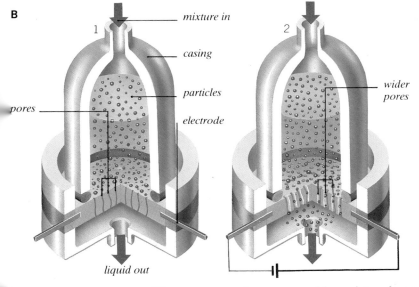

mixture in

casing

particles

electrode

pores

wider pores

liquid out

They shall not pass
*A gel is a soft aggregate of long-chain polymers. Some gels have been made that change shape, expanding or contracting in response to an applied electrical current. Such a gel could be used to make a variable filter [**B**]. This could consist of a barrier layer made of the gel, peppered with pores running from one face to another. When no current flows via the electrodes through the gel [1], the pores in the gel are too small to allow particles (red) to pass. But when current flows [2] the gel shrinks, with the result that the pores become larger and the red particles can pass.*

one another when it is a liquid, but locks them together as soon as a current is applied.

Electrorheological fluids are liquids that respond to electric current by changing viscosity. They contain minute particles in suspension: when a voltage is applied, the particles line up in columns, making the liquid "thicker." Such liquids make an ideal filling for automobile shock-absorbers because ride quality can be changed from "soft" to "hard" simply by varying the applied current.

The term *ceramic* covers a wide variety of nonmetallic materials, the most familiar of which is clay pottery. Advanced ceramics share the properties of pottery, being extremely hard and heat resistant, and are put to use where these attributes are desired. Tungsten carbide, for example, is used to make drill bits and grinding tools; and ceramics such as silicon nitride could one day be used to make engines that run "hotter" and therefore more efficiently than today's metal designs. Like pottery, however, ceramics are brittle – a drawback that is likely to be overcome by further research.

Another promising area of materials research focuses on substances that can be made to change their properties in response to external stimuli. An *electrorheostatic* fluid, for example, changes state from liquid to solid when an electric current is passed through it. It is an ideal substance to fill an automatic clutch because it allows the clutch plates to spin independently of

Super materials
Superconductors are materials that conduct electricity without any resistance, and thus without energy loss. It has been known for many years that alloys of the metal niobium behave as superconductors, but only when cooled by liquid helium to 3 K.

Recently, a more practical ceramic superconductor (yttrium-barium-copper oxide, YBCO) was discovered. This material superconducts at 95 K, a temperature that can be achieved by cooling in relatively inexpensive liquid nitrogen. In theory, YBCO superconductors make possible super-efficient motors and allow electricity to be carried over hundreds of miles without energy loss. Unfortunately, YBCO is brittle and almost impossible to draw into wires, so shaping it into useful devices is difficult.

Connections: Chemical Engineering 184 Steel Manufacture 188 Shaping Metals 190 Polymers 192 Principles 212 216 218 226 232 246

Life Science
How biotechnology is applied

Living cells are miniature chemical factories, carrying out a multitude of complex transformations and syntheses throughout their lives. Biotechnology gives us the means to harness their remarkable abilities on an industrial scale, to make products as diverse as vaccines and detergent. Although the word *biotechnology* is today associated with sophisticated genetic techniques, it covers a wide range of processes, some centuries old. Brewing, tanning, and sewage treatment, for example, are all industries that depend on the properties of living organisms.

Biotechnology is a multidisciplinary science that draws on a knowledge of biology, chemistry, physics, and engineering to exploit the properties of living cells. Today, much of the technology centers around the large-scale cultivation of microbes, such as bacteria and fungi. These microorganisms grow and divide at an incredible rate: 1 oz of bacteria could, under optimal conditions, grow to a mass of well over 3 tons in a single day. Moreover, many microbes have undemanding nutritional requirements, using simple raw materials such as glucose and starch as carbon sources, and yeast and soybean extracts as nitrogen sources. This means that they can be grown quickly and relatively cheaply in *bioreactors* – giant vats that hold up to 250,000 gallons.

Microscopic harvests

Occasionally, the cells themselves are the desired "products": for example, some fungi and bacteria are grown as animal feed supplements, and others are cultured because they are the natural enemies of crop pests. More commonly, however, the microbes are grown for the valuable chemicals they produce. Microorganisms produce a vast range of compounds, known as secondary metabolites. These are by-products, not essential for the organism's growth, and are generally produced at a late phase of a microbial culture, when the cells have stopped growing and dividing rapidly. By manipulating the culture conditions, the microorganism can be induced to manufacture large amounts of a particular secondary metabolite. For example, the mold *Penicillium* manufactures the antibiotic penicillin when the nitrogen content of its growing medium is reduced to below a certain level.

Other products obtained from industrial-scale microbial cultures include vitamins, such as B12 and B2; pigments, such as ß-carotene used as a yellow colorant in the food industry; flavoring compounds; simpler chemicals, such as amino acids; other organic acids; acetone; and alcohol for both consumption and use as an alternative fuel (gasohol).

Some bacteria manufacture carbohydrate polymers, such as xanthan gums and dextrans. In the living cell these molecules function as "larders" – stores of chemical energy. But they also have commercial value as edible thickeners, stabilizers, and fillers in the food and pharmaceutical industries.

Immune industry
*Today, biotechnologists are able to exploit the properties of antibodies [**A**] – molecules that are part of an animal's immune system. Antibodies are large Y-shaped molecules made when a specialized blood cell, called a B cell, encounters an* antigen – *a molecule that it recognizes as being foreign. The antibody works by binding to the antigen, "marking" it for destruction by other parts of the immune system.*

Antibodies are produced in response to many millions of different antigens: each type of

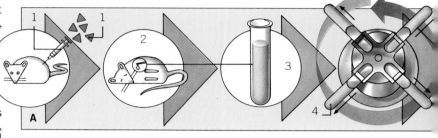

A

Microengineering (right)
The bacterium Escherichia coli *(here false-colored green/blue) is often used in genetic engineering experiments. It makes a convenient host for foreign genes because it occurs naturally and (usually) harmlessly in the human intestine. Under favorable conditions, it can divide once every 20 minutes, rapidly multiplying the quantity of the gene's valuable product.*

Genetically engineered bacteria are put to work in unlikely locations – cleaning up oil spills and purifying copper ore – as well as in the laboratory or pharmaceuticals plants.

Cellular catalysts

The ability of living organisms to carry out chemical syntheses and conversions depends entirely on the enzymes they contain. An enzyme is a catalyst that speeds up the rate of a biological reaction many thousands of times.

Any one cell contains hundreds of different types of enzyme, each one catalyzing one type of reaction. A complex chemical transformation like the synthesis of an antibiotic has numerous intermediate steps, each one requiring a different enzyme. Living cells are ready-made "packages" of the right enzymes, and so are ideal for complex syntheses. But some simple biochemical transformations, such as the conversion of the sugar glucose to fructose (which is sweeter, more soluble, and is much used in the confectionery industry), require only one enzyme, in this case glucose isomerase. This type of process is often carried out using the isolated enzyme or a partly purified cell extract.

Bioreactors (above right)
On the industrial scale, useful cells are cultured in large tanks, or bioreactors. The interior of the tank is usually stainless steel to resist attack by acids produced by the multiplying cells. Nutrients and other materials are fed into the reactor through valve-operated pipelines, while probes monitor the pressure, temperature, acidity, and degree of agitation of the culture.

Many microorganisms are aerobic, needing air to survive: air is filtered and pumped into the base of the reactor. It bubbles through the mixture and is released through a valve at the top.

Connections: Chemical Engineering 184 Brewing 198 Pharmaceuticals 200 Genetic Engineering 202 204 206 Principles 220 222 232

antibody binds specifically to one antigen, ignoring all others, and each type of antibody is made by one particular B cell.

The high specificity of antigen/antibody reaction makes antibodies valuable to scientists because they can be used to detect traces of chemicals that would be impossible to detect by conventional techniques.

To be usable, a single type of antibody must be made in large quantities. This is done by cloning a single B cell, and making it "immortal" by biological tinkering. The cell is then capable of producing monoclonal antibodies.

To produce a monoclonal antibody, a mouse is first injected with the "target" antigen [1]. Its spleen (containing many B cells) is removed [2], liquidized [3], and spun in a centrifuge [4], isolating a discrete layer of B cells [5], each of which produces a different type of antibody [6]. The B cells, however, would not survive long outside the body: to make them able to divide indefinitely in cell culture, they are fused [7] with "immortal" myeloma cells [8]. The hybrid cells [9] are then diluted and put on to a many-welled plate [10]. Here they are treated with the "target" antigen.

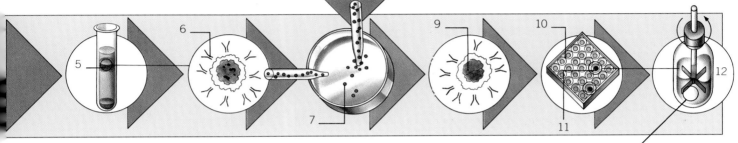

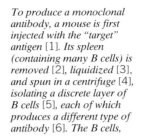

Enzymes are themselves a commercially important end-product of biotechnology. Tons of protein-digesting enzymes (proteases) are produced each year from microorganisms for use in detergents (the so-called biological powders). Bacterial amylases (starch-digesting enzymes) are also used in the manufacture of a wide range of foods.

Growing immunity

Although more demanding in their culture requirements than most microbes, animal cells have been cultured on a large scale since the 1950s, chiefly to grow viruses for vaccine production. Live vaccines, such as the oral polio vaccine, contain a special strain of the virus that induces immunity but does not cause the disease. Viruses can only multiply inside living cells, and are produced in the quantities required for vaccine manufacture from infected tissue cultures.

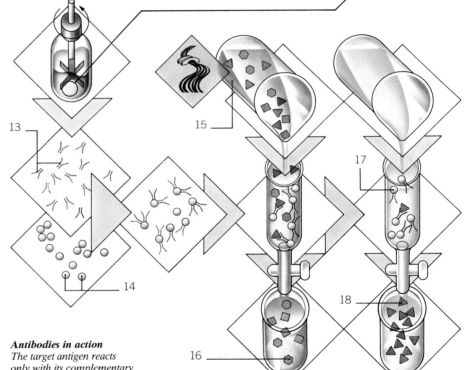

Antibodies in action
The target antigen reacts only with its complementary antibodies [11]. The B cell producing these antibodies is removed and cultured in a bottle [12], where it multiplies millions of times.

The antibodies [13] produced by the cloned B cells can be used in an immunosorbant column to separate out tiny quantities of the target antigen from mixtures. The antigen may, for example, be a pollutant of river water or a trace chemical in the human body that is diagnostic of a particular disease.

For use in the immunosorbant column, the antibodies are "glued" to resin beads [14], and used to fill a glass column. A mixture of chemicals suspected of containing the antigen is poured through the column [15]. Molecules of antigen (triangles) bind to the immobilized antibodies and are retained in the column. Molecules that cannot combine with the antibodies pass through the column [16]. The column is then washed with a solution that breaks the bonds between antigen and antibody [17], producing a purified solution of the antigen [18].

Bottled Biotechnology

How microbes are put to work in the brewing industry

Beer was first brewed in Babylonia more than 7000 years ago. For millennia its production was a mysterious business; the principles of brewing were poorly understood and the techniques jealously guarded by the brewers' guilds. But the identification of yeast as the agent of fermentation in the 19th century transformed brewing into a science. Today, computer-controlled breweries produce a stunning 23 billion gallons of beer every year, and the introduction of genetically engineered strains of yeast is soon to transform this most ancient of biotechnologies.

Beer, wine, and leavened bread are all produced by the action of yeast – a microscopic single-celled fungus. Like nearly all plant and animal cells, yeast cells *respire* sugars, chemically "burning" them to provide energy for growth and reproduction. If supplied with ample oxygen, yeast respires *aerobically,* yielding large amounts of energy, and water and carbon dioxide gas (CO_2) as waste products. However, in an oxygen-depleted environment the yeast switches to *anaerobic* respiration – or fermentation – growing more slowly and producing carbon dioxide and ethanol (alcohol) as "waste" products. In brewing, sugars derived from grain are fermented to give alcohol; in baking, the CO_2 gas liberated during fermentation rises through the dough to give bread a light texture.

Special brew

Modern brewing is an elaborate step-by-step process, central to which is the fermentation of *maltose* – a sugar extracted from germinated, roasted grains of barley. Fermentation is carried out in cylindrical vats, up to 65 ft high and 33 ft in diameter, by one of two species of yeast – *Saccharomyces cerevisiae* or *S. uvarum.* These naturally occurring microorganisms have been chosen for the flavors they produce and for their tolerance of alcohol (which allows strong beers to be brewed). The flavor of beer is also influenced by the type of water used, by the addition of hops (which contain bitter oils and resins), and by the way the beer is conditioned.

The distinctive character of the two main types of beer – ales and lagers – results from differences in the application of brewing biotechnology. Traditional ales are produced by fermentation that lasts six days, at a temperature of around 60°F, by strains of *S. cerevisiae* that float at the top of the brew. In contrast, lager yeasts are strains of *S. uvarum,* which sink to the bottom of the fermentation vat. These yeasts ferment at around 50°F and much more slowly, taking up to three weeks.

Soon after fermentation, beer is put into wooden casks or, more often, metal tanks, for *conditioning,* during which the traces of yeast left in the brew ferment any residual sugars. This secondary fermentation enhances flavor and produces CO_2, which gives beer its sparkle. Lagers are conditioned for longer than most ales and at near-freezing temperatures. The low temperature helps to precipitate the proteins that otherwise would cloud the beer.

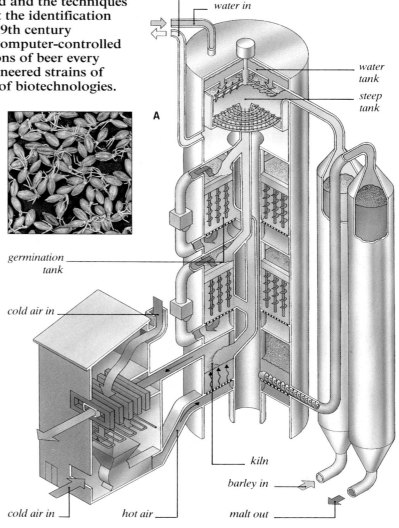

waste CO_2 out

water in

water tank

steep tank

A

germination tank

cold air in

kiln

barley in

cold air in   hot air   malt out

An intoxicating industry
The basic raw materials for brewing are malt (a source of sugar), hops (which provide bitter flavor), yeast (the key component in brewing, which ferments the sugar into alcohol) and, of course, water. But even before reaching the brewery (left), these materials have already been intensively processed. Water is treated to remove certain minerals and add others, while malt is prepared from grains of barley (a process called malting). The barley is soaked in water to start its germination: this activates the enzymes that convert the starch in each grain into the fermentable sugar maltose.

Connections: Food Preparation 90   Chemical Engineering 184   Biotechnology 196   Genetic Engineering: Lower Organisms 202   Principles 220   222   232

Copper kettle (right)
Kettles for boiling the sweet wort used to be made from copper, and were known as coppers. Today's coppers, however, are most often made from stainless steel. Traditionally, dried hop cones were added to the wort, but today, powdered compressed hops are used.

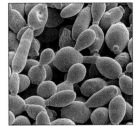

Maturation period
All beer is matured (or conditioned) after fermentation to improve its flavor and to clear any suspended solids. Mass-produced beers are conditioned in large tanks of several thousand gallons [**F**]. Ales are stored for about one week, but lagers are conditioned for up to a month. The beer is filtered before being poured into kegs, bottles, or cans. The beer is flash-heated (pasteurized) to protect it from spoilage by any live microorganisms. Carbon dioxide is injected into the container to exclude any air and prevent oxidation and spoilage.

grist

yeast in

waste CO_2 out

water in

hops in

B

C

D

E

masher

sweet wort

steam jacket

steam in

cooling jacket

beer out

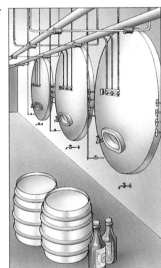

F

Today, malting often takes place in a tower maltings – essentially a series of tanks arranged vertically [**A**].
In the uppermost tank, barley is steeped in water for around two days. It is then transferred to two lower levels. Here, it is turned frequently by augers, and air is circulated through it. Aeration ensures that the grain germinates after a few days (top left). As it germinates, enzymes are released that begin to break down starch into sugar. Other enzymes in the grain produce amino acids, which later nourish the yeast.
Once the enzymes are fully activated, the malted barley passes into the kiln.

Here, air heated to around 176 °F is passed through for 24 hours. This stops further germination and roasts the malt, giving additional flavor. Some tower maltings are 130 ft high: computer control enables them to handle 22-ton batches of barley and produce malt of consistently high quality.
In the brewery, dried malt is milled – ground up into a coarse flour called the grist. This is mixed with water at

a temperature of 152 °F and mashed in a vessel called the mash mixer [**B**]. After mashing, all the starch in the grain has broken down, and the resulting sugar is dissolved in the hot water. This sugary solution, known as the wort, is separated from the grain in a lauter tun [**C**], where the grain is raked and sprayed with

water. The wort is then piped into a kettle [**D**] that holds over 7000 gallons. Hops (above) are added to the kettle, giving a characteristic bitter flavor, and the mixture is boiled by passing steam through a jacket at the base of the kettle. Boiling prevents the growth of bacteria that would spoil the brewing

process, inactivates the barley enzymes, and also extracts bitter and aromatic flavors from the hops.
Once the wort has cooled, yeast (above right) is added and the mixture ferments in a stainless-steel vessel [**E**] for several days. Fermentation produces the desired product – alcohol – as well as carbon dioxide (which is vented through the top of the vessel), and heat. A cooling jacket (made up of pipes carrying cold water) keeps the brew at optimal temperature. In lager manufacture, the yeast remains at the bottom of the tank, though convection currents carry some yeast to the top of the tank.

Magic Bullets
How new drugs are developed

Despite the formidable armory of compounds available to physicians, new drugs are always needed. Many conditions remain untreatable, and pathogens eventually become resistant to drugs used against them. But the design of new drugs is a huge challenge. Even when a promising compound has been identified it takes, on average, twelve years of research and an investment in excess of US$100 million before it reaches the pharmacy shelf. The goal of devising "magic bullets" – potent drugs without side effects – has been made easier by new advances in biotechnology.

For millennia people have used herbal extracts to treat illness and disease, and today about half of our most useful drugs are derived from chemicals originally extracted from plants and microorganisms. Bacterial and fungal invaders are combated by antibiotics made by microorganisms, while plants provide the basis for the heart drugs digitalis and digitoxin and the narcotic opium and its derivatives. The natural world is still being scoured for novel drugs. Programs such as that of the US National Cancer Institute collect and screen plants for anticancer activity and for activity against HIV, the cause of AIDS. But of the thousands of new products screened, few show any effect, and even fewer come to clinical trial.

Most "new" drugs are adaptations of existing compounds. Chemists tinker with the structure of a known drug of proven efficacy to produce a new compound with enhanced characteristics, such as increased activity or lower toxicity.

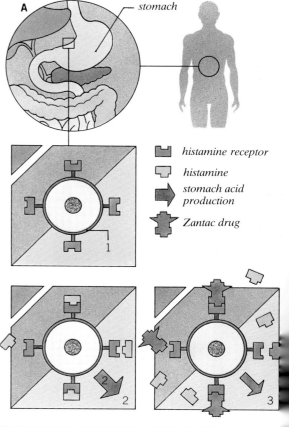

stomach

A cure for ulcers
*The antiulcer drug Zantac, made by the British company Glaxo, is the best selling pharmaceutical of all time [**A**]. It works by affecting the histamine receptors on parietal cells – those that line the stomach [1]. These cells produce stomach acid (to aid digestion) when histamine binds to receptors on the cell surface [2]. Excess acid production causes ulcers, because the acid attacks the lining of the stomach. Zantac blocks up the receptors, preventing the binding of the histamine ligand [3]. Acid secretion diminishes and the ulcers have a chance to heal.*

▢ histamine receptor
▢ histamine
➡ stomach acid production
➡ Zantac drug

Natural resources
Most drugs available today were found either by chance or by the systematic screening of millions of synthetic and natural chemicals. Living organisms provide many important pharmaceuticals. For example, the Madagascan periwinkle (far right) is the

source of the anticancer drug vincristine, and soil bacteria of the genus Streptomyces (right) provide us with numerous antibiotics. Genes for some drug compounds can now be transplanted into microorganisms capable of synthesizing the drug in large quantities.

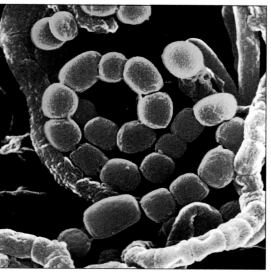

yews in the northeastern forests of Oregon and Washington state, the necessary level of destruction could not be sustained for long. The problem was overcome by isolating protaxols – chemical precursors of taxol – from renewable resources such as the needles and twigs of the Pacific yew and other, more common, yew species. These protaxols could be converted into taxol in the laboratory.

In other cases, the problem of limited supply was solved by bringing the plant into cultivation (as with the opium poppy) or by devising a complete laboratory synthesis.

Kill or cure?
Harvesting drugs from natural sources sometimes brings medical interests into conflict with environmental concerns. For example, the anticancer drug taxol comes from the bark of the Pacific yew tree (left) where it is present in minute quantities. The bark from a 100-year-

old tree yields only enough taxol for a single dose, and the tree is killed when its bark is stripped. In 1991, it was estimated that some 40,000 trees would be required to provide enough material to treat 12,000 patients in clinical trials. Although there are an estimated 28 million Pacific

Genetic medicines
*In a human cell [**B**], information stored in the form of the DNA double helix [1] is transcribed into the messenger molecule RNA [2] and then translated into a protein [3]. Genetic drugs work by preventing the synthesis of an unwanted protein. Short lengths of DNA, complementary to the gene that codes for the unwanted protein, are introduced into the cell [4]. They bind to the relevant gene forming a three-stranded triplex [5]. This stops the gene from being translated into RNA [6], and therefore prevents protein synthesis [7].*

Active ingredients

Drugs can act in a number of different ways. Some kill pathogens outright, or stop their growth, by inactivating their *enzymes* – the catalysts that build and maintain the invading cells. The antibiotic penicillin, for example, binds to an enzyme that constructs the thick protective cell wall of an invading bacterium. The disabled enzyme cannot complete the assembly of its wall, and the bacterium bursts under its own internal pressure and is killed.

Other drugs act on human rather than pathogenic cells, controlling or modifying their function. Human cells are covered with *receptors* – molecular "locks." When the appropriate "key," or *ligand,* binds to the receptor, the cell is induced to behave differently by, for example, producing more or less of a particular protein. Ligands may be hormones, neurotransmitters, or other types of molecules, and their binding to receptors is an essential mechanism by which the body regulates its internal environment – its control over heart rate, blood pressure, cholesterol level, nerve function etc. Drugs that work on human cells block receptors on the surfaces of specific tissues. In doing so, they prevent ligands from binding to the receptors and eliciting the normal responses from the cells. This intervention in the body's physiological management can produce useful effects. For example, the ligand histamine binds to receptors on cells of the stomach lining, inducing them to release stomach acid. Overproduction of acid (which causes ulcers) can be controlled by the drug Zantac, which blocks the attachment of the histamine molecules.

Hi-tech drug hunt

Biotechnological techniques allow receptors to be used directly in the search for new drugs. Receptors on a target cell are isolated, duplicated in great numbers, and then exposed to a wide range of chemicals. Those chemicals that do not latch on to the receptors are very unlikely to be good drugs; those that do bind may have drug activity and are subjected to further tests. This technique greatly accelerates the screening of possible drug compounds.

A still more sophisticated approach uses the techniques of X-ray crystallography and computer graphics to build up a three-dimensional picture of a receptor protein. The computer model is then used to "design" a molecule that will bind to the receptor and block the attachment of a ligand.

Although the targets of conventional drugs are cell receptor proteins, some recent research has been directed at nucleic acids – compounds such as DNA and RNA. The idea is to stop a cell from making an unwanted product by blocking the gene that codes for the compound concerned. Genetic medicines may one day be able to attack the viral and genetic diseases for which no effective treatments exist today.

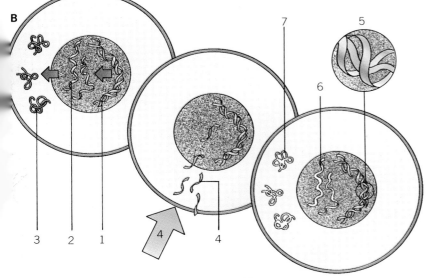

B

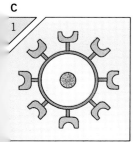

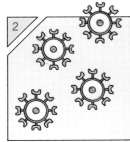

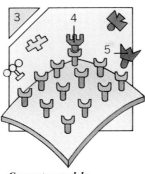

C

creening for cures
*eceptors – the cell proteins rgeted by drugs – are used irectly to identify new ctive compounds [**C**]. a cell that overproduces particular receptor [1] first isolated and cloned ?]. The cloned cells are en exposed to a variety chemicals [3]. Those that*

bind to the receptors [4,5] may be valuable drugs, and are tested further. The problem with this approach is that it is difficult to identify the receptor responsible for a particular cell response. Cancer cells in particular seem to lack any unique receptors that a drug could target.

Computer models
This computer system (right) displays a model of AZT – a drug used to treat AIDS. AZT works by inhibiting the action of an enzyme that enables the AIDS virus to subvert the human immune system.

Manufactured Microbes

How simple organisms are manipulated by genetic engineers

Biotechnology is set to become as great a force for change in the 21st century as the physical sciences have been in the 20th. New molecular engineering techniques allow today's biologists to alter a living organism at its most fundamental, genetic, level, and create "tailor-made" plants, animals, and microbes with novel, and highly valuable, properties. Bacteria transformed into "factories," that mass-produce human proteins for pharmaceutical use, are among the first products of the multibillion-dollar genetic engineering industry.

Animals or plants of a single species may differ widely in what they look like and what they can do. This *natural variability* has been exploited for centuries by farmers, who improved their stock by breeding from animals with desirable characteristics, such as high muscle bulk or copious milk production. However, the boundaries of natural variation are limited by the underlying genetic makeup of the species. Animals, for example, are never able to photosynthesize because, among many other things, they lack the genes that direct the manufacture of the green pigment chlorophyll.

Genetic engineering has freed biotechnologists from their dependence on organisms that occur naturally or that can be produced by conventional breeding techniques. Individual genes can now be extracted from one species (and even altered by delicate biochemical tinkering) and transferred into another species, to produce an organism with new, useful properties.

Tailored bacteria
*To splice a human gene into a bacterial cell [**A**], a pure sample of DNA is prepared from the human cells that contain the required gene [1]. DNA-cutting restriction enzymes [2] are used to snip the DNA into thousands of small pieces [3]. Each piece contains many genes: however, only one piece (colored yellow) contains the desired gene.*
Plasmid vectors are prepared by extracting all the DNA from bacteria and separating plasmid DNA [4] from chromosomal DNA [5]. The plasmids are then treated with restriction enzymes, to "open out" their circular shape [6].

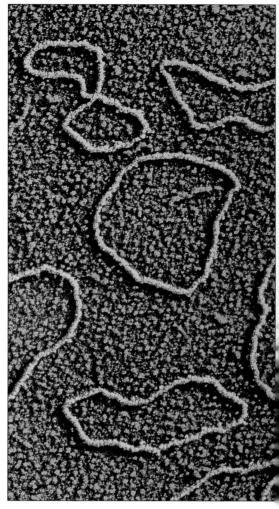

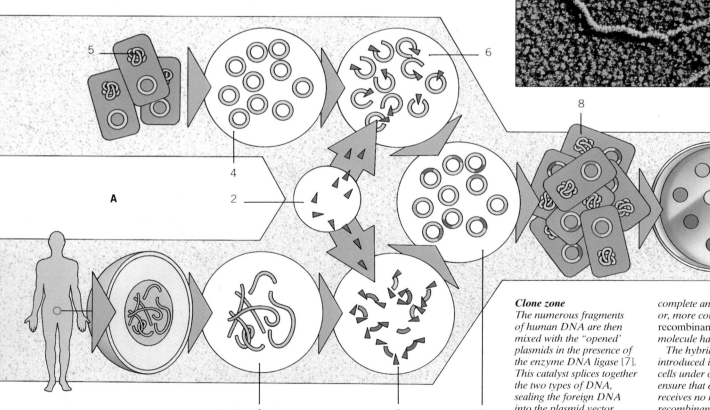

Clone zone
The numerous fragments of human DNA are then mixed with the "opened" plasmids in the presence of the enzyme DNA ligase [7]. This catalyst splices together the two types of DNA, sealing the foreign DNA into the plasmid vector. Gene splicing is now complete and a hybrid or, more correctly, recombinant *DNA molecule has been prepared.*
The hybrid DNA is introduced into bacterial cells under conditions that ensure that each bacterium receives no more than one recombinant plasmid [8]. The mixture of recombinant...

Connections: Chemical Engineering 184 Biotechnology 196 Brewing 198 Genetic Engineering 202 204 Principles 218 220 222

Plasmids (left)

Genetic engineers use plasmids as vectors to carry foreign genes into host bacterial or fungal cells. Plasmids are small rings of DNA found in most bacteria. They are less than one hundredth the size of the bacterium's main chromosome and can replicate independently of the main chromosome: this means that many copies of the same plasmid can be present in one cell. These features make them very useful and also relatively easy to manipulate.

Because of its small size, a plasmid treated with a highly specific restriction enzyme may be "snipped" at only one point (a large chromosome would be cut into many fragments). Foreign DNA can then be spliced into the open circle and the ends joined up again. Given the correct chemical environment, bacterial cells will take up the composite plasmids. The plasmids will then replicate inside their new hosts and express the foreign genes they contain.

Shown here, magnified 70,000 times, are plasmids from the common gut bacterium Escherichia coli. *Artificial color has been added for clarity.*

Drugs from bugs

The techniques for genetic manipulation were the fruit of decades of fundamental biological research. They were first used commercially in the early 1970s to splice genes for valuable human proteins into bacterial DNA. By culturing the transformed bacteria and allowing them to multiply rapidly, large quantities of the proteins could be produced and harvested. Although bacteria are still widely used as "protein factories", because they can be made to "accept" foreign genes relatively easily, new developments in genetic manipulation mean that genes can now be spliced into human and yeast cells. These more advanced cells are better at making human protein, and are used today to produce numerous proteins for pharmaceutical use, including insulin for treating diabetes, and hormones for treating growth defects. Previously, these proteins had to be laboriously and expensively extracted from human or animal tissue.

Human growth hormone, for example, could only be extracted in very small amounts from the pituitary glands of human cadavers. In the past, this procedure led to some patients receiving hormone contaminated with the agent of brain disease. Now, human growth hormone is produced in culture by engineered cells and is free from this fatal contaminant. Similarly, insulin used to be extracted from the pancreatic tissue of pigs and cows. Although animal insulin can be used to treat diabetes, it provokes immunity in patients over the long term, making it less effective and leading to undesirable side effects. Cultured yeast cells carrying transplanted viral genes are also used to make viral proteins needed to make vaccines such as that recently developed for hepatitis B.

From genes to proteins

Physically, a gene is a short segment of a long molecule of deoxyribonucleic acid – DNA. Most organisms contain DNA, which acts as a stable information store in each cell, governing its development and function. A typical human cell contains a total of 13 feet of DNA packaged into 46 separate chromosomes and enclosed in a nucleus [1]. This length of DNA holds around 100,000 genes. A chromosome is composed of two strands of DNA wound around each other into a double helix. Each strand is a chain of chemical subunits. Genetic instructions are encoded in the order of these chemical subunits along the strand. There are only 4 "letters" in the genetic code (denoted by the colors yellow, blue, red, and green).

Genes work by directing the manufacture of protein molecules. Cells are made of different types of protein; and specialized proteins called *enzymes* carry out all cellular functions that characterize life. Each gene is a set of instructions – in chemical code – that tells the cell how to assemble a particular protein.

When a cell "needs" to make a protein, a temporary copy of the relevant gene is first made in the form of a molecule of ribonucleic acid, or RNA [2] . This molecule acts as a *messenger*, carrying coded instructions from the DNA out of the nucleus to "protein factories" [3] within the cell. Here, the instructions are read by T-shaped RNA molecules [4]. These pick up amino acids [5] (the building blocks of a protein) and assemble them in the order specified by the sequences of coded "letters" on the messenger RNA. The chain of amino acids builds up to form a protein molecule [6].

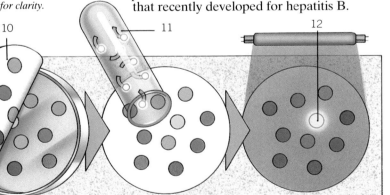

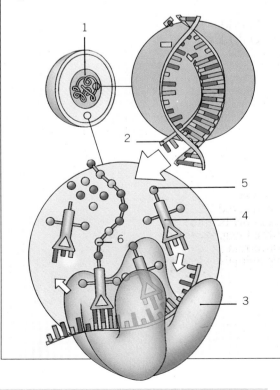

acteria is diluted and ultured on a plate covered ith solid nutrient medium], where the bacteria apidly multiply. Each acterium grows into a olony of millions of dentical individuals. But nly one of these cloned olonies contains the desired uman gene. To identify *this colony, the plate is blotted with a type of filter paper [10]. A sample of each colony adheres to the paper, which represents a faithful copy of the pattern of colonies on the plate. The blot is treated with a probe [11] that only binds to the desired human gene: the probe is tagged with a* *fluorescent marker, which shines when illuminated by ultraviolet light [12]. The bacterial colony containing the cloned gene has been identified on the blot. Its corresponding colony on the plate can then be picked out and cultured on a large scale to produce a particular human protein.*

Designer Living

How genetically engineered organisms are put to use

In 1989, a mouse made legal history by becoming the first animal to be covered by a US patent. The genetically engineered *oncomouse* was implanted with a gene that causes cancer, making it extremely valuable to scientists studying the disease in humans. Aside from its use in medical research, genetic manipulation has an ever-increasing number of applications: crop plants can now be genetically "designed" to resist attack by pests; patients with hereditary diseases can be treated with gene therapy; and genetic fingerprinting techniques are used to help catch criminals.

The genetics industry is emerging from its infancy. Whereas its first products, such as insulin and growth hormone, were derived from bacteria "reprogrammed" to produce human proteins, today's genetic engineers have more refined techniques at their disposal. These allow direct control over the inherited properties of higher plants and animals.

Genes to feed the world
Crop plants are constantly under threat from viral disease because most viruses are not killed by chemical sprays. Genes for virus resistance are sometimes found in wild relatives of crop plants, or even in unrelated species. In the past, plant breeders would cross a productive crop with a range of its wild relatives, hoping to incorporate the resistance gene. This "hit-and-miss" approach was difficult and time-consuming, sometimes taking decades to complete. Genetic engineering offers a way around this problem by taking a useful gene from one species and splicing it into the DNA of the crop species to produce a *transgenic* plant.

Transgenic plants, carrying bacterial genes for insecticidal proteins, have already been made. The whole living world is now being scoured for useful genes to make crop plants more productive and stress-resistant.

Animal engineering
In general, plants are more amenable to genetic engineering than animals. This is because a whole plant can be regenerated from leaf, stem, or root cells grown in culture. Cells genetically modified in culture grow into mature plants carrying the new gene in all their cells, and are capable of passing this gene to their offspring.

Animal cells fall into two broad categories: *germ-line* cells, which are the sperm and ova – the male and female reproductive cells; and *somatic* cells, which make up the rest of the body, including organs, muscle and bone-marrow. Transgenic animals that can pass their engineered characteristics on to their offspring can only be made by injecting a gene into a fertilized egg. At present, this procedure is highly uncertain, so the genetic engineering of domestic animals is less advanced than that of plants. In addition, animals respond unpredictably to manipulation: for example, pigs given extra copies of their own growth hormone genes grow more quickly, but also suffer from undesirable side effects.

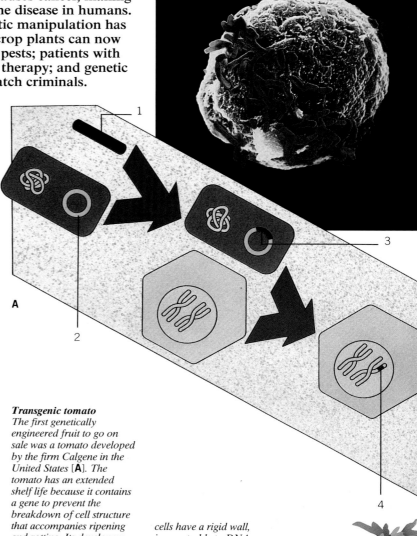

A

Transgenic tomato
*The first genetically engineered fruit to go on sale was a tomato developed by the firm Calgene in the United States [**A**]. The tomato has an extended shelf life because it contains a gene to prevent the breakdown of cell structure that accompanies ripening and rotting. Its developers also claim that this gives it a better flavor than conventional tomatoes because it can be left to ripen longer on the plant before being picked.*

The new tomato does not contain a gene introduced from another species, but an extra, modified copy of one of its own genes – one that directs cells to make the enzyme polygalacturonase (this catalyses the breakdown of a plant's cell walls during ripening). The additional copy of the gene is modified to be a "mirror image" of the native gene.

Introducing the modified gene into the tomato is no easy matter, because plant

cells have a rigid wall, impenetrable to DNA.

The gene [1] is therefore first spliced into a loop of DNA (or plasmid) carried by the bacterium Agrobacterium tumefaciens [2]. This pathogenic bacterium is a natural genetic engineer: it infects a plant and transfers a portion of its own DNA into the plant cell [3]. This foreign DNA is incorporated into the plant's own chromosomes [4]. The affected plant cells are cultured, induced to multiply [5], and then they differentiate to form plantlets [6]. Transferred into soil, the plantlets grow into mature plants [7] carrying the new traits.

Connections: Chemical Analysis 142 Polymers 192 Pharmaceuticals 200 Genetic Engineering 202 206 Principles 218 222

Targeted transfer (left)
The most common way of introducing genes into plant cells is to use the bacterium Agrobacterium tumefaciens as a vector. Agrobacterium naturally infects many plants, causing the disease crown gall, by splicing its own genes into the plant's chromosomes. For genetic engineering purposes, agrobacteria have been developed in which the disease-causing genes have been deleted. The "disarmed" bacteria can still infect plant cells with their own (or foreign) genes but do not cause crown gall. Agrobacterium (red) is here shown attacking a cell of a tobacco plant (yellow).

Genetic fingerprints

Apart from identical twins, no two human beings have exactly the same genes (or exactly the same DNA). In 1985, British scientists devised a way of analyzing these differences and presenting them as DNA "fingerprints," unique to the individuals from whom they were taken.

Fingerprinting is widely used in forensic science to match blood or semen samples to a suspect in cases of murder and rape. Until 1989, it was considered to be unassailable evidence of identity in a court of law, but errors in analysis have since removed this degree of certainty.

The second major use of fingerprinting is to investigate family relationships. Children inherit some DNA from their mother and some from their father. They therefore share DNA with brothers, sisters, and other close relatives. DNA fingerprinting is routinely used in cases of disputed paternity, but it has also had some more unusual applications. In 1992 it was used to confirm that five skeletons exhumed from a grave outside Ekaterinburg were those of Tsar Nicholas II of Russia, his wife and three daughters. A blood sample of a distant relation, Prince Philip, the Duke of Edinburgh, provided the means of identifying the members of the family.

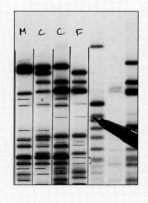

Relative values
Fingerprinting techniques examine the differences between pieces of DNA called minisatellite DNA. This DNA is found at sites between genes (a gene is a length of DNA that carries the instructions for making a protein). It is used in fingerprinting because it is highly variable. The chance of two unrelated individuals possessing exactly the same minisatellite regions is very small, so DNA fingerprinting provides a reliable test of identity. The minisatellite region is cut into fragments using restriction enzymes, catalysts that "snip" DNA at particular sites. The resulting fragments are separated according to size by a technique called gel electrophoresis, in which the fragments (which carry a small electric charge) are driven through a gel by an electric field. This produces a "ladder" of DNA, in which each "rung" corresponds to a fragment of different size. The ladders produced by DNA of different individuals can be compared; the differences can be readily detected and the people distinguished.

The bands in the fingerprints shown here are marked M for mother, C for child, and F for father. Both children share some bands with each parent, proving that the individuals from whom the samples were taken are indeed related. A lack of correspondence indicates no relationship.

Stopping the rot
In the cells of a normal tomato [8], the polygalacturonase gene [9] is "read" and transcribed into a messenger molecule called mRNA [10]. This in turn is translated into an enzyme [11], which attacks the plant's cell wall [12], causing it to break down.

The genetically altered plant is identical to the normal tomato. However, within its cells, both the polygalacturonase gene [13] and its mirror image [14] are transcribed into messenger RNA molecules. Because these molecules are complementary in structure, they lock together [15], thereby blocking the translation of the mRNA. The cell-wall-degrading enzyme is therefore not formed, and the tomato lasts longer before rotting.

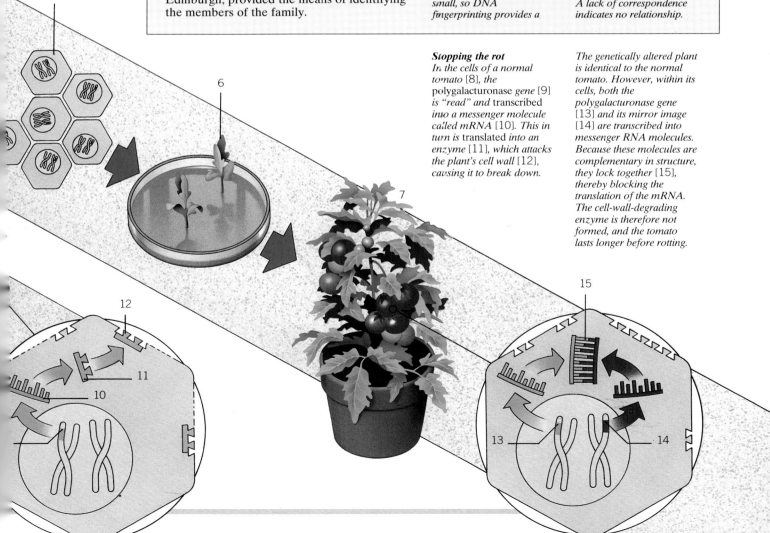

DNA Doctors
How gene therapy works

The growth, development, and functioning of the human body is governed by the 100,000 or so genes present in every cell. A fault in just one of these genes can cause a life-threatening disease, and indeed one infant in every hundred is born with a genetic defect. More than 4000 genetic diseases are known and, at present, most have no effective treatment. However, the prospects for dealing with these disorders have improved greatly with the first trials of gene therapy – a process that uses DNA technology to introduce normal genes into affected cells.

A human body is made up of trillions of individual cells. If an individual suffers from a disease caused by a defective gene, this gene is present in all cells of the body. Fortunately, for gene therapy to work, it is not necessary to introduce a normal gene into all of these cells. This is because many genes are only *expressed* (switched on) in particular types of tissue. For example, the gene which, when faulty, causes the disease hemophilia is expressed mainly in liver cells. So, at present, the aim of gene therapy is to introduce "good" genes into those tissues most seriously affected by the disease.

Biotechnological cures

The early successes of gene therapy have been in the treatment of the rare genetic disease, *severe combined immunodeficiency* (SCID). This is caused by a defect in the gene for the enzyme *adenosine deaminase* (ADA), the absence of which is fatal to white blood cells. Without functional white cells, the individual has little protection against infection and usually dies in early childhood. White blood cells are produced by the bone marrow, and if a compatible donor can be found, a marrow transplant can cure the defect. But where there is no donor, gene therapy now offers an alternative. In gene therapy, white cells are extracted from a SCID sufferer, a good copy of the ADA gene is introduced, and the functioning cells are then transfused back into the body.

The gene – physically a short segment of DNA – may be introduced into the cells in one of three ways. The white cells may be chemically treated to make them permeable to DNA. This technique has a poor success rate, the DNA only being taken up by one cell in around 10,000. Alternatively, the DNA may be injected using a microscopic pipette: this procedure, although more reliable, is extremely laborious. The third, and most promising, technique uses genetically engineered viruses to deliver packages of DNA into the target cells.

Although the treatment of SCID by repairing white cells is a successful strategy, the respite it gives is only temporary. This is because mature white cells have a limited lifespan and the treatment must be repeated every few months. White cells arise from small numbers of self-renewing *stem cells* in the bone marrow. Genetically modifying these cells rather than the white cells is the best hope for the long-term cure of SCID (see main illustration).

human DNA
viral protein coat
viral RNA
packing sequences
splicing sequences
normal ADA gene
RT enzyme

How retroviruses work
One of the challenges of effective gene therapy is to deliver a functioning gene accurately into a cell that has a faulty gene. This is done by exploiting the properties of special viruses, called retroviruses [A].

A retrovirus consists of a protein coat [1] enclosing a strand of the genetic material RNA [2], and an enzyme called reverse transcriptase (RT) [3]. First, the virus infects a human cell [4], penetrating its cell membrane. The RT "reads" the information in the single-stranded RNA and converts it into double-stranded DNA [5]. Human chromosomes are made of

DNA, so at this stage the virus has effectively "translated" its own genetic instructions into the language of the human cell. At both ends of the viral DNA are "splicing sequences" that allow the viral DNA to incorporate itself into one of the human chromosomes [6] in the

Whereas gene therapy is still a distant prospect for most disorders, DNA technology is already widely used diagnostically. DNA testing can identify carriers of many genetic diseases, and may soon be able to detect people's susceptibilities to common conditions, such as breast cancer, heart disease, and diabetes, which are believed to be genetically based. Such DNA tests depend on identifying, isolating, and cloning the gene involved in its normal and defective forms. These forms are then compared with test DNA from an adult or fetus. Diagnostic testing has been successful in reducing the incidence of inherited diseases such as thalassemia and Tay-Sachs disease. An ambitious international initiative – the Human Genome Project – aims to map the position and reveal the DNA sequence of all the genes in a human cell (collectively called the genome). This information will allow many more diseases to be screened before birth.

nucleus. The viral DNA can now lie dormant in the human cell. Some time later the viral DNA becomes active. It directs the human cell to make the components of the viral protein coat [7] and duplicates of the viral RNA [8].

Special "packing sequences" on the viral RNA then package the entire viral RNA strand into the fully assembled protein coat [9]. Completed viruses leave the cell (often killing in the process) and go on to infect other human cells [10]. Retroviruses are known to be responsible for many human diseases, including AIDS and some forms of cancer.

Connections: **Pharmaceuticals** 200 **Genetic Engineering** 202 204 **Principles** 218 220 222

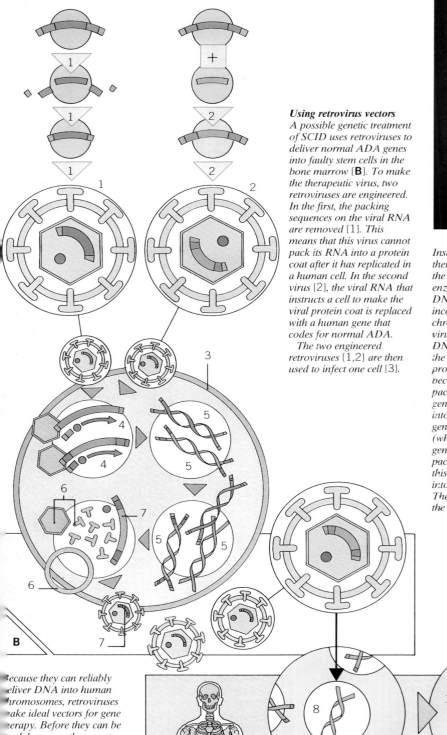

B

Using retrovirus vectors
*A possible genetic treatment of SCID uses retroviruses to deliver normal ADA genes into faulty stem cells in the bone marrow [**B**]. To make the therapeutic virus, two retroviruses are engineered. In the first, the packing sequences on the viral RNA are removed [1]. This means that this virus cannot pack its RNA into a protein coat after it has replicated in a human cell. In the second virus [2], the viral RNA that instructs a cell to make the viral protein coat is replaced with a human gene that codes for normal ADA.*

The two engineered retroviruses [1,2] are then used to infect one cell [3].

Inside the cell, both convert their RNA into DNA using the reverse transcriptase enzyme [4], and both viral DNA fragments are incorporated into human chromosomes [5]. When the viruses come to replicate, DNA from virus 1 directs the cell to make the viral protein coat [6]. But because this virus has no packing sequences, its genetic material cannot get into the protein coat. The genetic material of virus 2 (which includes the ADA gene), however, does have packing sequences, and it is this RNA which is packaged into the viral coat [7]. Therapeutic viruses carrying the ADA gene leave the cell.

Human genome *(above)*
The nucleus of each human cell contains over thirteen feet of DNA packaged into 46 chromosomes. The "letters" of the genetic code correspond to the molecular subunits, or bases, linearly arranged on DNA molecules. Human DNA has more than three billion of these bases.

The international Human Genome Project aims to determine the sequence of all the DNA in humans. Even with the enormous advances in DNA technology over the past decade, it is estimated that the project will take many years to complete and cost billions of dollars.

Because they can reliably deliver DNA into human chromosomes, retroviruses make ideal vectors for gene therapy. Before they can be used, however, they must be genetically modified in two important ways. First, the desired human gene must be implanted into the genetic material of the virus. This is accomplished using standard techniques for DNA manipulation. Second, the virus must be "defused" so that once it has delivered its genetic payload to the human chromosome, it can no longer replicate itself. This makes the virus "safe" because it cannot go on to infect further cells.

C

SCID treatment
*Therapeutic viruses produced by the above process are used to infect stem cells removed from the bone marrow of a SCID sufferer [**C**]. These cells [8] possess faulty genes for the enzyme ADA [9]. Infection by the retrovirus equips the cells with a good copy of the gene [10], in addition to the existing faulty copy. The treated cells are then returned to the patient's bone marrow [11]. White blood cells subsequently produced by these stem cells have a fully functioning ADA gene, and can play their normal role in the patient's immune system.*

*The paths of subatomic particles following a high-energy collision can be traced in
a bubble chamber. The faster or heavier the particle, the straighter its track.*

Principles

Introduction

The Properties of Matter

Mechanics

Electricity and Magnetism

Waves

Nuclear Physics

Introduction

This book is all about science and technology, but neither of these things is easy to define. The root meaning of science is *knowledge*, but today science is generally taken to mean a particular type of knowledge produced by systematic and experimental study of the natural world. The root meaning of technology is *practical art or craft*; but again, the term is now generally taken to stand for the practical applications of science.

The relationship between science and technology has not always been as close as it is today. This is partly because science is far younger than technology. Technology is as old as the earliest inventions – things like bone and stone implements, fire-making, and simple metal extraction. In fact, technology may be even older than humankind, because some other animals have been shown to use simple tools. Chimpanzees, for example, have been observed to strip leaves from branches and use the bared twigs to "fish" for termites in termite mounds.

Many of the earliest technologies were developed without very much scientific understanding of how or why they worked. The wheel, for example, was developed long before scientists understood the laws of motion and friction; lenses were made by people who knew little about the nature and properties of light; and explosives were invented many centuries before the birth of modern chemistry.

Although the Greeks developed a form of science several centuries before the birth of Christ, modern science did not emerge until the 16th and 17th centuries. At this time, science contributed relatively little to the development of technology; but new technologies certainly contributed to the rise of science. Thus, the Italian mathematician and astronomer Galileo Galilei made many of his most important discoveries using the newly invented telescope; the English natural philosopher Robert Hooke discovered plant cells with the aid of a simple microscope; and the English physiologist William Harvey developed his revolutionary theory of the circulation of the blood by likening the heart to a mechanical pump. Even as late as the mid-19th century, technology was still contributing more to science than science to technology. For example, new steam engines gave scientists the models with which to develop a scientific understanding of energy and energy conversion.

The harnessing of science to the production of new technologies only began in earnest in the late-19th century. In Germany, for example, the creation of professional laboratory science led to the development of the chemical industry; and at the same time, the development of the germ theory of disease in France and Germany supported the search for new ways of preventing and treating potentially fatal illnesses such as cholera, diphtheria, and tuberculosis.

Since the turn of the century, the relationship between science and technology has grown ever closer. The new nuclear physics of the 1920s and '30s inspired both the creation of the atom bomb during the Second World War and the development of civil nuclear power in the late-1940s and '50s; the emergence of information science from the pioneering work of code-breakers and cyberneticists during the Second World War inspired the rapid development of new computer technologies in the postwar period; and the discovery of the structure of the genetic material DNA in 1953 heralded a new era, not just in genetics, but in agricultural, industrial, and medical biotechnology.

Nowhere is the convergence of science and technology clearer than in the explosion of medical science and technology in the 20th century. For most of its history, medicine has been pure technology – a practical art or craft generally lacking in any firm scientific basis. In our century, for the first time, scientific research has become the basis for the vast majority of medical innovations. Even here, though, the relationship between science and technology is still two-way: new scientific understanding of diseases like diabetes or high blood pressure has guided the search for new drugs and treatments; and at the same time, new technologies like magnetic resonance imaging and DNA analysis have pushed forward the boundaries of scientific research.

Today, it is hard to tell where science stops and technology starts. In fields such as artificial intelligence and biotechnology, fundamental science and state-of-the-art technology are developing hand in hand. Though some areas of science (such as cosmology and evolutionary biology) seem largely irrelevant to technology, more and more scientific research is now conducted with at least one eye on potentially useful applications.

The following pages explore the fundamentals of science that underlie most technologies – from Archimedes' principle, which accounts for the buoyancy of ships and balloons, to the mathematical language of logic, which is drawn upon by all computer technologies.

Atoms and Elements
The building blocks of matter

The notion that all matter is made up of atoms is central to modern scientific thought. Yet atoms are so small – with diameters of less than one nanometer (a billionth of a meter) – that they are beyond the reach of direct experience. This means that their structure, and indeed their very existence, can only be considered in terms of abstract "models."

Over the years, models of atomic structure have become increasingly sophisticated, to the extent that they now provide a firm basis for the understanding of all physical and chemical phenomena. This understanding has led to the development of countless technologies, from microcomputers and lasers to solar panels and nuclear reactors.

This image, taken by a scanning tunneling electron microscope (STEM) shows the electric "signature" of a film of silver atoms (left). The STEM consists of an extremely fine "stylus" charged to a very high voltage. As the stylus is drawn across an object, its charged tip draws electrons away from the atoms on the surface. The small tunneling current that flows can be detected, measured and used to build up an image of the atomic pattern of the object. The STEM, however, provides no information about the internal structure of atoms.

Visualizing the atom

An individual atom can never actually be *seen*, no matter how powerful the microscope used to examine it. This is because the wavelength of light, to which the human eye is sensitive, is hundreds of times greater than the diameter of an atom: light, in effect, is too "coarse" to be able to pick out any detail in the atom's structure. Nevertheless, it is possible to generate *representations* of atoms: for example, an instrument called the scanning tunneling microscope measures the shape and size of the electric field around an atom; and a procedure called X-ray crystallography can measure the distance between atoms. These techniques strongly suggest that atoms are discrete units.

It is now known that atoms are not indivisible entities but are made up of even smaller particles – protons, neutrons, and electrons. The existence of these subatomic particles was first suggested by the behavior of cathode rays – energetic "rays" emitted from a negatively charged electrode (cathode) in a vacuum tube. At the end of the 19th century, the British physicist J. J. Thomson found that these rays were in fact fast-moving particles, each negatively charged and with a mass of around 9×10^{-31} kg. The particle that Thomson had discovered was the electron. Because atoms carry no net electrical charge, Thomson postulated that the tiny, negatively charged electrons were embedded in a positively charged "sponge" that made up the rest of the atom.

Thomson's model of the atom was superseded in 1911, when the New Zealand physicist Ernest Rutherford showed that an atom's positive charge was concentrated in a dense *nucleus*. The nuclear model of the atom supposes that the light electrons are in *orbit* around a heavy nucleus. The distance between the electrons and nucleus is more than 10,000 times the diameter of the nucleus, and the atom is therefore made mostly of empty space.

*In a cathode-ray tube [**A**] "rays" emitted by a negatively charged electrode [1], are accelerated by a positively charged electrode [2] before striking a fluorescent screen [3]. Here, they form a visible glowing spot. Application of an electric field across the tube [4] causes the rays to be deflected. The direction and degree of deflection [5] suggests that the rays are in fact tiny negatively charged particles, or electrons, with a mass 2000 times less than that of an atom. These observations are expressed in Thomson's model of the atom, in which electrons [6] are embedded in a large positive "sponge" [7].*

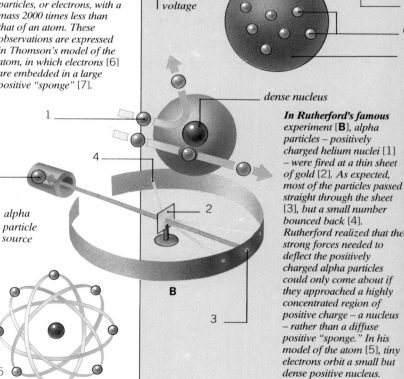

dense nucleus

*In Rutherford's famous experiment [**B**], alpha particles – positively charged helium nuclei [1] – were fired at a thin sheet of gold [2]. As expected, most of the particles passed straight through the sheet [3], but a small number bounced back [4]. Rutherford realized that the strong forces needed to deflect the positively charged alpha particles could only come about if they approached a highly concentrated region of positive charge – a nucleus – rather than a diffuse positive "sponge." In his model of the atom [5], tiny electrons orbit a small but dense positive nucleus.*

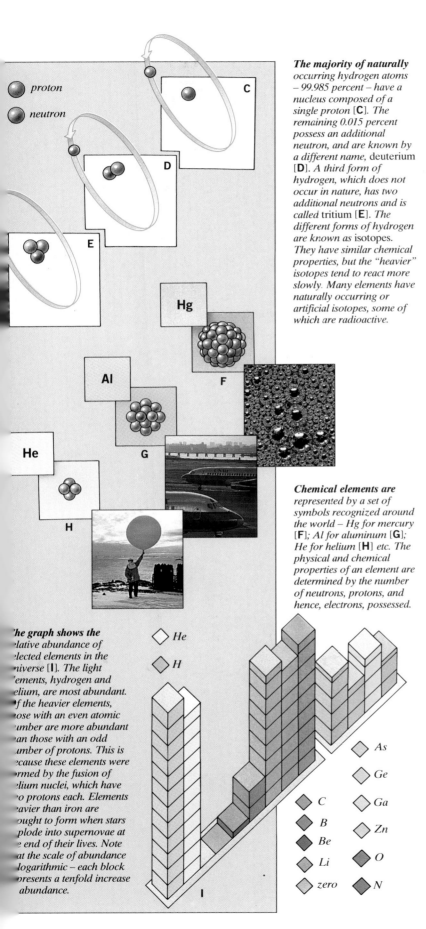

proton

neutron

The majority of naturally occurring hydrogen atoms – 99.985 percent – have a nucleus composed of a single proton [C]. The remaining 0.015 percent possess an additional neutron, and are known by a different name, deuterium [D]. A third form of hydrogen, which does not occur in nature, has two additional neutrons and is called tritium [E]. The different forms of hydrogen are known as isotopes. They have similar chemical properties, but the "heavier" isotopes tend to react more slowly. Many elements have naturally occurring or artificial isotopes, some of which are radioactive.

Chemical elements are represented by a set of symbols recognized around the world – Hg for mercury [F]; Al for aluminum [G]; He for helium [H] etc. The physical and chemical properties of an element are determined by the number of neutrons, protons, and hence, electrons, possessed.

The graph shows the relative abundance of selected elements in the universe [I]. The light elements, hydrogen and helium, are most abundant. Of the heavier elements, those with an even atomic number are more abundant than those with an odd number of protons. This is because these elements were formed by the fusion of helium nuclei, which have two protons each. Elements heavier than iron are thought to form when stars explode into supernovae at the end of their lives. Note that the scale of abundance is logarithmic – each block represents a tenfold increase in abundance.

Protons and neutrons

Water is a compound. It is composed of atoms of hydrogen joined to atoms of oxygen and can be decomposed into these constituents by passing an electrical current through it, or by appropriate chemical treatment. By contrast, oxygen and hydrogen cannot be split into simpler substances by conventional treatment. Oxygen and hydrogen are elements: an atom can be described as the smallest possible particle of an element that still exhibits that element's chemical behavior. There are more than 100 elements, which combine in different ways to give millions of different compounds.

One element differs from another in many ways – some elements are metals, others nonmetals; some solids, others gases; some highly reactive, others inert. Elements have these different properties because they differ at the atomic level. Atoms of hydrogen, for example, all have one electron, the negative charge of which is balanced by a single positively charged *proton* in the nucleus. Atoms of oxygen have eight electrons, balanced by eight protons. Atoms of aluminum have thirteen electrons and thirteen protons, and so on. Although, in most atoms, the number of electrons is equal to the number of protons, the identity of an atom is determined by the number of protons in its nucleus, rather than the number of electrons in orbit around it. The number of protons in the nucleus is known as the atomic number.

Hydrogen, the simplest element, has one proton, and an atomic number of 1. The next simplest element, helium, has two protons and hence the atomic number 2. But helium atoms are four times heavier than ordinary hydrogen atoms, not twice as heavy, as might be expected. The discrepancy in mass is made up by neutrons – nuclear particles which have the same mass as protons, but are electrically neutral. Helium has two neutrons and two protons: it is said to have an atomic mass of 4.

The origin of the elements

The universe is believed to have begun around 20 billion years ago in a single event of unimaginable ferocity – the Big Bang – in which an incredibly dense ball of subatomic particles exploded outward. Formation of the elements began seconds after this event. Hydrogen, the nucleus of which is made of a single proton, was the first element to be formed, and is still by far the most abundant in the universe. In the heat following the Big Bang, hydrogen nuclei fused to produce helium. These two elements together account for about 99 percent of all atoms.

Millions of years later, hydrogen and helium began to coalesce into stars and galaxies. The temperature within the stars rose to the point that nuclear fusion could begin: fusion in the stars produced still more helium, and as the temperature rose even further, progressively heavier elements were formed by fusion of larger nuclei.

Atoms, Electrons, and Energy

The electrical basis of chemical bonding

Electrical charge is a fundamental property of matter. It originates within each atom: protons fixed in the atomic nucleus carry a positive charge, while the electrons in the atom's outer layers give a balancing negative charge. Unlike protons, electrons can move away from their atomic "hosts" when supplied with relatively small amounts of energy, and can form associations with other atoms. These associations constitute chemical bonds. The chemical behavior of any element is therefore determined by the number and arrangement of its electrons.

Today's models of atomic structure and electronic behavior allow chemists to predict accurately how atoms will behave in chemical reactions.

Excited electrons

By night, city streets around the world are illuminated by the glow of sodium lamps. Unlike conventional lightbulbs, which give out white light (a mixture of frequencies), sodium lamps emit light of a characteristic yellow color (and therefore of specific frequency). This simple observation provides an important clue to the way in which electrons are organized in atoms.

A sodium lamp is made of a sealed glass tube filled with sodium vapor at very low pressure. Applying a high voltage between two electrodes in the tube makes the gas glow. This is because the voltage causes electrons in the gas atoms to gain energy and become "excited." As these excited electrons fall back to their original, lower, energy state (known as the *ground state*), they emit their excess energy in the form of light. The frequency of the light emitted depends on how much energy the electrons lose when falling from the higher to the lower energy level. The unique frequency of the light from a sodium lamp means that all the electrons fall between the same two fixed energy levels, and in doing so emit a uniformly sized packet of energy, called a *quantum*.

This type of evidence forced a refinement of Rutherford's early model of the atom. Instead of orbiting the nucleus in an unspecified way, electrons could be thought of as occupying concentric "shells" around the nucleus, the electrons in each shell all possessing very similar amounts of energy. Electrons in shells nearest the nucleus have less energy than those in shells farther out: and movement between shells involves the input or output of a discrete packet, or quantum, of energy. Although electrons in one shell are all at a similar energy level, no two electrons in an atom have precisely the same amount of energy. Subtle differences in an electron's momentum, magnetic properties, and direction of spin ensure that each one is at a unique energy level.

Neon lights, which produce a characteristic red light, are often used in advertising displays (right). Each one comprises a glass tube filled with low-pressure neon gas. Electrons in the neon atoms are stimulated by a high voltage, and emit red light as they "relax."

A neon atom [A] has ten electrons, two in its first shell, and eight in its second. Electrical stimulation promotes one of its electrons to a higher energy shell [B]. Moments later, the electron falls back to its original shell [C]. Like any moving electric charge, it emits electromagnetic energy, here in the form of light. Because the difference

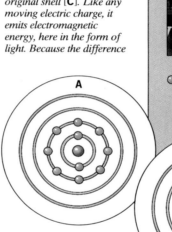

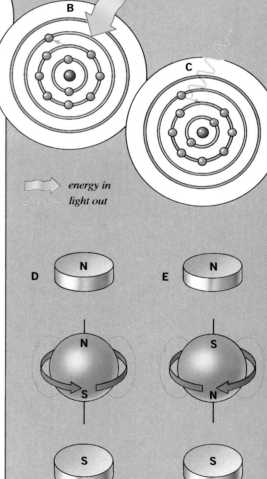

○ *electron*

in the energy levels of successive shells is fixed, stimulated electrons in all the neon atoms in the tube all fall through the same distance, and all emit light of the same wavelength (and therefore color).

⇨ *energy in*
~ *light out*

The precise energy level of *any electron in an atom is described in terms of four quantum numbers – values that give its distance from the nucleus, momentum, magnetic properties, and the direction of its spin.*

For example, the spin quantum number tells us whether the electron spins from left to right [D] or from right to left [E] when placed in a magnetic field. The energy level of an electron that spins from left to right is slightly different from one that spins right to left. Each electron in an atom has a unique combination of the four quantum numbers, which describes its energy level.

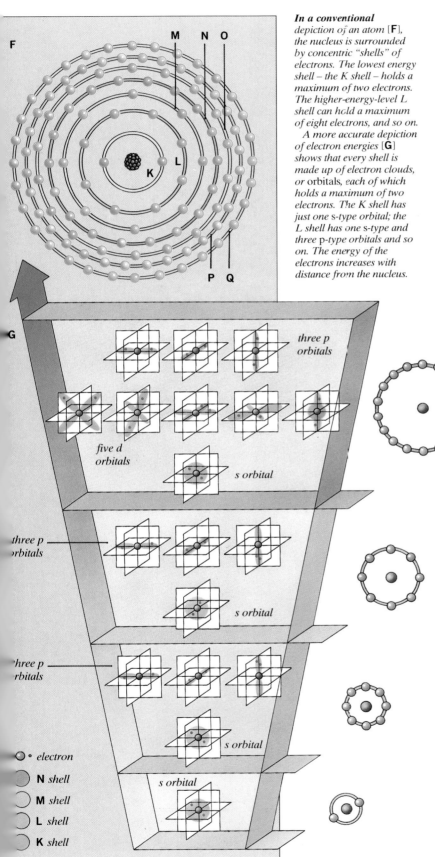

F

M N O

K **L**

P Q

G

three p
orbitals

five d
orbitals

s orbital

three p
orbitals

s orbital

three p
orbitals

s orbital

s orbital

○ • electron

◯ **N** shell

◯ **M** shell

◯ **L** shell

◯ **K** shell

In a conventional depiction of an atom [**F**], the nucleus is surrounded by concentric "shells" of electrons. The lowest energy shell – the K shell – holds a maximum of two electrons. The higher-energy-level L shell can hold a maximum of eight electrons, and so on.

A more accurate depiction of electron energies [**G**] shows that every shell is made up of electron clouds, or orbitals, each of which holds a maximum of two electrons. The K shell has just one s-type orbital; the L shell has one s-type and three p-type orbitals and so on. The energy of the electrons increases with distance from the nucleus.

Shells, orbitals, and atomic structure

In some cases it is useful to think of electrons orbiting the atomic nucleus in concentric shells (designated K,L,M,N,O,P,Q), with electrons in one shell having similar amounts of energy, and shells increasing in energy level with distance from the nucleus. However, a full explanation of atomic bonding requires a still more sophisticated model of atomic structure.

In this model, electrons are represented not by solid spheres occupying circular orbits, but by fuzzy "clouds" around the nucleus. This is a closer approximation to the real behavior of electrons because, even though the energy possessed by an electron is well defined, its position is not. The best that can be done is to calculate the *probability* that an electron with a given energy will be in any one place at any one time – this probability can be shown visually as a cloud, or *orbital*. The shape and size of the orbital depends on the energy of the electrons it contains, but any one orbital can contain a maximum of two electrons. Some orbitals are spherical; others, dumbbell-shaped; still others have highly complex three-dimensional shapes.

The electron shell nearest the nucleus – the K shell – contains a single spherical *s* orbital, and can therefore contain a maximum of two electrons. The second shell – the L shell – consists of a spherical *s* orbital and three dumbbell-shaped *p* orbitals, oriented at 90° to one another. The *p* orbitals have slightly higher energy than the *s* orbital. The L shell therefore contains a maximum of eight electrons. The M shell also has one *s* and three *p* orbitals – a maximum of eight electrons. The fourth shell can take up to eighteen electrons: this is because in addition to *s* and *p* orbitals, it possesses five *d* orbitals. The sixth and seventh shells also have *f* orbitals, and can accommodate up to 32 electrons.

In an unexcited (ground state) atom, electrons occupy the lowest available orbitals. In an atom of neon, for example, which has a total of ten electrons, the *s* orbitals of the first and second shells, together with the *p* orbitals of the second shell are fully occupied, and no other orbitals are used.

When an element bonds with another to form a chemical compound, it is the interaction of the atoms' *outermost* electron shell that holds the atoms together and determines the properties of the compound. The significance of this is that elements with similar electron configurations, particularly in their outermost shells, tend to have similar chemical and physical properties. For example, neon, with ten electrons, has fully occupied *s* and *p* orbitals in its outermost shell, as does argon, with 18 electrons. Both elements are gases at room temperature, and both are inert, forming compounds very unwillingly. The similarities between elements arising from similarities in their electronic configuration are expressed in the Periodic Table.

The Periodic Table

A useful way of classifying the elements

The idea that the world around us is made of a few fundamental substances is an old one. The ancient Greeks believed that all matter was composed of just four "elements" – earth, fire, air, and water. Today, chemists recognize more than 100 elements – chemical building blocks that cannot readily be broken down into smaller units. Nature provides 88 of these elements; the remainder are unstable, existing only fleetingly in the laboratory.

In the late 19th century, the Russian chemist Dmitri Mendeléev realized that when the elements are ordered from the lightest to the heaviest, certain properties recur at definite intervals: this *periodicity* is most conveniently expressed in the form of the Periodic Table.

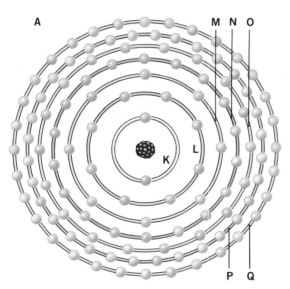

A catalog of elements

Atoms of each element have a characteristic number of protons within their nucleus – one for hydrogen, six for carbon, up to 103 for the element lawrencium. This number is called the atomic number, and in a neutral atom it is always matched by the number of electrons in "orbit" around the nucleus. Although the atomic number identifies an atom, it is the number and arrangement of its electrons that governs its chemical properties.

In the Periodic Table, the elements are arranged from left to right and from top to bottom in order of ascending atomic number. The further organization of the table into rows and columns – or *groups* and *periods* – identifies similar arrangements of electrons in different elements: for example, the elements of group 1 – lithium, sodium, potassium etc. – all have a solitary electron in their outermost shell and so have similar chemical properties.

The Periodic Table is divided into periods – horizontal rows running from left to right – and groups – vertical columns running from top to bottom. Each element has a unique "address" in the table, given by its group and period. The table is a useful visual way of correlating the properties of an element with its electronic structure.

Electrons in an atom can orbit the nucleus in as many as seven concentric "shells," designated K,L,M,N, O, P, and Q [**A**]. The innermost K shell holds a maximum of two electrons; the L and M shells, 8 each; the N and O shells, 18 each; and the P and Q shells, 32 each.

The period that an element falls into tells us which of the seven possible electron shells are occupied in the element concerned. In hydrogen and helium (which have one and two electrons respectively) only the K shell is occupied. These elements fall into the first period. The eight elements of the second period (lithium to neon) make use of L and K shells. Third period

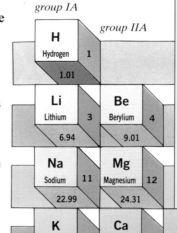

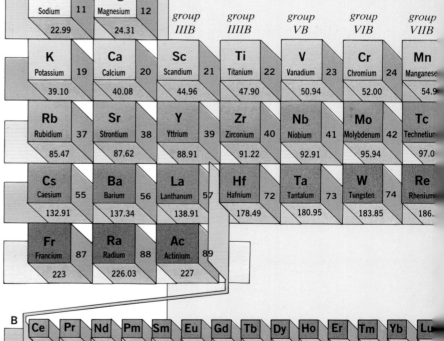

group IA	group IIA						
H Hydrogen 1 1.01							
Li Lithium 3 6.94	**Be** Berylium 4 9.01						
Na Sodium 11 22.99	**Mg** Magnesium 12 24.31	group IIIB	group IIIB	group VB	group VIB	group VIIB	
K Potassium 19 39.10	**Ca** Calcium 20 40.08	**Sc** Scandium 21 44.96	**Ti** Titanium 22 47.90	**V** Vanadium 23 50.94	**Cr** Chromium 24 52.00	**Mn** Manganese 54.9	
Rb Rubidium 37 85.47	**Sr** Strontium 38 87.62	**Y** Yttrium 39 88.91	**Zr** Zirconium 40 91.22	**Nb** Niobium 41 92.91	**Mo** Molybdenum 42 95.94	**Tc** Technetium 97.0	
Cs Caesium 55 132.91	**Ba** Barium 56 137.34	**La** Lanthanum 57 138.91	**Hf** Hafnium 72 178.49	**Ta** Tantalum 73 180.95	**W** Tungsten 74 183.85	**Re** Rhenium 186.	
Fr Francium 87 223	**Ra** Radium 88 226.03	**Ac** Actinium 89 227					

B | Ce | Pr | Nd | Pm | Sm | Eu | Gd | Tb | Dy | Ho | Er | Tm | Yb | Lu |

C | Th | Pa | U | Np | Pu | Am | Cm | Bk | Cf | Es | Fm | Md | No | Lr |

Two series of elements – the lanthanides [**B**] and actinides [**C**] – stand apart from the normal order of the Periodic Table. The lanthanides, with two electrons in their outer shell, were once thought to be scarce, hence their collective name, the rare earths. The actinides – rare apart from uranium and thorium – are all radioactive.

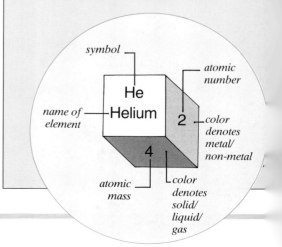

symbol

name of element

He Helium

atomic number

2

4

atomic mass

color denotes metal/ non-metal

color denotes solid/ liquid/ gas

elements use the K,L, and M shells, and so on.

Within a single period, trends are observed in the properties of the elements. Each period starts with a very reactive metal. Moving to the right, successive elements are progressively less reactive and less metallic, until a highly reactive nonmetal is reached. Each period closes with a noble gas – a very unreactive element.

Vertical columns (groups) are occupied by elements with similar properties. The members of the groups designated

IA to VIIIA are called representative elements. Those groups designated IB to VIIB contain elements known as the transition metals. Elements in a representative group, for example IA, resemble one another more closely than they resemble elements in the equivalent transition group (IB).

The number of the group tells us how many electrons are present in the outermost shell of the atom concerned. This determines what sort of bonds the atom forms, and how strong they are.

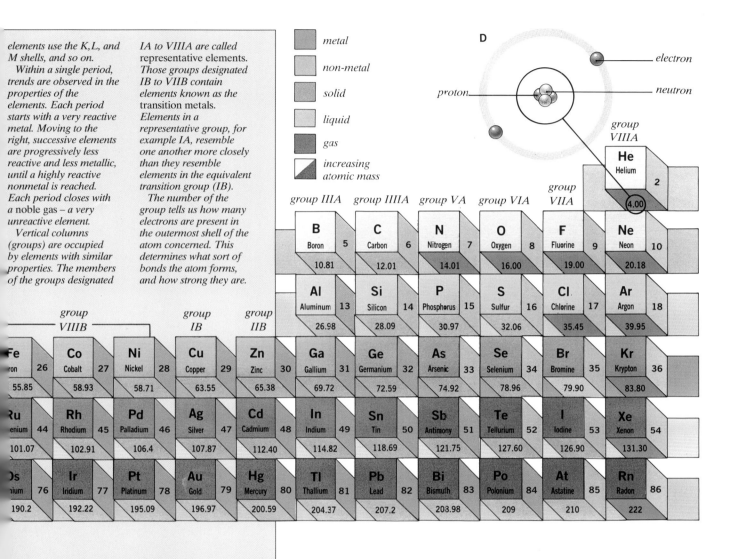

Group IA
All are highly reactive metals that are never found in a free state in nature. Their atoms have a lone electron in the outer shell that is easily lost, so these atoms readily become positive ions. Group I elements react violently with cold water, producing hydrogen gas and a metal hydroxide (an alkali that neutralizes acids).

Group IIA
These elements are also metals, though less reactive than those of Group I. This is because they must lose two outer electrons to form positive ions (cations).

Group IIIA
These elements show a marked shift away from metallic character. The first member of the group, boron, is a metalloid that does not form simple ionic compounds and is totally unreactive toward water. Other members of this group, such as aluminum, can lose their three outer electrons, or lose just one of these electrons, leaving behind a relatively stable pair of electrons in the outer shell. Group III elements can also form covalent compounds by sharing their outer electrons with other atoms.

Group IIIA
The first three members – carbon, silicon, and germanium – are metalloids that form covalent bonds. Only lead and tin lower in the group show metallic character. This is because their outer electrons are shielded from the attractive force of the nucleus by numerous shells. The valence electrons of these elements can therefore be lost with greater ease.

Group VA
These elements bond covalently, but the first member, nitrogen, also forms negative ions (anions) by accepting three electrons. Compounds of the nitrogen ion are called nitrides.

Group VIA
The elements sulfur, selenium, and tellurium can all accept two electrons to form anions, as well as bonding covalently with a large number of nonmetals.

Group VIIA
Called the halogens, these are reactive nonmetals, never found naturally in elemental form. With seven electrons in their outermost shell, they readily accept one more to achieve a stable electron structure.

Group VIIIA
The noble gases exist as single atoms, with little tendency to bond because their outermost shells are filled and have no need to accept or donate electrons.

Atomic number – which is equivalent to the number of protons in the nucleus of an atom – increases from left to right, and from top to bottom in the Periodic Table. Atomic mass – the sum of the mass of protons, neutrons and electrons in an atom – increases in a similar fashion. An atom of helium, for example [D], has an atomic number of two (it has two protons) and an atomic mass of four – made up of two protons and two neutrons (electrons have negligible mass). Atomic mass is not always a whole number, because it is often the average of two or more atomic forms, or isotopes, of the element concerned.

Chemical Bonding
The forces that hold atoms together

Solitary atoms are rarely found in the natural world. Matter, whether solid, liquid, or gas, is usually made up of atoms bonded to one another to form molecules. A molecule can consist of two or more atoms of the same element, or atoms of two or more different elements held together by chemical bonds.

The number of molecules that can be built up from the 88 naturally occurring elements – the atomic building blocks – is vast: millions of compounds are known today, and millions await discovery. Despite this enormous diversity, molecules are held together by two main types of chemical bond – ionic and covalent – both of which depend on electronic interactions between the atoms involved.

Stable atoms and ionic bonds

Late in the 19th century, more than 100 years after the discovery of the elements nitrogen and oxygen, British chemists identified a new group of gaseous elements. These elements – named helium, neon, argon, krypton, xenon, and radon – had eluded discovery for so long because they were all odorless, colorless and nonflammable, and because they did not react with any other substance. The behavior of these gases – collectively known as the noble gases because they "shun" the other elements – tells us much about the role of electrons in chemical bonding.

In an atom, electrons are organized in concentric "shells" around a positively charged nucleus. Each shell can hold only a certain number of atoms. A bond between two atoms arises when electrons in the outermost, or *valence* shells of the atoms interact. In all the noble gases, the valence shell is full, containing eight electrons, and cannot accommodate any more electrons. This condition makes the atoms unusually stable.

In forming chemical bonds, atoms "seek" to achieve the stability of noble gases by replicating their electron configuration. To end up with the desired eight electrons in its valence shell, an atom can gain electrons from another atom, or lose electrons to it. In doing so, both atoms involved become *ions* – one has a surfeit of electrons (a negatively charged *anion*) while the other has a deficit (a positively charged *cation*). The resulting positive and negative ions are bound together by electrostatic attraction. This constitutes an *ionic bond*.

Compounds held together by ionic bonds tend to be solid at room temperature and usually have a regular crystalline structure. This is because each positive ion in the solid is surrounded on all sides by negative ions; and each negative ion is surrounded by positive ions. The net result is a crystal, the geometry of which is determined by the sizes of the ions involved.

Ionic compounds include table salt, limestone, and the ores of many important metals (left).

Some elements are more likely to participate in ionic bonds than others [A]. A sodium atom, for example, has just one electron in its valence shell [1]. It loses this electron relatively easily, forming a positive ion that has the noble gas structure of eight electrons in its outer shell [2]. Chlorine [3] also readily takes part in ionic bonds. It has seven electrons in its valence shell, and gains an electron to achieve noble gas structure, forming a negative ion [4].

When sodium and chlorine are brought together, the sodium atom donates its valence electron to the chlorine atom, forming the ionic compound sodium chloride (common salt).

The tendency of two atoms to form an ionic bond depends on the ease with which one atom gives up its valence electron (its ionization energy) and the ease with which the other accepts these electrons (its electron affinity). These graphs show ionization energy [B] and electron affinity [C] of elements from sodium to argon. Sodium (Na) has a low ionization energy and chlorine (Cl) has a high electron affinity

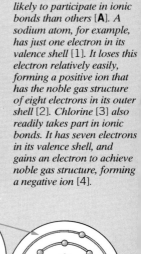

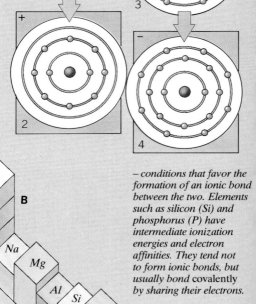

– conditions that favor the formation of an ionic bond between the two. Elements such as silicon (Si) and phosphorus (P) have intermediate ionization energies and electron affinities. They tend not to form ionic bonds, but usually bond covalently by sharing their electrons.

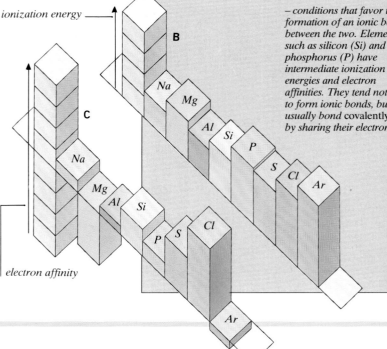

ionization energy

electron affinity

Covalent bonds

Atoms can achieve noble gas structure by forming covalent bonds. Just as in ionic bonds, the participating atoms end up with valence shells filled to capacity with electrons. However, in covalent bonds this is achieved not by gaining or losing electrons outright, but by sharing electrons with other atoms.

Each electron shell around the atomic nucleus comprises one or more orbitals – areas of space within which the electrons are most likely to be found. Each orbital represents a particular energy level, and its shape (some are circular, some dumbbell-shaped, and some have more complex configurations) is determined by its energy. An orbital can contain a maximum of two electrons, hold just one electron, or be totally unoccupied.

In a simple covalent bond between two atoms, one orbital from the first atom overlaps one orbital from the second, and the electrons contained in the orbitals are shared between the two atoms. Usually the two orbitals each contain just one electron, so each participating atom "gains" one electron in its valence shell. This is the case with the element fluorine (F), which forms the fluorine molecule (F_2). Each fluorine atom has seven electrons in its valence shell: by bonding covalently with another fluorine atom, and sharing an electron pair with it, both atoms achieve the desired *octet*, or set of eight, electrons in the valence shell.

Two atoms held together by one electron pair in this way are said to be joined by a *single bond*. But in many compounds, atoms share two pairs of electrons, forming a *double bond*, or three, making a *triple bond*. For example, the element nitrogen (N) is abundant in the earth's atmosphere as the simple molecule N_2. The nitrogen atom has five valence electrons: it shares three of these with another nitrogen atom. The resulting molecule is therefore held together by three electron pairs – a triple bond.

Organic geometry

Carbon is the central element in more than eight million known compounds. Many of these are essential biological molecules, and some – plastics, fuels, pesticides, and detergents – are key products of the technological age. These carbon, or *organic*, compounds are held together by covalent bonds. The great number of carbon compounds comes about because the element is very flexible in its bonding, able to form single, double, and triple covalent bonds, and to link up with other carbon atoms into long chains.

In part, carbon's flexibility comes from the ability of its electron orbitals to *hybridize*. A solitary carbon atom has a total of six electrons: two in its inner, spherical *s* orbital, two in its outer *s* orbital and two in dumbbell-shaped *p* orbitals. Within one atom, these orbitals can "merge" in different ways, allowing carbon to bond in a variety of configurations.

Hydrogen atoms bond with one another to form H_2 molecules [D]. A covalent bond is formed by the overlap of the spherical s orbitals of the two hydrogen atoms. As a result, the valence shell of each hydrogen atom (which can accommodate only two electrons) is filled, and the hydrogen achieves the stable electronic configuration of the noble gas helium. Hydrogen also bonds covalently with fluorine [E]. In this bond, the s orbital of the hydrogen atom overlaps with one of the three dumbbell-shaped p orbitals of the fluorine atom.

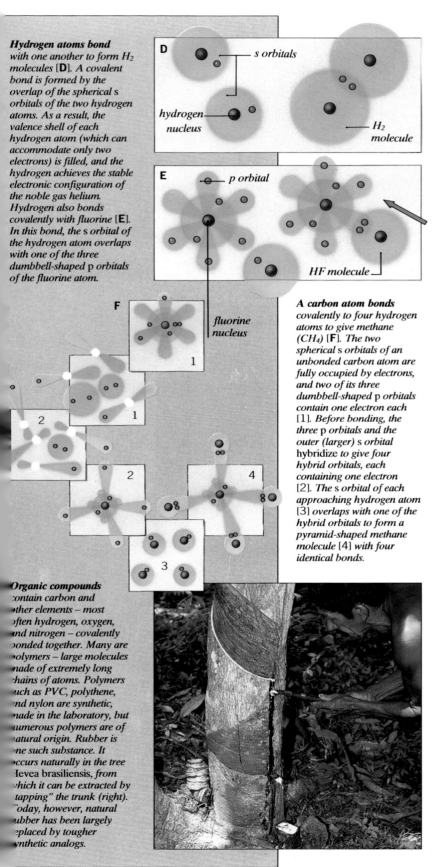

A carbon atom bonds covalently to four hydrogen atoms to give methane (CH₄) [F]. The two spherical s orbitals of an unbonded carbon atom are fully occupied by electrons, and two of its three dumbbell-shaped p orbitals contain one electron each [1]. Before bonding, the three p orbitals and the outer (larger) s orbital hybridize to give four hybrid orbitals, each containing one electron [2]. The s orbital of each approaching hydrogen atom [3] overlaps with one of the hybrid orbitals to form a pyramid-shaped methane molecule [4] with four identical bonds.

Organic compounds contain carbon and other elements – most often hydrogen, oxygen, and nitrogen – covalently bonded together. Many are polymers – large molecules made of extremely long chains of atoms. Polymers such as PVC, polythene, and nylon are synthetic, made in the laboratory, but numerous polymers are of natural origin. Rubber is one such substance. It occurs naturally in the tree Hevea brasiliensis, from which it can be extracted by "tapping" the trunk (right). Today, however, natural rubber has been largely replaced by tougher synthetic analogs.

Energy and Chemical Reactions

Controlling chemical interactions

An iron nail exposed to the elements soon becomes covered in rust (iron oxide), formed by the reaction of iron with water and oxygen. This reaction occurs spontaneously, without the need for external energy, but it also takes place extremely slowly. In contrast, a match can lie inert for years until it is struck, whereupon it releases a burst of light and heat energy. The amount of energy released (or consumed) during a reaction, and the rate at which the reaction proceeds are often of as much practical importance as the type and amount of reaction product. For example, fuel is burned for the heat it produces rather than for the products of combustion, which are not particularly desirable.

Heat in or heat out

Energy can assume many different forms, which may be converted into one another with varying degrees of difficulty. One conversion that we all carry out daily is that which changes chemical energy into heat – namely, burning fuel to produce carbon dioxide and water.

Chemical energy is stored in the bonds that hold molecules together. When burning a fuel, the bonds of the *reactant* molecules – oxygen and the fuel itself – store more energy than the bonds of the *products* – carbon dioxide and water. As the reaction proceeds, the "excess" energy is given out to the surroundings in the form of heat and light. Any reaction – not only combustion – that gives out heat is termed *exothermic*. The opposite situation is that of an *endothermic* reaction, in which the products are at a higher energy state than the reactants, and energy must be taken in from the surroundings for the reaction to proceed.

Controlling reaction rate

The rate at which a reaction proceeds depends fundamentally on the identity of the reactants involved: zinc, for example, reacts more quickly with hydrochloric acid than with acetic acid. However, a number of other factors – heat, concentration, and use of catalysts – can significantly affect reaction rate and are used by chemists to control chemical syntheses.

The molecules in a reaction mixture are constantly in motion, regularly bumping into one another. Reaction occurs only when molecules of reactant collide with one another at sufficient speed (and therefore with enough energy). Heating the reaction increases the speed at which molecules move and therefore raises the probability of a "successful" collision. Reaction rate therefore increases. The same effect is achieved by increasing the concentration of the reactants, which collide more often because they are "crowded" together more closely.

In an endothermic reaction [**A**] *the reactants (green) have less chemical energy than the products (yellow). To make the reaction proceed, energy must be supplied. Production of iron from its ore (right) is an endothermic process: heat is supplied to it by the reaction of carbon and oxygen.*

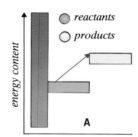

○ *reactants*
○ *products*

energy content

A

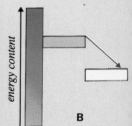

energy content

B

In an exothermic reaction [**B**] *the reactants (green) have more chemical energy than the products (yellow). As the reaction proceeds this energy is liberated as heat. The combustion of crude oil (left) is a highly exothermic reaction that gives out so much energy that it is difficult to stop.*

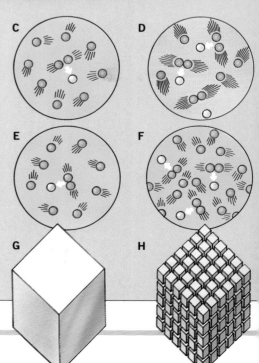

At low temperatures, few collisions [**C**] *between reactant molecules (green) are sufficiently energetic to bring about a chemical transformation. Elevating the temperature* [**D**] *makes the molecules move faster, and more higher-energy collisions occur, forming product (yellow) more rapidly. Increasing pressure (for gaseous reactants) or concentration (for liquids) from a low* [**E**] *to a high* [**F**] *value gives more collisions in a given time and so also increases rate. An analogous process for solid reactants* [**G**] *is powdering the solid: this makes a greater surface area available for reaction* [**H**].

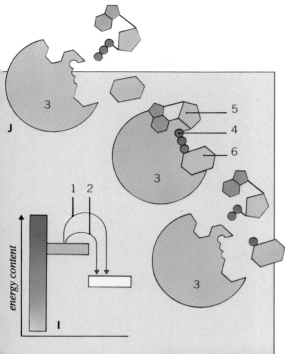

A chemical reaction needs an input of energy, called activation energy, to set it going [**I**]. *This energy can be represented as an energy "hill" or barrier* [1] *that must be ascended by the reactants before reaction occurs. The use of a catalyst lowers the hill to be surmounted* [2] *and so reduces activation energy.*

Enzymes [**J**] *only bind to certain reactants – this "lock and key" specificity means that different enzymes can control different reactions in the same cell. Here, a respiratory enzyme* [3] *catalyzes the transfer of an atom of phosphorus* [4] *from an ATP molecule* [5] *to a glucose molecule* [6].

Catalysts and activation energy

Molecules react only when they collide with one another. Mere collision between two reactants, however, is in itself not enough to ensure that they will undergo chemical change. The reactants must possess more than a critical amount of energy – the *energy of activation* – for reaction to occur: otherwise, they will just "bounce off" one another unchanged. Supplying energy to the system increases the number of reactant molecules that possess more than this threshold amount of energy, and so makes the reaction proceed more quickly.

Every chemical reaction has its own, characteristic energy of activation. Those with a low energy of activation proceed at observable rates at room temperature: those with high activation energies (such as the combustion of a match) proceed very slowly or not at all at room temperature. This is why a match does not spontaneously burst alight, but needs to be struck: this action provides the activation energy for combustion to begin. Once begun, the reaction produces copious energy (it is exothermic) and provides its own activation energy, making it self-sustaining.

The rate of a chemical reaction can be increased by using a *catalyst* – a substance that lowers the activation energy of the reaction. Most catalysts work by temporarily binding to reactant molecules and holding them in the "correct" position for reaction to occur. The products are then released, and the catalyst itself is ready to accept another charge of reactant. Catalysts are found in most industrial syntheses, including the manufacture of fuels, plastics, and ammonia. All life depends on specialized biological catalysts called *enzymes*. These molecules greatly speed up vital cellular reactions: moreover, they are remarkable for their specificity: each one catalyzes only a narrow range of reactions. The human body contains around 30,000 different enzymes.

Reversible reactions

In some chemical reactions, the reactants are entirely converted into products. In others, however, the products apparently react with one another to remake the original reactants. In these *reversible* reactions, there comes a point at which the rate at which the products are made equals the rate at which the reactants are reformed. At this point, known as *equilibrium,* it appears as though the reaction has come to an end because the proportion of reactants to products remains constant. In fact, the equilibrium is *dynamic* – maintained by constant and simultaneous forward and backward reactions. This fact is exploited by chemists, who usually wish to maximize yield of the products. Increasing the rate of the forward reaction relative to the rate of the backward reaction (for example, by increasing the concentration of the reactants) causes the equilibrium to shift so that more product is made.

The role of a catalyst is to speed up the rate of a reaction. Without catalysts many important industrial reactions would occur at uneconomically slow rates. Many industrial catalysts are solids, especially metals such as platinum and iron, though some are liquids or gases (the manufacture of sulfuric acid by the lead chamber process, for example, is catalyzed by gaseous nitrogen oxides).

Some reactions are catalyzed by zeolites – complex compounds of aluminum and silicon. These molecules (right) have "pores" that trap molecules of reactant.

The reversible reaction between hydrogen (H_2) *and iodine* (I_2) *gives the products hydrogen iodide* (*HI*) *and heat* [**K**]. *If more reactants are added, the equilibrium of the reaction is disturbed (the balance tips to the left)* [**L**]. *By an important principle called Le Châtelier's principle, a disturbed chemical equilibrium tends to right itself. This is achieved by producing more hydrogen iodide and heat to balance the scales* [**M**]. *Similarly, if equilibrium is disturbed by removing heat* [**N**], *Le Châtelier's principle dictates that more heat (and therefore more hydrogen iodide) is made* [**O**].

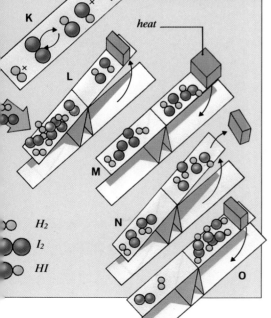

Carbon Chemistry

The diversity of organic compounds

The nonmetallic element carbon makes up less than one percent of the earth's crust, yet it forms 35 times as many compounds as all the other elements of the Periodic Table combined. The great diversity of carbon compounds includes plastics, fuels, fertilizers, synthetic fibers, and pesticides as well as virtually all the molecules that characterize life. Carbon compounds are sometimes referred to as *organic,* because of their close association with living organisms. Indeed, 18th-century chemists believed that carbon compounds were possessed of a special vital force. This notion was laid to rest in 1828 by the German chemist Friedrich Wöhler, who made urea – an organic compound – from inorganic lead cyanate and ammonia.

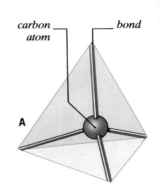

carbon atom ____ bond

A

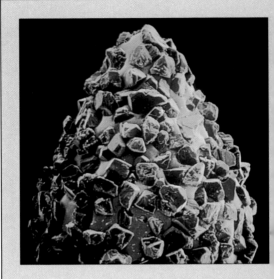

The atomic building block

Carbon is a member of group IIII of the Periodic Table, meaning that it has four electrons in its outermost (valence) shell. To achieve the "desired" stable electronic configuration of a noble gas (a full valence shell containing eight electrons), carbon readily forms *covalent bonds,* sharing electrons with other atoms, notably carbon and hydrogen (yielding *hydrocarbons,* many of which are valuable fuels), and oxygen, nitrogen, and chlorine. Atoms of carbon form strong covalent bonds with one another. This gives them a uniquely strong tendency to link up into extremely long chains, a phenomenon called *catenation.* This is true of the naturally occurring forms of the element – diamond and graphite – which consist of billions of atoms bound together into covalent crystals, and of the organic compounds of carbon. Many of these are *polymers* – giant molecules containing thousands of linked carbon atoms.

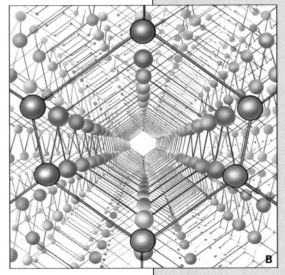

Carbon has four valence electrons and forms four single covalent bonds. The bonds (which are areas of negative charge) repel one another equally. This mutual repulsion causes the bonds to occupy positions around the atom that are as far apart as possible: geometrically, this gives a tetrahedral form [A]. In diamond, this tetrahedral bond pattern is repeated billions of times as each atom bonds to four others in an infinite three-dimensional array [B]. Because it is held together by many strong covalent bonds, diamond is very hard – ideal for the tips of dentists' drills (above).

Molecular permutations

The huge number of natural and synthetic carbon compounds is only partly accounted for by the carbon atom's near-unlimited capacity to bond with itself. Carbon compounds exist in a variety of conformations: they can be linear or ring-shaped, and many exist in *isomeric forms.* (Isomers are molecules with the same numbers of atoms bonded in different arrangements.)

Carbon atoms in a chain can also be linked together by different types of bond. The simplest of these is the single bond, in which one carbon atom shares one of its four valence electrons with another atom. Atoms linked by a double bond share two pairs of electrons, and those joined by a triple bond share three electron pairs. The number of permutations presented by changing the chain length, atomic arrangement, and frequency and distribution of single, double, and triple bonds in a carbon compound is almost limitless.

Compounds of carbon and hydrogen (hydrocarbons) that contain only single bonds between carbon atoms are called alkanes. *The smallest of these are methane (CH_4) [C] and ethane (C_2H_6) [D], but alkanes can have up to several thousand carbon atoms. Hydrocarbons with double bonds are called* alkenes *(the simplest is ethene, C_2H_4) [E] and those with triple bonds are known as* alkynes *(the simplest is ethyne, C_2H_2) [F]. Alkanes tend to be the least reactive of the three classes because they have no "spare" bonds between carbon atoms. For this reason they are known as* saturated *compounds.*

The three isomers of the alkane pentane [G] have the same chemical formula (C_5H_{12}) but different arrangements of atoms give them different chemical and physical properties. The larger the compound, the greater the number of possible isomers.

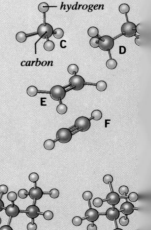

hydrogen

carbon

C

D

E

F

G

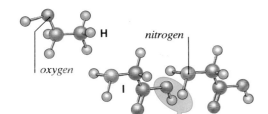
oxygen H nitrogen I

*Functional groups greatly affect the characteristics of an organic compound. Ethanol [**H**], for example, is a liquid that is soluble in water, whereas its parent hydrocarbon ethane is an insoluble gas.*

*Reactions between functional groups drive many important biological reactions. Amino acids, for example, are characterized by two functional groups – a carboxyl and an amino. Bonds between amino acids result from a condensation reaction (one that involves the loss of water) between the two types of group [**I**]. Thousands of amino acids link together to form a protein such as insulin (left).*

*The basic hexagonal structure of benzene [**J**] is the starting point for an enormously large number of derived compounds. In some of these, one or more hydrogen atoms are substituted with other groups to produce compounds like TNT, aspirin, and scent molecules (right). In others benzene rings are fused together to form polycyclic compounds. The best known of these is naphthalene, present in mothballs, but many of these compounds are important biological molecules. The largest tend to be carcinogenic – capable of inducing cancer in humans and other animals.*

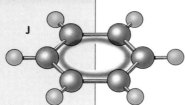

J

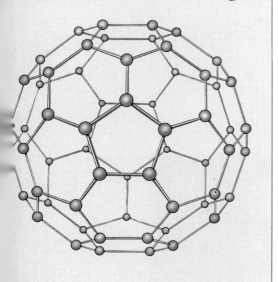

*The buckminsterfullerene molecule resembles a soccer ball [**K**], which has led to the use of its more familiar name "buckyball."*

The molecule is a spherical cage with an open interior large enough to contain other atoms. The chemical possibilities of such novel structures are currently being explored.

Despite its unusual geometry, bonding in the buckminsterfullerene molecule is fairly conventional. The bonding orbitals of all the constituent carbon atoms are hybridized; delocalized bonds – bonds spread out over many carbon atoms – hold the molecule together.

Group influences

The properties of an organic molecule are influenced not only by the length and arrangement of its carbon chains and the types of bonds it contains, but also by the presence or absence of *functional groups*. These are chemical subunits that replace one or more hydrogen atoms in a hydrocarbon and significantly alter its properties. The influence of these groups is such that organic compounds can be organized according to which groups they possess. *Alcohols* are compounds in which a hydroxyl (OH) group substitutes for a hydrogen atom. Methanol (wood alcohol), ethanol (the alcohol present in intoxicating drinks), and ethylene glycol (used as antifreeze) are all members of this chemical family. Similarly, *carboxylic acids* (such as citric acid) all possess the COOH group, *amines* all have an NH_2 group, and so on. Organic compounds may possess several functional groups of the same or different type.

Aromatic compounds

Among the most important of organic molecules are the *aromatics,* so called because they are often fragrant themselves or extracted from aromatic substances, such as cinnamon and wintergreen. All the members of this large family of compounds, discovered in the 1830s, are based on the structure of the "parent" hydrocarbon molecule benzene.

Benzene has the chemical formula C_6H_6. Its six carbon atoms are joined together in a regular hexagon, in which a single bond joins each carbon to its two neighbors. Another single bond joins each carbon to a hydrogen atom. The fourth bond of each atom (carbon has four bonding electrons) is *delocalized* – spread out over all six carbon atoms. This makes the benzene ring more chemically stable than similarly sized linear hydrocarbons, and greatly alters the types of reactions in which it can participate.

Spherical carbon

Carbon occurs in two naturally occurring forms, or allotropes – diamond and graphite. Both substances are made of carbon atoms: the difference between them lies in the way the atoms are linked together. In 1985, chemists discovered a third allotrope of carbon, formed by vaporizing graphite with a high-power laser. Analysis of this new allotrope showed it to be a spherical molecule made up of 60 carbon atoms, arranged in hexagonal and pentagonal "panels." The highly symmetrical molecule was named buckminsterfullerene in honor of the architect R. Buckminster Fuller who used geodesic domes resembling the molecule in his designs for buildings. Other fullerene molecules with 70 and 76 atoms have since been discovered and applications for their unique architecture are now being sought. Studies have already proved that these molecules can act as high-temperature superconductors and catalysts.

Matter, Heat, and Energy
Energy, temperature, and the behavior of particles

Matter, whether solid, liquid, or gas, is made up of tiny particles (molecules or atoms) in constant motion. Whenever we measure the temperature of a substance – be it on the Celsius, Fahrenheit, or (more scientific) Kelvin scale – we are actually measuring the speed at which its component particles are moving. In a gas, the particles are separated by expanses of empty space and move quickly in all directions; in a solid, they are bound closely together by attractive forces, but can still vibrate about their nearly fixed positions. Particles in a liquid are held together by weaker forces, and can move more freely past one another. In each case the motion of the particles is driven by the heat energy contained by the material.

Lively molecules
In 1827, the Scottish botanist Robert Brown used a microscope to observe pollen grains suspended in water. The grains moved – not slowly and predictably as might be expected if carried by convection currents in the water, but following rapidly changing zigzag paths. This irregular movement (which came to be known as *Brownian motion*) remained a mystery until 1905, when Albert Einstein explained the phenomenon using *kinetic theory*.

Kinetic theory assumes that molecules are always in motion. In a gas or liquid, where they are not bound in one position, the molecules move around, randomly colliding with one another. Over time, this movement distributes the molecules evenly in the available space. Brownian motion is therefore caused by water molecules colliding with the pollen grains. This constant buffeting sets the grains in motion, and causes random changes in their direction.

Conduction, convection, and radiation
Heating a metal bar causes the atoms nearest the heat source to vibrate faster. As they knock into their neighbors, these atoms pass on the increased vibration, and in this way heat energy is *conducted* through the substance. Metals are generally better conductors than other solids. This is because they have "free" electrons that can carry heat energy from hot to cold areas.

Liquids and gases are poorer conductors. This is because forces between particles are weaker than in solids, so it takes longer for the vibrations to be transmitted to all parts of the fluid. However, heat can move through a gas or liquid by a different process called *convection*, in which regions of warm fluid move in bulk.

Heat can also be transferred by *radiation* – electromagnetic waves given out by all warm bodies. Unlike conduction and convection, radiation needs no medium and can carry heat through a vacuum.

Brownian motion is caused by random collisions [**A**] between pollen grains [1] and the invisible water molecules around them [2]. It is also apparent in gases: air molecules, for example, constantly buffet particles of smoke, the irregular movement of which can be seen under a microscope. Air molecules also buffet invisible chemicals – such as scent molecules – making them spread out over a wide area. This process, known as diffusion, lets us detect odors from a distance. The feathered antennae of the male Atlas moth (above), allow it to detect the scent of a female 5 miles away.

Heating water in a pan [**B**] makes its particles move farther apart. The hot water [1] becomes less dense than the cold water above it [2], and rises to the surface, transferring heat as it goes. It then cools and sinks. The process of heat transfer by bulk movement in a fluid is called convection.

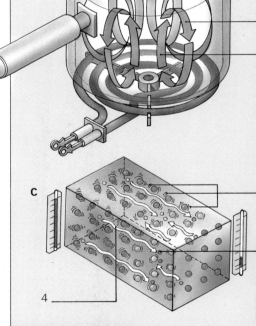

Conduction [**C**], *unlike* convection, involves heat tranfer by the transfer of vibrations [3] (and moving electrons [4]): the material itself does not move. Radiative heat can travel through a vacuum: satellite are often covered in reflective foil to deflect this heat (left).

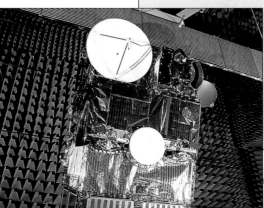

At atmospheric pressure, *solid carbon dioxide (CO₂) does not melt when heated. Instead it sublimes, passing directly from the solid to the gas phase. Sometimes called dry ice, it is used as a stage effect to produce "clouds" made of CO₂ gas (right).*

D

1 atmosphere pressure

0.006 atmospheres pressure

water

ice

steam

1

2

3

32 °F   212 °F

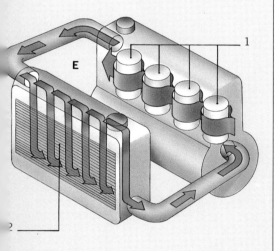

E

1

2

For any substance, *the relationship between temperature, pressure, and the three states (or phases) of matter – solid, liquid, and gas – can be expressed in a phase diagram. The phase diagram for water [D] shows that increasing temperature at a pressure of one atmosphere causes melting at 32 °F [1] and boiling at 212 °F [2]. But below a pressure of 0.006 atmospheres, heating causes ice to convert directly into steam without first passing through the liquid phase [3], a phenomenon known as sublimation. While water sublimes at low pressure, carbon dioxide does so at atmospheric pressure.*

Almost all car engines *are water-cooled [E]. Water flows around the hot engine cylinders [1] and to the radiator [2], where its heat is dissipated. Water is an ideal inexpensive coolant because it absorbs a lot of heat with relatively small increases in temperature and because it flows readily.*

Pouring water on a fire *(right) helps extinguish the flames in more than one way. Initially, the cold water absorbs energy from the combusting materials: the energy is used both to raise the temperature of the water to boiling point, and then to supply the latent heat of vaporization and convert the water into steam. It takes more than two million joules of energy to convert a kilogram of water at 212 °F to steam at the same temperature. The clouds of steam formed by the vaporation of the water then help to keep atmospheric oxygen from reaching the fire, and so choke out the flames.*

Pressure, temperature, and states of matter

When a pan of water is heated to its boiling point, some of the water molecules gain enough energy to form bubbles of vapor. These rise through the liquid, and break the surface, passing into the atmosphere as steam. For this to happen, the pressure in the vapor bubbles must be the same as, or greater than, the pressure of the atmosphere bearing down on the liquid's surface. The transition between liquid and gas therefore depends on external pressure: this is why water boils at less than 212 °F on a mountain peak, and at more than 212 °F in a pressure cooker.

The transition between solid and liquid also depends on pressure. For most substances, raising the external pressure increases the temperature at which melting occurs because it "squeezes" molecules of the solid together. Water, however, exhibits the opposite behavior because, unusually, it expands as it melts.

Stored heat

Compared to most other substances, water is difficult to heat up – a lot of energy must be put into the liquid before its temperature rises. Water is said to have a high *specific heat capacity* – a quantity defined as the amount of heat energy needed to raise the temperature of one kilogram of a substance by one Celsius degree. For water, this value is 4200; for alcohol 2450; for glass 840; and for copper 390. This means that the energy needed to boil a kettle of water (approximately 1 kg) could heat a 1 kg lump of copper to more than 1450 °F.

But just as water absorbs a lot of energy as its temperature rises, so it must lose a lot of energy when it cools. That is why it takes so long for a hot-water bottle to cool down, and why coffee stays hot for longer than thick soup. The non-water contents of the soup – the fats and solids – have a lower heat capacity than water, and cool down with smaller outputs of energy.

Hidden heat

Even when a solid has been heated up to its melting point, it needs extra energy to make it melt. For example, ice at 32 °F takes in a surprisingly large amount of heat (330,000 joules per kilogram) from its surroundings before it becomes water at 32 °F. This heat is known as the *latent heat of fusion*. It is needed because the molecules in a solid occupy relatively fixed positions in a crystal lattice. To melt the solid, the attractive forces between the molecules must be overcome, separating them into the more random motion of the liquid. Latent heat of fusion explains why snow lingers so long, even after the air temperature has risen well above "freezing point," and why ice is extremely efficient at cooling drinks.

Similarly, extra energy must be supplied to convert a liquid into gas: this *latent heat of vaporization* explains why the evaporation of sweat helps keep the body cool.

Solids

Properties and applications

As molten salt is cooled, its molecules pull one another into place and become symmetrically fixed within a vast repeating latticework – a crystal. Many solids possess such a regular crystalline structure. This is clearly visible in snowflakes, diamond, and quartz, but most metals and rocks are also composed of crystals, although these are too small to be seen by the naked eye. In other solids, such as glass and the mineral opal, the component atoms are fixed in more random arrangements: these solids are said to be amorphous. The properties of a solid depend not only on the nature and arrangement of its component atoms or molecules, but on the types of bond that hold them together.

Covalent solids

Covalent molecules such as methane (CH_4), carbon dioxide (CO_2), and hydrogen chloride (HCl) are held together by strong internal bonds, formed when their constituent atoms share electrons. But these strong covalent bonds act only *within* the molecules: their influence does not extend to adjacent molecules. For this reason the molecules of covalent compounds can easily be separated. This is reflected in the properties of covalent solids, which tend to be weak and usually melt at low temperatures. (Methane melts at –296.5 °F.)

However, certain covalent compounds, such as diamond and quartz, have markedly different properties, being extremely hard with high melting points. These compounds are not made up of small, discrete molecules: instead, each diamond or quartz crystal is in effect a giant molecule, in which each atom is joined to all its neighbors by covalent bonds.

Ionic solids

Silver bromide (a component of photographic film), potassium nitrate (used in explosives), and table salt are *ionic* solids. All are arrays of positive and negative *ions* – atoms that have lost their electrical neutrality by losing or gaining electrons. In table salt, for example, negatively charged chlorine ions attract positively charged sodium ions and vice versa. The result is a highly regular and repeating structure – an ionic crystal.

The electrostatic forces holding the ions in place are strong, so ionic solids tend to be hard and melt at high temperatures (table salt melts at 2575 °F). Even though they are composed of charged particles, ionic solids are poor conductors of electricity because the component ions are immobilized in the crystal lattice. Only when molten or in solution do the ions become free to move. Ionic liquids or solutions can therefore conduct electricity.

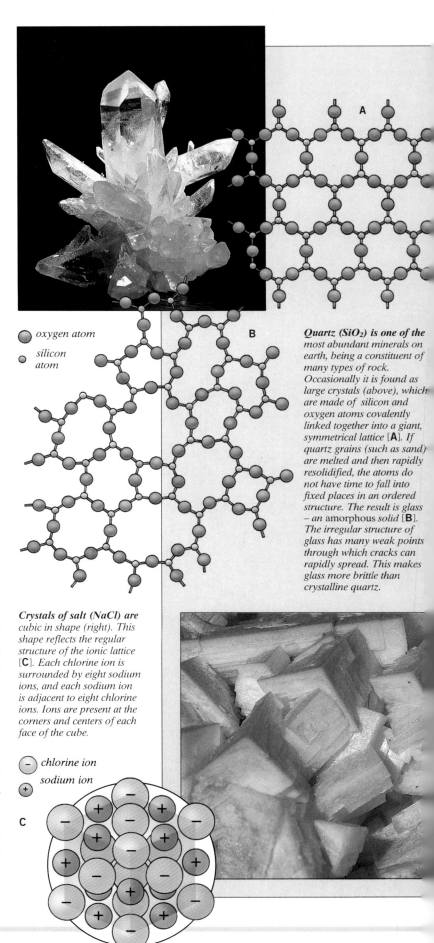

oxygen atom

silicon atom

Quartz (SiO_2) is one of the most abundant minerals on earth, being a constituent of many types of rock. Occasionally it is found as large crystals (above), which are made of silicon and oxygen atoms covalently linked together into a giant, symmetrical lattice [A]. If quartz grains (such as sand) are melted and then rapidly resolidified, the atoms do not have time to fall into fixed places in an ordered structure. The result is glass – an amorphous solid [B]. The irregular structure of glass has many weak points through which cracks can rapidly spread. This makes glass more brittle than crystalline quartz.

Crystals of salt (NaCl) are cubic in shape (right). This shape reflects the regular structure of the ionic lattice [C]. Each chlorine ion is surrounded by eight sodium ions, and each sodium ion is adjacent to eight chlorine ions. Ions are present at the corners and centers of each face of the cube.

chlorine ion

sodium ion

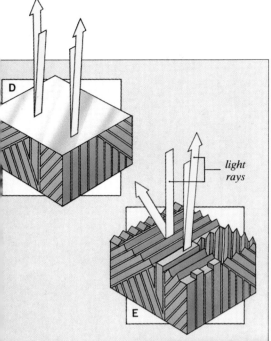

light
rays

Metals are lustrous (reflect light) *because their free or* delocalized *electrons (which are responsible for the bonds that hold metal atoms together) readily absorb and then reemit light energy. A sheet of metal can be polished to a highly reflective finish* [D], *but treating the sheet with dilute acid (a process called acid etching) gives the metal a matte finish* [E]. *This is because etching corrodes away the smooth surface, exposing the boundaries of the individual crystals that make up the metal. These crystals, each around 0.25 mm across, are oriented in different directions and so scatter light.*

Copper – which was first discovered and used by humans in the late Stone Age (around 8000 BC) – is extremely malleable and can be worked into intricate, decorative shapes even when cold (right).

Although copper occurs in native form, most metals are produced by heating their naturally occurring ores in the presence of a reducing agent, such as charcoal. When the molten metal first solidifies, it has a coarse crystalline structure. Working the metal by repeated hammering reduces the size of its component crystals and improves the strength and durability of the material.

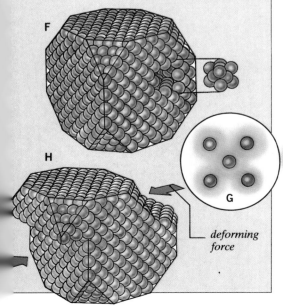

deforming force

A crystal of copper [F] *is a close-packed array of positively charged copper ions "glued" together by a cloud of free or* delocalized *electrons* [G]. *When a metal is hammered* [H], *layers of atoms within the crystals slide over one another: the atoms readily form bonds with their new neighbors.*

— metal ion

— delocalized electrons

Metallic solids

Around three-quarters of the elements of the Periodic Table are classed as metals. Most abundant in the earth's crust are aluminum, iron, sodium, potassium, and magnesium. The majority occur naturally as *ores* in combination with nonmetallic elements, but a few of the less reactive metals, such as silver, gold, and platinum are found in *native* (uncombined) form.

Although they differ widely in their chemical reactions (sodium, for example, reacts violently with water, whereas gold and platinum are inert), metals share a suite of physical properties. They are good conductors of both heat and electricity; they have relatively high melting points; they are *lustrous*, reflecting light of any wavelength; they are *malleable*, capable of being beaten into shape, and *ductile* (they can be drawn out into wire). These properties can be largely attributed to the metallic bond that holds atoms within a metal together.

Crystals and the metallic bond

Like ionic solids, metals are crystalline – their atoms occupy predictable positions in a regular lattice. But unlike ionic solids, which are made up of at least two different types of ion – one positive and one negative – metal crystals contain just one type of atom. Because all the atoms in a metal crystal are the same, they normally pack together closely in fairly simple repeating patterns. Metallic elements are therefore usually very dense.

Metal atoms typically have few valence electrons (those electrons that occupy the outermost shell). In a metal crystal these electrons are lost relatively easily. They become *delocalized* – detached from their parent atom and shared between all the atoms within the crystal – leaving behind positively charged metal ions. These ions tend to repel one another, but are held together by the "glue" of negatively charged delocalized electrons.

Because the delocalized electrons can move freely within the metal crystal, and from one crystal to another, metals are good conductors of electricity. And when one end of a metal bar is heated, electrons move away from the hot end, carrying energy with them. This helps make metals good conductors of heat.

The character of the metallic bond also explains the ductility and malleability of metals. In an ionic solid, such as salt (sodium chloride), a positive ion must always be surrounded by negative ions and vice versa. If this order is changed, by for example hammering a salt crystal, the repulsive forces between positive ions brought into close proximity to other positive ions cause the crystal to shatter. In a metal crystal, however, there is nothing to distinguish the bonds between any two atoms – the bonds are *nondirectional*. This means that when the metal is hammered, bonds between atoms are broken and then reform without destroying the integrity of the crystal.

Liquids
Their properties and applications

We are familiar with many different liquids – water being the most obvious – but we also know oils, alcohols, and even one metal, mercury. It is easy to say what all these substances have in common: they have a definite volume but not a fixed shape, and they flow, assuming the shape of their container. However, an exact scientific definition turns out to be much harder to make. Matter, in all its forms, is made up of atoms or molecules (groups of atoms). A liquid is between a solid and a gas in character: like a solid, the molecules are fairly closely packed, with strong forces between them. However, like a gas, the individual components can readily move past one another and are not confined to one area.

Surface tension
The molecules in a liquid tend to attract one another – this force is known as *cohesion*. A molecule deep within a container of liquid is attracted by all of its neighbors: because it is pulled in all directions at once, it experiences no net force. However, a molecule at the surface of the liquid has no neighbors above it and so experiences a net downward force. The downward pull on all the surface molecules creates a kind of tension over the surface of the liquid that acts like the the rubber skin of a balloon. It is this *surface tension* that gives water droplets their spherical shape.

Just as molecules of a liquid can attract one another, they can also be attracted by molecules of other substances. The attractive force between unlike materials is called *adhesion*. It is adhesion that makes water "wet" other materials, and that makes it climb up the side of a glass to form a visible curve, or *meniscus*.

The special properties of water
For most substances it is true to say that the solid form is denser than the liquid. A significant exception to this rule is water, the solid form of which – ice – floats on water rather than sinking in it. Water behaves in this way because its molecules (which consist of one atom of oxygen and two of hydrogen) are *polar* – the two hydrogen atoms carry a slight positive charge, and the single oxygen atom carries a slight negative charge. Because of this polarity, there is a weak attraction between hydrogen atoms of one water molecule and the oxygen atom of an adjacent molecule: this is called a *hydrogen bond*. These bonds influence the way in which ice crystallizes – its lattice is "looser" than it would be without hydrogen bonds – and it is for this reason that ice is less dense than water. Water's unusually high boiling point also results from hydrogen bonding because extra heat must be applied to weaken these bonds.

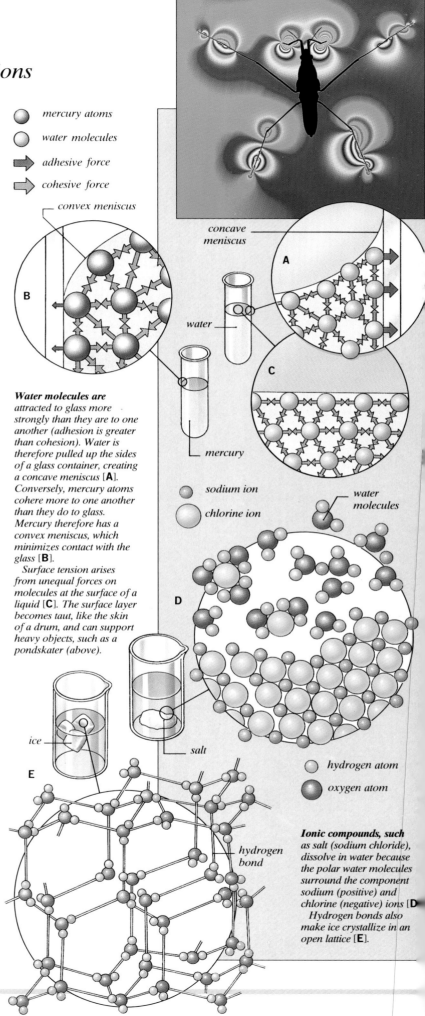

mercury atoms
water molecules
adhesive force
cohesive force

convex meniscus

concave meniscus

A
B
water
C
mercury

sodium ion
chlorine ion
water molecules

D

ice
salt
E

hydrogen atom
oxygen atom

hydrogen bond

Water molecules are attracted to glass more strongly than they are to one another (adhesion is greater than cohesion). Water is therefore pulled up the sides of a glass container, creating a concave meniscus [A]. Conversely, mercury atoms cohere more to one another than they do to glass. Mercury therefore has a convex meniscus, which minimizes contact with the glass [B].

Surface tension arises from unequal forces on molecules at the surface of a liquid [C]. The surface layer becomes taut, like the skin of a drum, and can support heavy objects, such as a pondskater (above).

Ionic compounds, such as salt (sodium chloride), dissolve in water because the polar water molecules surround the component sodium (positive) and chlorine (negative) ions [D]. Hydrogen bonds also make ice crystallize in an open lattice [E].

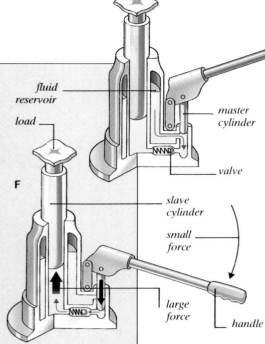

Pressure is force per unit area. Pascal's principle means that a small force applied to a small area can be transmitted to become a large force over a large area. This is how a hydraulic car jack works [F].

The handle is a lever that the operator uses to push down on a small-diameter master cylinder. Hydraulic fluid transmits this pressure to the wider slave cylinder, which pushes up on the car with greater force. A one-way valve traps the hydraulic fluid under the slave cylinder, and allows the fluid above to be refilled from a reservoir.

fluid reservoir

load

F

master cylinder

valve

slave cylinder

small force

large force

handle

An automobile battery charge indicator [G] depends on Archimedes' principle. It contains a green plastic ball [1], trapped in a Y-shaped, perforated tube [2], immersed in the battery acid [3].

When the charge is low, the density of the battery acid decreases. The volume of acid displaced by the ball weighs less than before. This gives correspondingly less upthrust on the ball, which sinks out of view of the observer [4].

In a fully charged battery, the acid has a high density. Upward force on the ball increases, and the ball floats [5], becoming visible in the viewing window.

viscous liquid in a pipe [H] does not flow at uniform rate away from given starting position]. At the sides, a viscous fluid, such as oil, "sticks" the pipe, forming a static boundary layer. The speed flow increases progressively farther into the pipe, with a maximum at the center [2]. A low-viscosity fluid, such as water, flows a more uniform rate across the pipe [3].

The viscosity of a liquid creases as its temperature es: for this reason, oil owing in long continental pelines (right) is often ated. This makes its nsportation more onomically viable.

H

562

Liquids under pressure

A diver swimming beneath the sea feels pressure because of the weight of water bearing down on his body. The degree of pressure depends on the density of water and the diver's depth – each thirty-foot increase adds pressure equivalent to that of another earth atmosphere. Mercury is more dense than water: the pressure three feet down a column of mercury is the same as that forty feet under water.

Liquids are almost incompressible – their volume hardly changes however much they are squeezed. When pressure is applied to an enclosed liquid, it is transmitted evenly to all parts of the liquid. This principle was first formulated by the French mathematician Blaise Pascal in 1652. It explains the action of *hydraulic* machines – ones that use liquids to transmit pressure and amplify forces. Automobile brakes, automobile jacks, and mechanical excavators are all hydraulic devices.

Archimedes' principle

When an object is immersed in a liquid, it takes up space that was previously occupied by that liquid. Archimedes' principle states that the object will experience an upward (or *buoyant*) force equal in size to the weight of liquid that it displaces. The principle was formulated by the Greek philosopher Archimedes sometime around 250 BC. It explains the buoyancy of objects in all fluids, gases as well as liquids. It is Archimedes' principle that accounts for the buoyancy of a balloon or airship.

If two solid balls of the same size, one made of iron, the other wood, are fully immersed in water, the iron ball sinks whereas the wooden ball bobs up to the surface. The buoyant force on each ball is the same, because each displaces an equal weight of water. The iron ball is denser than water and so weighs more than the buoyant force; the wooden ball weighs less than the buoyant force, and so rises to the surface.

Viscosity

Starting an automobile on a cold day can be difficult because the engine is hard to turn over. This is because the lubricating oil becomes much thicker, or more *viscous* at low temperatures. The viscosity of a liquid is a measure of its "stickiness," or the ease with which it can be made to flow. Water, for example, has a low viscosity, whereas honey has high viscosity.

The Trans-Alaskan pipeline carries oil hundreds of miles through temperatures as low as –40°F. It is surrounded by heating elements that raise the oil temperature to 140°F. At this temperature, the lower viscosity of the oil produces great savings in pumping costs.

At extremely low temperatures some elements become *superfluids* – liquids with no viscosity at all. Helium becomes a normal liquid at –449°F. If it is cooled another two degrees, however, it becomes a superfluid, flowing even through microscopic cracks.

Gases

Their properties and applications

The earth's atmosphere is a sea of air more than 60 miles deep. Its constituent gases – mainly nitrogen, oxygen, and water vapor – have a strong tendency to spread into the vacuum around the earth, but are held close to the planet's surface by gravitational attraction. Their behavior is governed by a few simple laws that relate the pressure, temperature, and the volume occupied by a gas. These laws, first set forth in the 17th century, are drawn upon in the design of countless devices, from engines, pumps, and refrigerators, to barometers and balloons. Today, the gas laws can be explained in terms of kinetic theory – the random and independent movement of billions of separate gas molecules.

Boyle's law and Charles' law

In the course of one hour a person on the earth's surface inhales and exhales about 195 gallons of air. But this apparently large volume of air can be squeezed into a tank small enough to be carried on the back of a scuba diver. This is because air is a gas, with vast expanses of empty space between its molecules. The amount of gas that the tank can accommodate depends on how closely together the molecules are squeezed, and therefore on the pressure used to force the gas into the tank. The relationship between the pressure and volume of a gas is expressed in Boyle's law (named after the 17th-century Irish scientist, Robert Boyle). This states that the volume of a gas is *inversely* proportional to its pressure: this means that halving the pressure of a gas results in a doubling of its volume, and vice versa.

This relationship holds only for a gas in which the individual molecules are completely independent of one another. Such *ideal* gases, however, do not exist in practice: in a *real* gas, there are small but significant attractive forces between molecules, especially at high pressure, when the molecules are close together. One consequence is that halving the pressure of a gas does not quite double its volume. But at low pressure, gases deviate very little from Boyle's law, and their behavior is close to ideal.

The relationship between gas temperature and volume is also a simple one, described by by Charles' law (named after the French physicist Jacques Charles). This states that the volume of a fixed mass of gas is directly proportional to its temperature (when measured on the Kelvin scale). So doubling the temperature of a gas also doubles its volume. This effect is put to work in an internal-combustion engine. Within each of the engine's cylinders, burning fuel causes a gas to expand and push on a movable piston, which in turn drives the rotation of the wheels via a crankshaft.

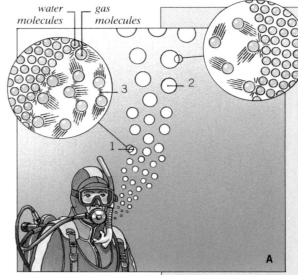

water molecules — gas molecules

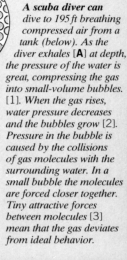

A scuba diver can dive to 195 ft breathing compressed air from a tank (below). As the diver exhales [A] at depth, the pressure of the water is great, compressing the gas into small-volume bubbles. [1]. When the gas rises, water pressure decreases and the bubbles grow [2]. Pressure in the bubble is caused by the collisions of gas molecules with the surrounding water. In a small bubble the molecules are forced closer together. Tiny attractive forces between molecules [3] mean that the gas deviates from ideal behavior.

B

pressure / volume

The graph shows how *the volume of an ideal gas changes as pressure is increased [B]. It assumes, however, that the temperature of the gas remains constant as it is pressurized. In reality, heating of the gas would occur, disturbing the regular shape of the curve.*

The movement of a piston *in a car engine is cyclical [C]. First it moves upward, compressing a gaseous mixture of fuel and air [1]. Forced into a smaller volume, the gas warms up (by Charles' law) [2] and the molecules move more rapidly. A spark plug [3] ignites the compressed gas, and energy is released as it burns. The resulting rise in temperature makes the molecules move faster still [4] and collide more often with the walls of the cylinder and with the piston. The gas exerts an increased pressure, pushing back on the piston, which turns a crankshaft. The piston then begins to move up once again [1].*

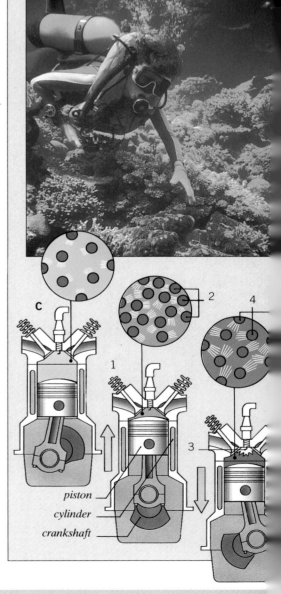

piston

cylinder

crankshaft

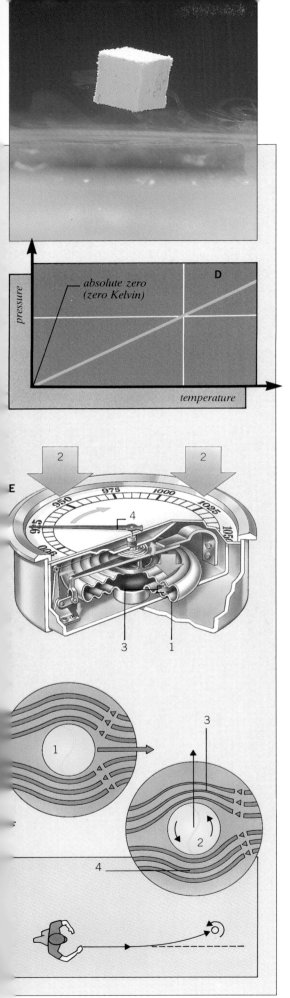

Absolute zero cannot be measured directly. Instead it is derived by plotting the pressure of a gas against temperature on a graph [D]. The direct relationship between the two values is represented by a sloping line: extending this line (extrapolating) until it hits the temperature axis gives a theoretical value at which a gas exerts zero pressure. This value is absolute zero. Near this temperature, materials behave oddly. Some are superconductors, carrying electric current without any resistance. An electromagnet made out of such a material can lift an iron block without continual input of energy (left).

The exact pressure exerted by the atmosphere depends on location, altitude, temperature, and weather. Indeed, variations in atmospheric pressure, measured using an aneroid barometer, can be used to track and predict changes in the weather.

The barometer [E] contains an evacuated metal chamber, supported by ribbed walls [1]. As air pressure increases [2] the chamber walls are pushed inward [3]. A system of gears and levers amplifies this movement and translates it into the movement of a pointer [4] on a dial, in this case calibrated in millibars.

A baseball pitcher can make a ball curve as it flies though the air [F] by spinning it on its axis as it is thrown. When thrown normally [1], the pressure on both sides of the ball is the same and it flies straight. If, however, the ball is spun on its axis [2], it "drags" some of the air next to it around with it as it spins. This increases the speed of air on one side of the ball [3] and decreases it on the other [4]. Because of the Bernoulli effect, the air on one side is at lower pressure than the air on the other side. The ball is pushed sideways into the area of low pressure, curving as it moves forward.

The Kelvin scale and absolute zero

Although temperature is commonly measured on the Celsius and Fahrenheit scales, the Kelvin scale has greater scientific significance. The two familiar scales are built around arbitrarily chosen temperatures – the freezing and boiling points of water (at a pressure of one atmosphere). These temperatures are assigned the values of 0°C and 32°F, and 100°C and 212°F respectively, and the temperature difference between the two values is divided into equal units, called Celsius or Fahrenheit degrees. The Kelvin scale, in contrast, is based on a fundamental physical constant – an absolute zero point (equivalent to –273.15°C) below which a substance cannot be cooled. Absolute zero itself has never been reached by progressive cooling, and the value is derived theoretically.

The size of one Kelvin is the same as that of one Celsius degree, so the melting point of water is given as 273.15K.

Atmospheric pressure

Like any fluid, the atmosphere exerts a pressure in all directions, which depends on its depth and density. The gravitational pull of the earth means that the atmosphere is much denser near the surface than at higher altitudes: more than 50 percent of atmospheric gas is within 4 miles of the earth's surface.

At ground level, the atmosphere is at its densest and deepest. It presses on the human body with a force of 100,000 newtons (equivalent to a weight of 10 tons) per square meter of area. Fortunately, the human body exerts an equivalent, balancing, outward pressure. Atmospheric pressure can be measured in a number of different units: one newton per square meter (N/m^2) is equivalent to one *pascal* (Pa). One hundred pascals is equivalent to a *millibar*, one *bar* being the pressure exerted by the atmosphere at sea level at a temperature of 0°C and a latitude of 45°.

Fluid flow

A fast-flowing fluid – which may be either a gas or a liquid – exerts less pressure than a similar fluid that is static or moving slowly. This effect, first noted by Daniel Bernoulli in 1740, has many everyday manifestations. For example, a jet of air blown between two leaves of paper generates an area of low pressure between the two sheets: the higher – atmospheric – pressure around the sheets then pushes them together. For the same reason, two high-sided vehicles moving side-by-side along a highway can be dangerously drawn toward each other.

However, the most spectacular application of Bernoulli's principle is in the science of *aerodynamics*. The wings of an aircraft and blades of a helicopter are shaped so as to make air flow more rapidly over their top than over their bottom surfaces. This shape, known as an airfoil, causes a higher pressure to act on the underside of the wing, in this way producing *lift*.

Mixtures and Solutions

The ways that substances can mix together

Very few everyday substances are encountered in a pure, unadulterated form. Most are mixtures of elements or compounds, or solutions – molecular level associations between two substances, one of which, the *solute*, is dissolved in the other, the *solvent*. Blood is a life-giving solution of gas and nutrient molecules, and carries away the dissolved waste products of human metabolism. But blood is also a suspension – a liquid throughout which larger particles, such as white and red blood cells, are distributed. Tap water is a solution, which is given its taste by the material dissolved in it. And even most familiar metals, from dental amalgam to "gold" rings, occur in the form of solid solutions, or alloys.

Intimate attraction

For one compound (a *solute*) to dissolve in another (a *solvent*), the forces that naturally bind the atoms or molecules of the solute together have to be overcome. The atoms of the solvent must then surround those of the solute to produce a *solution*. Whether one substance can dissolve in another depends on the nature of the two materials. An ionic substance, such as table salt, is made up of positive sodium and negative chloride ions. It only dissolves in solvents that can surround its ions with other charged particles. Water is such a solvent because its molecules are *polar* – one side of the water molecule carries a slight positive charge, the other a slight negative charge.

Many covalent molecules, such as oils and greases, contain no charged particles. They therefore do not dissolve in polar solvents, but need non-polar or organic solvents. Dry cleaning fluid (trichloroethene) is one such liquid.

Crossing the divide

A plant depends on the properties of solutions to carry water from its roots to its leaves. Osmosis is a process that occurs when two similar solutions come into contact across a semi-permeable membrane – a barrier through which solvent, but not solute, molecules may pass. If one of the solutions has a higher concentration of solute than the other, solvent will travel from the weaker solution to the stronger until the difference is removed. The water in the cells of a plant root is a more highly concentrated solution of sugars and minerals than the water in the earth surrounding it. By osmosis, water is attracted through the semi-permeable skin of the root into the plant.

Artificial semi-permeable membranes are designed to let through only molecules of a certain size. When used in dialysis machines, these membranes remove toxic waste products from the blood of a patient with kidney failure.

A useful rule for deciding whether one substance will dissolve in another is "like dissolves like" [**A**]. Ionic liquids, made up of separate positive and negative ions, can thus dissolve in other ionic liquids [1]. For example, when molten alumina (ionic aluminum oxide) is electrolyzed into pure aluminum, it is first dissolved in cryolite, another ionic compound.

Polar *molecules, such as water, have areas of higher and lower electron density that act as tiny positive and negative charges. They can dissolve polar molecules* [2], *and ionic compounds* [3] *by surrounding negative ions with their positive ends, and positive ions with their negative ends. Non-polar compounds, such as oil, do not dissolve in polar solvents* [4].

Seawater is a solution of *sodium chloride and other substances* [**B**]. *In one ton of seawater* [1] *there are 40 lb of chloride ions* [2], *22 lb of sodium ions* [3] *and 11 lb of other salts* [4].

A dialyzer makes use of *the properties of semi-permeable membranes* [**C**]. *A patient's blood is separated from a solution by a membrane that allows waste products and glucose to pass, but blocks larger molecules. If the solution was of the same strength as the blood, equal numbers of molecules would be exchanged* [1]. *But in a dialyzer, the solution is weaker than blood, so there is a net flow of waste products and glucose into the solution* [2]. *A pump* [3] *draws blood from an artery, and after contact with the solution* [4] *it is returned to a vein. The used solution drains away* [5].

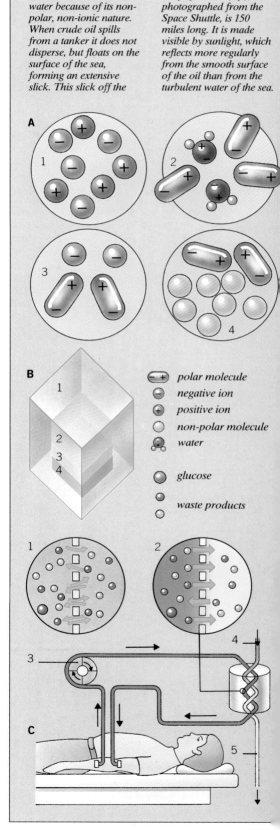

Oil does not dissolve in water because of its non-polar, non-ionic nature. When crude oil spills from a tanker it does not disperse, but floats on the surface of the sea, forming an extensive slick. This slick off the

Gulf of Oman (left), photographed from the Space Shuttle, is 150 miles long. It is made visible by sunlight, which reflects more regularly from the smooth surface of the oil than from the turbulent water of the sea.

- ⬭ polar molecule
- ⊖ negative ion
- ⊕ positive ion
- ◯ non-polar molecule
- water
- glucose
- waste products

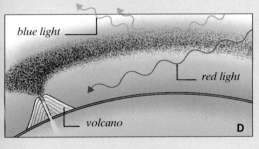

When a volcano erupts, it throws millions of tons of fine dust into suspension in the upper atmosphere [**D**]. As the particles are carried by winds they form a thin layer that encircles the globe (right). The dust is made evident by deep red sunsets visible for a long time after the eruption. These occur because the dust particles scatter blue light more effectively than red. The result is that blue light arrives at our eyes from all directions, while

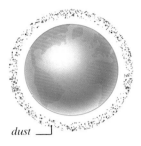

red light takes a more direct route through the atmosphere. Around sunset, the sun is low in the sky, and light must travel through a great thickness of the dust. So much blue light is scattered that by the time the light reaches the earth's surface the sun appears to be an intense red color.

Both face cream and margarine are emulsions – suspensions of tiny droplets of two immiscible liquids (oil and water) held together by an emulsifier [**E**]. This has long molecules with a charged, hydrophilic (water loving) head, and a non-polar, hydrophobic (water-hating) tail, which dissolves in oil. In a face cream, the emulsifier allows small droplets of oil to bind with water, which makes up the bulk of the product. Conversely, margarines are made from vegetable oils, but contain a small amount of water. The emulsifier allows the water to exist as stable small droplets in a "sea" of oil.

A pure metal is relatively soft because of its crystal structure [**F**]. Adjacent layers in the regular lattice can easily be pushed past one another by a cleaving force [1]. The metal can be strengthened by becoming the solvent in an alloy with another, solute metal. In a substitutional alloy [2], slightly smaller solute atoms directly replace some solvent atoms in the lattice. The size difference warps the crystal structure slightly, stopping layers from flowing smoothly past each other. Even smaller solute atoms can fit into the spaces between the solvent atoms, interlocking adjacent layers more tightly [3].

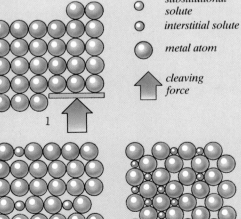

Suspensions and red skies
In a solution, particles on the scale of ions, atoms and small molecules, are completely intermingled and cannot be separated by a process as simple as filtration. *Suspensions* contain much larger particles than a solution, which can be filtered out or made to settle. Unlike solutions, suspensions have a cloudy appearance because the particles that they carry are large enough to scatter light.

Suspensions do not only form in liquids. The explosive eruption of a volcano releases vast quantities of very fine pumice dust into the atmosphere, where it can remain suspended for several years. The suspended dust can have a significant effect on the earth's energy balance, reflecting sunlight back out to space and so causing a small depression in world temperature. The dust also has a more pleasant effect as it scatters sunlight, causing vivid red sunsets for years after the eruption.

Mixing the immiscible
Oil does not dissolve in water. But if the two substances are brought together and shaken, tiny oil droplets form in the water. This is a type of suspension, called an emulsion, distinct from a solution because the droplets are much larger than individual molecules. Emulsions have a characteristic appearance. The mayonnaise-like substance seen in a car radiator when oil has leaked into the cooling water is an emulsion. And mayonnaise itself is an emulsion, though considerably more palatable.

Cosmetic creams are similar mixtures. The oil and water from which these creams are made tend to separate out into distinct layers. To prevent this unwanted separation, a third substance, an *emulsifier*, is added to the mixture. This works in a very similar way to a detergent, allowing "bridging" bonds to form between the polar water molecules and non-polar oil molecules in the emulsion.

Mixtures of metals
Metals have widely varying properties – gold is very stable, whereas potassium ignites on contact with water; lead is extremely dense and soft whereas aluminum has high strength and extremely low density. However, in many instances, different and more controllable attributes are needed, which can only be met by *alloys* – mixtures, or, more exactly, solid solutions, of the atoms of one or more *solute* metals in the crystal structure of a *solvent* metal.

Alloys are usually made to be stronger and more heat-resistant than the metals that they are based upon. But in some cases, weakness is a virtue. Pure bismuth has a melting point of 520°F, but mixed with lead, tin, and cadmium it becomes an alloy called Wood's metal, with a melting point of just 158°F. This is the ideal material for automatic sprinkler plugs, which are melted by the heat of a fire, allowing water to flow and extinguish the flames.

Energy on the Move
The laws that govern the transfer of heat

Thermodynamics is the science that describes the fundamentals of energy, temperature, and heat flow, and the nature of change itself. Its roots lie in nineteenth century musings on the nature of heat and matter, prompted by a need to describe the practical workings of steam engines. Its laws predict which changes are impossible and which are inevitable. Some events, such as a collision of two particles, can easily be reversed, as if on a film played backward, whereas other events cannot. And just as thermodynamics began where abstract philosophy met engineering, it now describes the economics and the efficiencies of everything from cars to power stations, as well as suggesting why time seems to flow in only one direction.

The zeroth law of thermodynamics
The first laws of thermodynamics to be formulated were named, unsurprisingly, the first and second laws. But, after the discovery of these rules it was realized that another, more fundamental law was needed to make the others rigorous. To give a logically numbered system, this new law had to be called the zeroth law – the one before the first.

The zeroth law seems to be a statement of the obvious: it states that a body is in thermal equilibrium with its surroundings if the amount of heat flowing into it is exactly matched by the heat flowing out. A thermometer placed in a patient's mouth takes two or three minutes to reach a stable reading, but once it does, the thermometer is in thermal equilibrium with the mouth. The zeroth law says that if two objects are in thermal equilibrium with a third object or system, they are in equilibrium with each other, and hence they are at the same temperature.

Balancing inputs and outputs
Every time a mixture of fuel and air inside an internal combustion engine cylinder explodes, an energy conversion takes place. Before the explosion, the energy takes a potential form, stored in the chemically unstable fuel. The explosion converts all of this potential energy into two other forms – heat (or internal energy) in the exhaust gases; and *work*, mechanical energy produced as the expanding gases drive the piston downward.

For any process taking place in an enclosed system, energy cannot be created or destroyed. All that can happen is that energy can be transferred from one form to another. This is a statement of the first law of thermodynamics, a fundamental rule of nature also called the law of conservation of energy. According to this, all the energy released when the fuel and air in an engine explode is converted into other forms, which can all be accounted for.

*The concept of thermal equilibrium is central to thermodynamics [**A**]. A can of drink in a refrigerator is in thermal equilibrium with its surroundings – the amount of heat flowing into the can is matched by the heat flowing out – and so it remains at a constant cool temperature [1]. But if the refrigerator door is opened, equilibrium is lost. More heat enters the can than leaves, and so the can warms up [2]. In time, it warms to room temperature, and regains thermal equilibrium. By the zeroth law of thermodynamics, two objects in equilibrium with their surroundings are also in equilibrium with each other. So, when two cans at the same temperature touch, they exchange equal amounts of heat [3].*

*The first law of thermodynamics states that energy is always conserved as it flows from one form to another [**B**]. When a burner is used to heat gas inside a cylinder [1], the increased pressure created pushes a piston outward, doing work, a mechanical form of energy. In an ideal situation [5] where no heat is lost to the surroundings, all the heat from the burner becomes work on the piston or increased internal energy – a measure of the speed with which individual gas molecules move around [2], spin [3], and vibrate [4].*

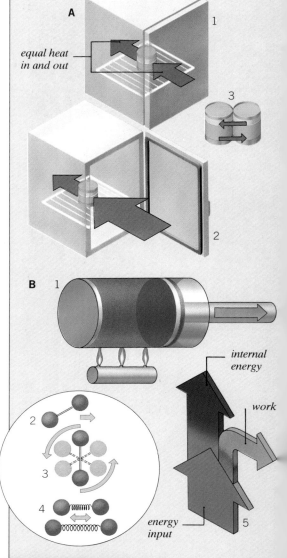

Thermodynamics grew up as a science in the 19th century, when methods were needed to quantify and then improve the efficiency of steam engines. The modern successor to the steam engine is the steam turbine (left), which takes superheated high-pressure steam and converts the thermal energy it contains into mechanical energy or work. This turbine is part of a set in a power station. It can turn up to 40 percent of the energy in the fuel burned into electrical energy. The laws of thermodynamics predict that it is impossible for any engine to convert 100 percent of its energy input into work.

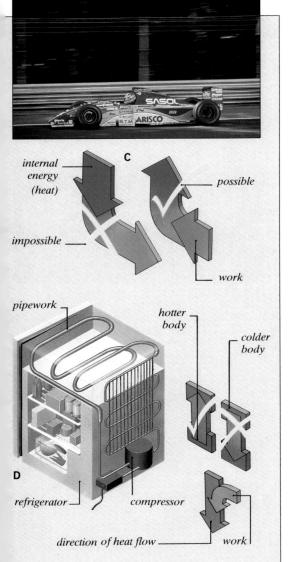

internal energy (heat)

C

impossible

possible

work

pipework

hotter body

colder body

D

refrigerator

compressor

direction of heat flow

work

A formula one racing car (left) is designed to convert chemical energy in its fuel into mechanical and then kinetic energy. The second law of thermodynamics states that it is impossible for the car's engine, no matter how sophisticated, to convert 100 percent of the energy of the fuel into work or kinetic energy [C] (left). But it is possible – in fact inevitable – for the work done to power the car forward to be entirely converted back into heat [C] (right). This could be heat in the air pushed out of the way by the car's body or, more evidently, the intense heat in the brake disks, here shown glowing red hot.

Heat will always naturally flow from a hotter body to a colder one, but for flow in the opposite direction, energy must be expended. In a refrigerator [D], this is supplied in the form of electricity. Inside the pipework loops of a refrigerator is a coolant which, as a vapor, absorbs heat from food and drink inside the refrigerator. It loses this heat in a series of steps in which it is compressed to a liquid, cooled to room temperature, and then expanded to a vapor once more. The work needed for this to happen comes from an electric compressor, which pumps coolant around the pipes.

The second law

Although it is possible to convert mechanical work entirely into heat, the second law of thermodynamics states that it is impossible to perform the opposite process – the conversion of an amount of heat entirely into work. In a real engine some energy is used to overcome frictional forces, some to heat the engine and its surroundings, and yet more is lost through leakage. But even in a completely idealized engine, where frictional and other losses have been eliminated, it is still impossible to convert all of the heat released by burning fuel into mechanical energy.

In the 19th century, a Frenchman, Nicolas Carnot, considered a cycle, or series of steps, that an idealized engine would have to pass through when working. He discovered that there was no possible way to avoid conducting a proportion of the heat released from the fuel to the surroundings of the engine.

Cool box

The second law of thermodynamics can be expressed in a number of different ways. One of these is the statement that heat, on its own, will only flow from a hotter body to a colder one. The law does not say that flow in the opposite direction is impossible, but that some additional input – namely work – is needed in order to make this happen.

A refrigerator is a type of machine that, in thermodynamics, is referred to as a heat pump. It works by extracting heat from the food and drinks inside it, and pumping this heat out into the hotter air surrounding the machine. In most cases the work needed to pump the energy against the direction in which it would normally flow comes from the motor that drives an electric refrigerator's compressor. But the result can be achieved through other means – many mobile homes and campers have refrigerators that cool food by burning gas!

Law and disorder

The second law of thermodynamics can also be expressed in terms of one of the most puzzling of physical quantities, *entropy*, a measure of the disorder of a system. Tanks full of hot and cold water are in a highly ordered, low entropy state. But if the tanks are emptied into a bath, the hot and cold water mix together, increasing their disorder and hence entropy. The second law implies that it is impossible for the entropy of a system ever to decrease. It either remains the same or increases. This means it is impossible for the bath of water ever to "unmix" itself into separate bodies of hot and cold water.

Unlike the three dimensions of space, the fourth dimension, time, seems to move in only one direction. The second law of thermodynamics shows that natural processes have a preferred direction, always moving forward through time. Any movement backward would be against the law – and hence impossible.

According to the second law of thermodynamics the amount of disorder or entropy in the universe is constantly increasing. Stars are highly ordered – hence low entropy – systems, stable for many billions of years. But at the end of their lifetimes, the most massive stars explode out into space in huge supernovae. The remnants of one such gigantic explosion, which was observed on the earth in AD 1054, are still visible as the Crab nebula (right). This is a huge, rapidly expanding cloud of dust and gas. It has much more disorder, and therefore higher entropy, than the star from which it originated.

Forces in Equilibrium
Balancing action and reaction

Each of the 1362-foot-high twin towers of the World Trade Center in New York is subject to two huge forces. One is gravity – the building's 300,000-ton weight pushing it down on to the earth; the other is an upward force supporting the building. These two forces exactly cancel each other out – the building is said to be in *equilibrium*. Architects have to ensure that every force in a building is balanced by another. This applies not only to the vertical forces of weight and support but also to the sideways pulls of the winds. When the forces on a body do not balance, an overall force acts and the body accelerates. The acceleration does not have to be be in a straight line – the body could also start to spin.

Staying aloft with balanced forces
As an airplane flies along, four forces act on it in different directions. Its *weight* pulls it downward while the *lift* from its wings pushes it upward. The *thrust* from its engines pushes it forward, while the *drag* from the air forced out of the way pulls it back. When the plane is in steady flight – moving forward at a steady speed without accelerating or decelerating – all four of these forces balance each other out and the plane is in *equilibrium*.

Like an aircraft, a bird can control its lift, thrust, and drag so that they are in balance for steady flight, or out of balance to accelerate and maneuver. A rising dove makes maximum use of its wings to provide lift greater than its weight, a swan decelerates to land by spreading its wings to increase drag. A plunging gannet folds its wings to minimize drag – in the same way a jet fighter swings its wings back to enable it to fly supersonically.

Cantilevers and the center of mass
All buildings, from tents to skyscrapers, must stand firm, and therefore have to be in equilibrium. This may be effected with a system of ropes or cables in tension, as in a tent or suspension bridge, or by distributing loads along arches. In addition it must be established that the ground is capable of providing an upward force equal to the weight of the structure, and that the floors and roofs are well supported.

One commonly used type of structure is the cantilever, a construction that permits large distances to be spanned. It comprises a beam supported by a pier not at its center of gravity, but a short way along its length. This gives the beam a tendency to topple, which is counteracted by firmly anchoring the "short" end. The cantilever is the basis of a simple type of bridge, formed from two cantilevers that meet over a river. It is also used in the design of the canopies of sports stadiums and hangars.

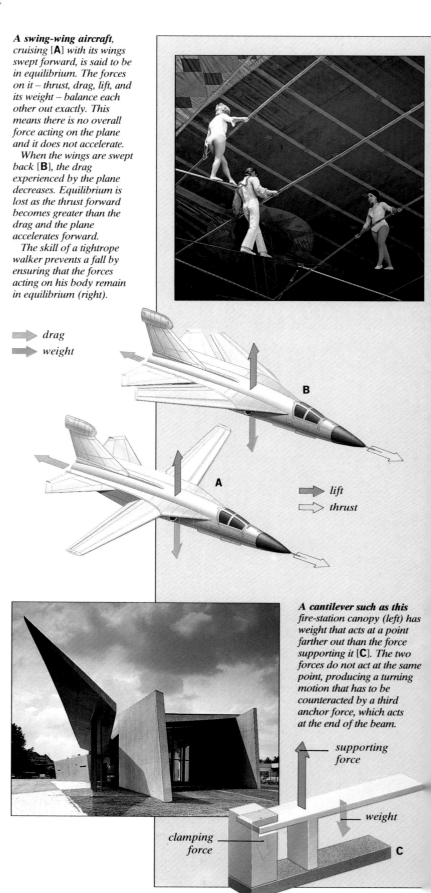

A swing-wing aircraft, cruising [A] with its wings swept forward, is said to be in equilibrium. The forces on it – thrust, drag, lift, and its weight – balance each other out exactly. This means there is no overall force acting on the plane and it does not accelerate.

When the wings are swept back [B], the drag experienced by the plane decreases. Equilibrium is lost as the thrust forward becomes greater than the drag and the plane accelerates forward.

The skill of a tightrope walker prevents a fall by ensuring that the forces acting on his body remain in equilibrium (right).

drag
weight

B

A

lift
thrust

A cantilever such as this fire-station canopy (left) has weight that acts at a point farther out than the force supporting it [C]. The two forces do not act at the same point, producing a turning motion that has to be counteracted by a third anchor force, which acts at the end of the beam.

supporting force

weight

clamping force

C

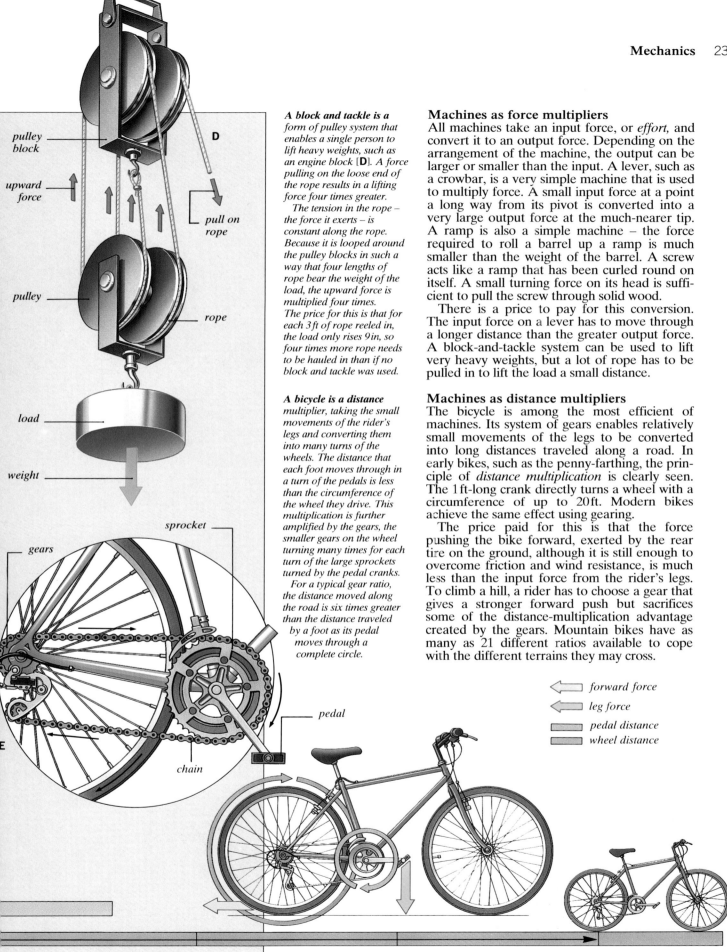

pulley block

upward force

pull on rope

pulley

rope

D

load

weight

sprocket

gears

pedal

chain

E

*A block and tackle is a form of pulley system that enables a single person to lift heavy weights, such as an engine block [**D**]. A force pulling on the loose end of the rope results in a lifting force four times greater.*

The tension in the rope – the force it exerts – is constant along the rope. Because it is looped around the pulley blocks in such a way that four lengths of rope bear the weight of the load, the upward force is multiplied four times. The price for this is that for each 3ft of rope reeled in, the load only rises 9in, so four times more rope needs to be hauled in than if no block and tackle was used.

A bicycle is a distance multiplier, taking the small movements of the rider's legs and converting them into many turns of the wheels. The distance that each foot moves through in a turn of the pedals is less than the circumference of the wheel they drive. This multiplication is further amplified by the gears, the smaller gears on the wheel turning many times for each turn of the large sprockets turned by the pedal cranks.

For a typical gear ratio, the distance moved along the road is six times greater than the distance traveled by a foot as its pedal moves through a complete circle.

Machines as force multipliers

All machines take an input force, or *effort,* and convert it to an output force. Depending on the arrangement of the machine, the output can be larger or smaller than the input. A lever, such as a crowbar, is a very simple machine that is used to multiply force. A small input force at a point a long way from its pivot is converted into a very large output force at the much-nearer tip. A ramp is also a simple machine – the force required to roll a barrel up a ramp is much smaller than the weight of the barrel. A screw acts like a ramp that has been curled round on itself. A small turning force on its head is suffi-cient to pull the screw through solid wood.

There is a price to pay for this conversion. The input force on a lever has to move through a longer distance than the greater output force. A block-and-tackle system can be used to lift very heavy weights, but a lot of rope has to be pulled in to lift the load a small distance.

Machines as distance multipliers

The bicycle is among the most efficient of machines. Its system of gears enables relatively small movements of the legs to be converted into long distances traveled along a road. In early bikes, such as the penny-farthing, the prin-ciple of *distance multiplication* is clearly seen. The 1 ft-long crank directly turns a wheel with a circumference of up to 20ft. Modern bikes achieve the same effect using gearing.

The price paid for this is that the force pushing the bike forward, exerted by the rear tire on the ground, although it is still enough to overcome friction and wind resistance, is much less than the input force from the rider's legs. To climb a hill, a rider has to choose a gear that gives a stronger forward push but sacrifices some of the distance-multiplication advantage created by the gears. Mountain bikes have as many as 21 different ratios available to cope with the different terrains they may cross.

forward force

leg force

pedal distance

wheel distance

Bodies in Motion

The behavior of objects on the move

For centuries, astronomers have been accurately predicting the motion of planets, and the orbits of satellites are known for years to come. But motion in space is relatively simple – uncomplicated by air resistance or friction. Such simplicity is often at odds with our own direct experience, in a world where friction and air resistance, with gravity, dominate the motion of everyday objects. Such great mathematicians as Galileo, Descartes, Newton, and Leibniz provided the principal techniques for analyzing motion, such as *idealizing* – imagining a situation stripped of complexities like friction. This tool must be carefully used to avoid oversimplification, but it can provide powerful predictions of real motion.

On track – displacement and distance

The movement of a racing car can be measured using two related, but distinct, quantities. The *distance* it travels along the track, given in yards is the more familiar one. A more exact measure is the car's *displacement* from the starting point. This is given as the length *and* the direction of the straight line between its initial and present positions. Such quantities, with a direction as well as a magnitude, are called *vectors* – other examples are force, acceleration, and velocity. Those quantities, like distance, that have only magnitudes are known as *scalars*. Thus velocity – the rate of movement in a particular direction – is the vector counterpart of speed.

An example of the use of vectors is in navigation. A ship or airplane navigator calculates the craft's position by adding together known distances traveled in specific directions. The result – the ship's *bearing* – is itself a vector and consists of a distance and a heading.

A path through the air – ballistics

As a motorcycle stunt rider flies through the air, he follows a curved *parabolic* path. Because his direction is constantly changing, his velocity also changes. When a body changes velocity it accelerates – in this case the acceleration is due to the force of gravity on the motorcycle and rider. The link between force and acceleration was formally established by Isaac Newton, whose first law states that a body will remain at rest or moving at a steady velocity unless it is acted upon by an external force.

Ballistics is the study of the paths of projectiles moving through a gravitational field, and was developed to allow medieval gunners to be able to predict where their cannonballs would land. Calculations are made easier if the slowing effects of air friction are ignored. However, from watching the path of a tennis ball when served, it is clear that this approximation is not a very accurate one.

As it crosses the finish line at the end of a two-hour race, a Grand Prix car's displacement relative to the earth is zero. But because the earth is constantly moving, the car's displacement relative to the sun will have changed.

racing car

start/finish line

distance

displacement

track

The progress of a racing car around a circuit [**A**] *illustrates the distinction between* displacement *and* distance traveled. *Partway through the lap* [1] *the car has traveled the distance of the curved path of the first corner. However, its* displacement *– distance in a specific direction – from the startline is smaller, being the length of the straight line between the starting point and the car. At the end of a lap the distance traveled is the lap length itself. But the displacement from the startline is zero* [2].

take-off ramp

landing ramp

weight

horizontal velocity

vertical velocity

The flight of a stunt motorcycle [**B**] *can be broken down into two separate motions, one horizontal and constant, the other vertical and constantly changing. This component is affected by the bike and rider's weight. At first the bike's vertical velocity is upward* [1]. *Gravity reduces this until midway through the flight, when vertical velocity is zero* [2]. *The continued action of gravity accelerates the rider downward, so that when he lands, his downward velocity is the same as was his upward velocity at takeoff.*

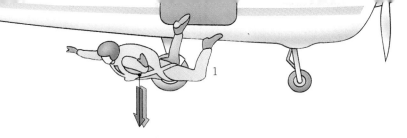

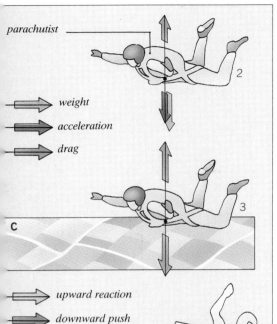

parachutist

weight

acceleration

drag

C

upward reaction

downward push

high jump bar

high jumper

rack

*When a skydiver jumps from a plane [**C**], her free fall is characterized by the constantly changing overall force on her body. At first only gravity acts, increasing her speed by roughly 22 mph each second [1]. However, almost instantly a second force, drag, comes into play. This upward force increases with the skydiver's speed, slowing her acceleration [2]. The drag builds up to a point where it exactly cancels out any further acceleration caused by the force of gravity [3]. From this point on, the skydiver's speed remains constant – she is said to have reached terminal velocity.*

*A high jumper relies on Newton's third law to clear a bar [**D**]. To make the jump successfully, he must leave the ground with an upward velocity of almost 18 mph. He attains this by bending his jumping leg and straightening it rapidly, pushing down on the ground as he does so. The result of this violent downward thrust is an equal and opposite reaction that carries the jumper upward.*

Inside a gearbox, the metal teeth of gearwheels constantly mesh in and out of each other (left). Although the metal surfaces appear smooth, investigation with a high-powered microscope reveals them to be covered with a succession of jagged bumps and grooves. It is this roughness that produces energy-wasting frictional forces. To overcome friction the gearbox is filled with oil. The high viscosity of the oil (its "thickness") enables it to build up a film that keeps metal surfaces separate and able to move easily past each other. Oil is not only a lubricant – as a bonus it also prevents corrosion.

Free fall and Newton's second law

Newton's second law states that the acceleration of a body is proportional to the force it experiences. Thus the huge force from a charge exploding in a gun barrel can accelerate a stationary shell to 5280 ft/s (3600 mph) in less than a tenth of a second. More strictly, this law states that a body's acceleration is proportional to the *total* force on it. This means that a body in equilibrium, where all forces cancel one another out, does not accelerate at all.

The acceleration of a body also depends on its mass. The engine of a racing car produces the same power, and hence force at its driving wheels, throughout a race. However, as it uses up its fuel, its mass decreases. This means that the car accelerates – and brakes (negative acceleration) – more quickly, so that the fastest lap times are set at the end of a race. For the same reason, the tires wear at a faster rate at the start of the race.

Reacting well

For every force there is an equal and opposite reaction. This is the simplest statement of the third of Newton's laws of motion. The principle can be seen in the recoil of a gun when it is fired, the forward push on the bullet being answered by the sharp backward "kick" of the gun itself. So-called recoilless guns make use of this backward reaction force to load the next bullet. The upward force pushing a rocket into space is the reaction to the downward force the rocket itself exerts on the stream of gas particles ejected at high speed from its nozzle.

The third law, like all of Newton's laws, is universal – the pull of the earth on the moon is exactly balanced by the moon's pull on the earth. Forward motion depends on the ground's providing a forward thrust in response to the backward push of a foot. This is a rule that applies equally to an athlete on earth or an astronaut on the surface of the moon.

Necessary friction

Almost every movement on earth relies on friction. Even standing up depends on sideways frictional forces that prevent our feet from sliding apart. Yet friction is also the enemy of energy-efficient motion. It results in a force that opposes motion, slowing objects to a standstill unless more energy is supplied. The energy lost to friction is converted into unwanted heat, and it causes wear on those surfaces in contact.

The wheel is one of the most successful ways of overcoming friction, although even this still needs a bearing where the moving axle rubs against a stationary part of the vehicle. Inside a bearing the moving and stationary parts are separated by oil, which acts as a lubricant. Even more successful are such vehicles as airplanes and hovercraft, which are only slowed by the friction of the air rushing past their structures. Magnetically levitated trains travel almost frictionlessly, held up by powerful electromagnets.

Work, Energy, and Power
Measuring activity

Work, energy, and power are words in everyday use, but to the physicist they have precise and distinct meanings.
Work is the product of force and distance. Engines do work because they exert forces that cause material to move. *Energy* is the ability to do work. The fuel of an engine can supply energy, as can a moving object, such as a flywheel. While work is being done and energy is supplied, the fuel is consumed or the flywheel slows down. *Power* is the rate of doing work. To sell early steam engines, their work-rates were compared with those of horses, giving rise to the term *horsepower*. This unit has now been replaced by the *watt*, allowing direct comparison between the power of objects as diverse as lightbulbs, lasers, and stars.

Work, power, and potential energy
When a force moves an object, work is done. A weightlifter, for example, does work lifting a barbell through a vertical distance. If he is unable to lift the weight, even though he strains and does work on his muscles, he does no work on the barbell. To be able to lift the weight, the weightlifter must expend energy. Energy is measured in *joules*, as is work, because the amount of work done is equal to the change in energy possessed by the barbell.

The raised barbell is said to possess *gravitational potential energy* – it has the potential to do work by virtue of its height above the ground. If it is let go the barbell will fall back to earth as a result of gravity, and do work on the floor. The work done will be a product of the gravitational force and the distance fallen: the greater the height of the barbell, the greater its potential to do work, and so the greater its gravitational potential energy.

The way it moves
A moving object, such as an automobile, is said to have *kinetic energy* because it will do work on anything with which it collides. This energy increases with speed: the faster an object moves, the more energy it has. Kinetic energy is equal to the work done to accelerate an object to a particular velocity; it is also equal to the work done to bring it to rest. An engine works to increase the speed and, therefore, the kinetic energy, of an automobile; a brake absorbs this energy, converting it into heat energy, and thereby slows the automobile down.

Kinetic energy is also dependent on mass. The greater the mass of a body, the greater its kinetic energy. A 175 lb athlete sprinting at 33 ft/s possesses 4000 joules (4 kJ) of kinetic energy. This is almost negligible compared with a 220-ton blue whale swimming at exactly the same speed, which possesses 10 million joules (10 MJ) of kinetic energy.

A jet fighter takes ten seconds to climb to 3300 ft but a propeller-driven aircraft of the same weight takes five times as long [A]. In climbing to the same height, the two planes do equal amounts of work, but the power output of the jet is five times greater. A weightlifter does work as he raises a barbell [B]. At the start and end of the lift [1,6], his legs push rapidly to provide maximum power. His weaker arm muscles [3] raise the weight slowly, with lower power. While the weightlifter maneuvers his body [2,3], [4,5], the barbell stays at the same height so no work is done on it.

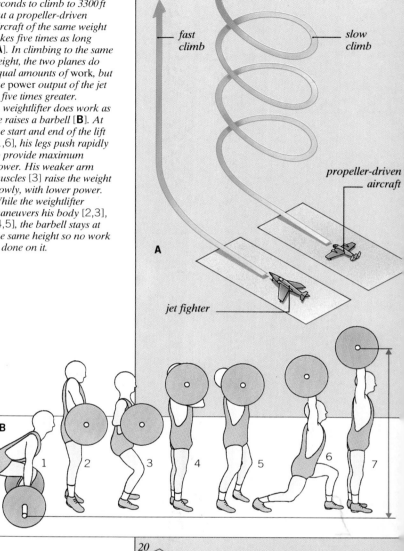

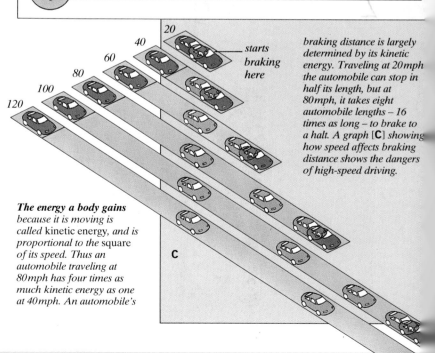

The energy a body gains because it is moving is called kinetic energy, *and is proportional to the square of its speed. Thus an automobile traveling at 80 mph has four times as much kinetic energy as one at 40 mph. An automobile's* braking distance is largely determined by its kinetic energy. Traveling at 20 mph the automobile can stop in half its length, but at 80 mph, it takes eight automobile lengths – 16 times as long – to brake to a halt. A graph [C] showing how speed affects braking distance shows the dangers of high-speed driving.*

*Piledrivers are commonly seen on building sites. As they work, energy undergoes a number of changes and conversions [**D**]. The energy is initially contained in the fuel [1] that is burned in the engine. The inefficiencies and frictional losses inside the engine mean that less than half of the input turns into useful work [2]. This is used to lift the weight, increasing its gravitational potential energy. The weight then falls, converting potential to kinetic energy as its speed increases [3]. It then strikes the pile, a sunken girder that is part of a building's foundations. The weight exerts a force and moves the pile a short distance into the ground [4]. However, only a small proportion of the energy available from burning fuel does useful work on the pile. Much of the energy heats the engine and its surroundings. As the weight falls, some energy is transferred to air molecules. On impact, the weight, pile, and ground all heat up slightly. All the energy contained in the fuel is converted to heat in the form of vibrations in the atoms of the engine, air, piledriver, pile, and ground.*

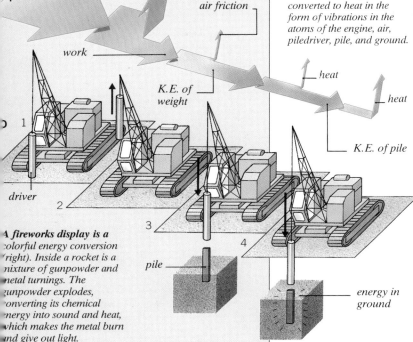

fuel energy

waste heat

air friction

work

K.E. of weight

heat

heat

K.E. of pile

driver

1

2

3

4

pile

energy in ground

A fireworks display is a colorful energy conversion (right). Inside a rocket is a mixture of gunpowder and metal turnings. The gunpowder explodes, converting its chemical energy into sound and heat, which makes the metal burn and give out light.

An automobile traveling at 30 mph plows aside 60 lb of air every second. As it does so, some of the kinetic energy of the automobile is transferred to the air molecules. If the air flows smoothly around the automobile, the amount of energy lost depends on the speed traveled. But if the air flow becomes turbulent, the energy used up depends on the square of the speed, so the energy loss becomes significant at high velocities. Automobile bodies undergo trials in wind tunnels (left) to smooth out the shape and so minimize energy loss. This increases the efficiency of the automobile and decreases fuel consumption.

Energy conservation

Energy is neither created nor destroyed, but is transferred from place to place and from one form to another. Plants convert light energy from the sun into carbohydrate fuel stores by the process of photosynthesis. After living organisms die they may eventually turn into fossil fuels – gas, coal, and oil. These fuels have *chemical potential energy*, which can be converted into heat energy when burned in a power station. The heat energy turns water into steam, which transfers its heat into the kinetic energy of a turbine. This in turn drives a generator that produces electrical energy for the national grid.

This energy flow starts with the sun. As the sun burns, it converts its mass (m) into energy (E), according to the equation $E = mc^2$, where c is the speed of light. Mass and energy are totally interconvertible, and physicists believe that the total amount of mass and energy in the universe has remained constant since the Big Bang.

Maximizing efficiency

The amount of useful work done by a machine is always less than the energy put into it. Some of the input energy is lost to the surroundings as heat, light, or sound. To keep the machine working, extra energy must be put in. The *efficiency* of a machine is the ratio of its useful energy output to its input, and is often expressed as a percentage.

An automobile typically has an efficiency of about 30 percent – most of the energy available from burning fuel simply heats the engine and other moving parts. This does not mean that 30 percent of the energy is available to be continuously converted into ever-increasing kinetic energy. Most of it is used to overcome friction between automobile and road, and to push air out of the way. Much time and money is spent trying to isolate and eliminate sources of energy loss. But, the perpetual-motion – or 100 percent efficient – machine is an impossible dream.

Circular Motion

The forces that govern orbits and oscillations

The relationship between force and acceleration is simple – when a force acts on a body, the latter accelerates in proportion. This is a simple statement of Newton's second law of motion.

When a steady force pushes on a moving object at right angles to its direction of motion, the path of the object becomes circular. This type of motion describes the movement of satellites in orbit around the earth. Similarly, if forces act in such a way as to make an object constantly accelerate toward one central point, a regular oscillation results. This type of motion – simple harmonic motion – describes the movement of a pendulum steadily swinging inside a clock, and of an object bobbing up and down on the sea.

Circular motion

When a force acts on an object, the object accelerates. Acceleration, in this context, does not just mean speeding up: it can equally mean that the object slows down (negative acceleration) or changes direction. An object moving in a circle is always changing direction, and is therefore constantly accelerating. It also follows that a force constantly acts upon it. This force is known as *centripetal force* (Latin for "center-seeking"). It always pushes the body toward the center of the circle, at right angles to its direction of travel. Without this force, the object would simply shoot off in a straight line.

A satellite in orbit around the earth moves at high speed. Without centripetal force provided by the gravitational pull of the earth, it would head off into outer space. Similarly, an object spun around at the end of a string follows a circular path. In this case, the centripetal force is provided by tension in the string.

Inertia and the myth of centrifugal force

Centrifugal force seems to be a very familiar concept. Passengers on a bus rounding a corner must hold on tight to counteract this apparent force, and a racing car driver tilts his head into bends for the same reason. But centrifugal force does not really exist. What is being experienced by the passengers and driver is actually a *lack* of centripetal force, an absence of the center-seeking push that would normally act to keep the body moving in a circle.

The natural tendency of a moving object is to continue moving in a straight line. This property – a reluctance to change direction or speed – is called *inertia*, and it increases with the object's mass. The inertia of the bus passenger's body makes him or her "want" to continue moving in a straight line in spite of the fact that the bus is rounding a corner. This tendency, which is detected by the passenger's senses, is given the common name centrifugal force.

Every circular motion is the result of a force that constantly pushes, and therefore accelerates, an object toward the center of a circle. In the case of a satellite [B] this force is gravity (or weight). The satellite's orbiting velocity is selected to exactly counteract the downward force, or centripetal acceleration [A], and therefore keep the satellite in a stable orbit. If the orbiting speed is too slow, the satellite spirals down to burn up in the earth's upper atmosphere. If it is too fast, the satellite drifts away from the earth. The satellite's speed is constantly adjusted by onboard thrusters.

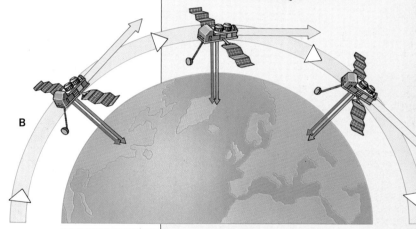

too fast
correct speed
too slow

weight | centripetal acceleration | velocity

A centrifuge is a piece of laboratory equipment designed to separate molecules from solution, or particles from a liquid. It is often applied in medicine to separate red and white blood cells from blood plasma [C]. A test tube of blood is placed into the centrifuge where it is spun at speeds of several thousand rpm. The denser red cells experience a greater centrifugal force than the lighter white cells. The red cells are pushed through the plasma, settling at the bottom of the tube. The white cells are buoyed up by the plasma and gather in a distinct white band farther up the tube.

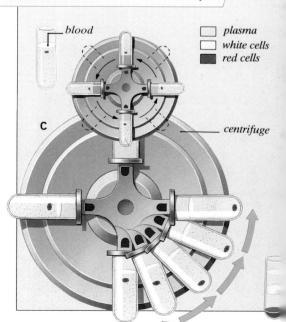

blood

plasma
white cells
red cells

centrifuge

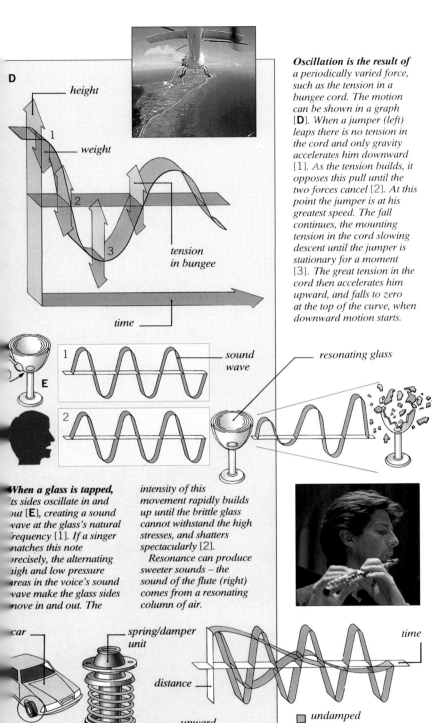

Simple oscillations

Just after a diver leaps off a diving board, the end of the board visibly bounces up and down with a regular period. This type of motion, called *simple harmonic motion* (SHM), occurs in a variety of situations. It is responsible for the regular swing of a clock pendulum, the oscillation of a bungee jumper on an elastic cord, and the regular motion of a wave.

The regular oscillations of SHM can be described as the movement of a body around a central midpoint. The acceleration the body experiences is proportional to its distance from the central point and is always directed toward it. Thus the force on the body is greatest when it is at the extreme ends of the motion.

The movement is repeated over a cycle with a regular frequency. It is this property that makes systems producing SHM ideal for use as timing devices – both pendulums and the quartz crystals in digital watches move in this way.

Vibration and resonance

Solid but flexible objects, such as diving boards, branches of a tree, or strings of a guitar, vibrate when they are hit or slightly displaced. If the physical dimensions and properties of the object stay the same, then the frequency of vibration is the same whenever the object is displaced in this way. This frequency is known as the *natural frequency* of the object.

Often, objects are made to vibrate by some other vibrating body. The shaking of an auto engine, for example, makes other parts of the vehicle vibrate. If the frequency of the vibration matches the natural frequency of a component, it goes into high-amplitude vibration. This is *resonance*. Larger auto components, such as body panels, have lower natural frequencies than those produced by the engine, and so do not resonate. Small parts, however, can rattle in an irritating way, sometimes resulting in the loosening of screws and bolts.

Damped oscillations

The pendulum of a grandfather clock requires a regular push at the same point in its cycle to carry on swinging indefinitely. Without this, the size or *amplitude* of its oscillations would gradually decrease until the bob stopped moving altogether. The movement is opposed or *damped* by friction, both air resistance and the forces in the bearing round which the pendulum turns. Similarly, a plucked guitar string gradually stops vibrating back and forth. It is being damped – energy is drawn from the string and turned into sound.

The graph (left) shows the different degrees of damping that can exist in a vibrating system. *Undamped*, or simple harmonic motion (red) continues indefinitely, whereas slightly damped motion (green) ceases after a few oscillations. In *critically damped* motion (blue), the system does not oscillate at all, but settles directly into its equilibrium position.

D

height

weight

tension in bungee

time

Oscillation is the result of a periodically varied force, such as the tension in a bungee cord. The motion can be shown in a graph [D]. When a jumper (left) leaps there is no tension in the cord and only gravity accelerates him downward [1]. As the tension builds, it opposes this pull until the two forces cancel [2]. At this point the jumper is at his greatest speed. The fall continues, the mounting tension in the cord slowing descent until the jumper is stationary for a moment [3]. The great tension in the cord then accelerates him upward, and falls to zero at the top of the curve, when downward motion starts.

E

1 ——— sound wave

2

resonating glass

When a glass is tapped, its sides oscillate in and out [E], creating a sound wave at the glass's natural frequency [1]. If a singer matches this note precisely, the alternating high and low pressure areas in the voice's sound wave make the glass sides move in and out. The intensity of this movement rapidly builds up until the brittle glass cannot withstand the high stresses, and shatters spectacularly [2].

Resonance can produce sweeter sounds – the sound of the flute (right) comes from a resonating column of air.

---

car

spring/damper unit

distance

upward movement

spring

holes

oil

time

☐ undamped
☐ light damping
☐ critical damping

A car mounted only on springs would resonate, bouncing up and down after it had passed a bump. The oil inside a shock absorber [F] can only pass slowly through holes in the top of the piston, damping out the oscillations and making the ride more comfortable for the passengers.

Static Electricity
Electrical charges and their interactions

Electrostatic charges cause dramatic flashes of lightning, and make pieces of plastic packaging cling to fingers and woolen sweaters crackle as they are pulled off. Despite their nuisance or amusement value, these effects offer an insight into the nature of electricity. The existence of both attractive forces, as in clinging plastic, and repulsive ones, those that make newly brushed hair stand on end, show that there must be two types of electrical property – positive and negative charge. These properties are rooted deep inside atoms, in the negatively charged electrons and positive protons. To a large extent any balance or imbalance in the numbers of these particles governs the chemical and physical behavior of a compound.

Loss and gain of electrons

Atoms are made up of nuclei of positively charged protons and neutral neutrons, surrounded by a "cloud" of negatively charged electrons. The protons in an atom lie deep in its nucleus and only fundamental processes like radioactivity can change their number.

Electrons, however, are much more loosely attached. High temperatures and high voltages can supply enough energy to allow them to "escape." Even friction can scrape some electrons off one material and onto another, creating a deficit of electrons in the first substance and a surplus in the other. The material with the excess of electrons in effect becomes negatively charged, whereas the material that lost these particles becomes positive. These two oppositely charged materials will attract each other, the force between them proportional to their difference in charge and decreasing with the square of their distance apart.

Induced charges and electric fields

An *electric field* surrounds a charged particle, in the same way that a magnetic field surrounds a magnet. The field can be represented by a series of lines of which the concentration at any point shows the magnitude of the force another charged particle will experience there.

Placing a positively charged object near another one made of a conducting material produces an *induced* charge on the conductor's surface. Electrons within the conductor are free to move and, because they are negatively charged, are attracted toward the positive object. Negative charge thus gathers on one side. This effect is used by spray guns to make pesticides stick to their targets, and explains why small pieces of paper can be picked up by a charged rod, irrespective of the sign of the charge. A charge of the opposite sign is induced in the surface of the paper, producing an attractive force that holds it to the rod.

The attractive forces between charged particles can be used to clean air in a room or office.

An electrostatic cleaner [**D**] contains a fan, which blows air past two electrodes, positive and negative. The dirty air first travels through a positive grid, an anode, *that gives particles in the air a positive charge by stripping off some of their negatively charged electrons. The air passes on through a* cathode, *or negatively charged grid, that attracts and traps the positively charged dirt. A charcoal filter then absorbs any remaining airborne particles and the cleaned air is returned to the room.*

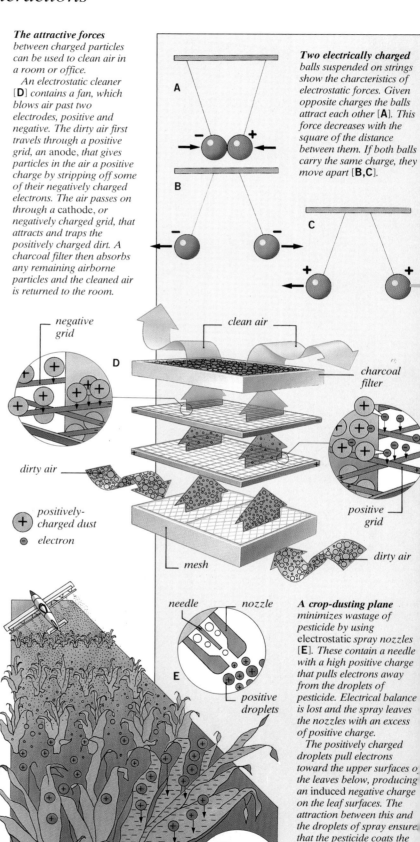

Two electrically charged balls suspended on strings show the charcteristics of electrostatic forces. Given opposite charges the balls attract each other [**A**]. This force decreases with the square of the distance between them. If both balls carry the same charge, they move apart [**B,C**].

negative grid

clean air

charcoal filter

dirty air

positive grid

positively-charged dust

electron

mesh

dirty air

needle   nozzle

E

positive droplets

A crop-dusting plane minimizes wastage of pesticide by using electrostatic *spray nozzles* [**E**]. These contain a needle with a high positive charge that pulls electrons away from the droplets of pesticide. Electrical balance is lost and the spray leaves the nozzles with an excess of positive charge.

The positively charged droplets pull electrons toward the upper surfaces of the leaves below, producing an induced *negative charge* on the leaf surfaces. The attraction between this and the droplets of spray ensure that the pesticide coats the leaves evenly and is not carried away by the wind.

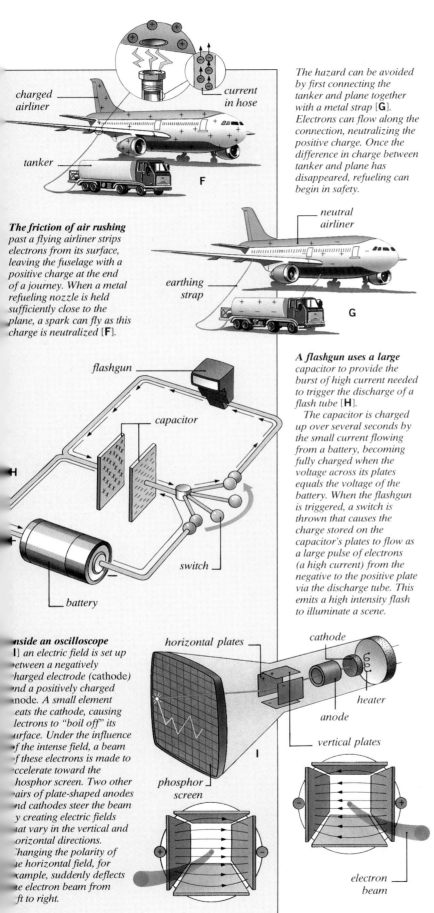

charged airliner

tanker

F

current in hose

The friction of air rushing past a flying airliner strips electrons from its surface, leaving the fuselage with a positive charge at the end of a journey. When a metal refueling nozzle is held sufficiently close to the plane, a spark can fly as this charge is neutralized [F].

The hazard can be avoided by first connecting the tanker and plane together with a metal strap [G]. Electrons can flow along the connection, neutralizing the positive charge. Once the difference in charge between tanker and plane has disappeared, refueling can begin in safety.

neutral airliner

earthing strap

G

flashgun

capacitor

H

switch

battery

A flashgun uses a large capacitor to provide the burst of high current needed to trigger the discharge of a flash tube [H].

The capacitor is charged up over several seconds by the small current flowing from a battery, becoming fully charged when the voltage across its plates equals the voltage of the battery. When the flashgun is triggered, a switch is thrown that causes the charge stored on the capacitor's plates to flow as a large pulse of electrons (a high current) from the negative to the positive plate via the discharge tube. This emits a high intensity flash to illuminate a scene.

Inside an oscilloscope [I] *an electric field is set up between a negatively charged electrode (cathode) and a positively charged anode. A small element heats the cathode, causing electrons to "boil off" its surface. Under the influence of the intense field, a beam of these electrons is made to accelerate toward the phosphor screen. Two other pairs of plate-shaped anodes and cathodes steer the beam by creating electric fields that vary in the vertical and horizontal directions. Changing the polarity of the horizontal field, for example, suddenly deflects the electron beam from left to right.*

horizontal plates

phosphor screen

cathode

heater

anode

vertical plates

I

electron beam

Making sparks fly

Electrostatic behavior was first investigated by rubbing rods of different materials to charge them. Frictional charging is an everyday occurrence – the rub of shoes on a nylon carpet can charge the wearer's body to the extent that a small electric shock is experienced when a piece of metal is touched.

If two bodies carry different overall charges, for instance if one is highly positive and the other negative, a *potential difference*, measured in volts, is said to exist between them. A current – a movement of charged particles – will seek to flow between the bodies to neutralize this. An *insulator*, such as air, normally prevents this, but if the potential difference is very great, or the bodies very close, its insulating properties break down. When this happens, sparking, or *arcing*, occurs, for example, between the highly charged lower layers of a thundercloud and the relatively neutral earth.

Capacitors, charge storage, and current

A capacitor is a simple electrical device consisting of two metal plates separated by a gap across which no electrons – and hence no direct current – can flow. However, when a capacitor is connected to a power supply a surplus of electrons builds up on one plate and a deficit on the other. Electrostatic attraction between the two plates holds the charge in place, and the capacitor can therefore slowly store up large amounts of electrical charge.

The two types of electrical charge (positive and negative) were distinguished from each other as early as the 19th century. Through an accident of history, it became the convention (and still is) to say that a current flows from regions of positive to regions of negative charge. In fact, and rather confusingly, current in a circuit is actually carried by negatively-charged electrons moving from regions of negative to regions of positive charge.

Tracing a pulse

Just as an electric field surrounds a charged particle, so a field exists between two metal plates carrying opposite charges. If a charged particle (such as an electron) enters this field, it is attracted to one of the plates and repelled by the other. This behavior is behind the function of numerous devices, from oscilloscopes and television tubes to huge particle accelerators.

In an oscilloscope a high voltage (corresponding to an intense electric field) between two plates is used to accelerate a beam of electrons toward a phosphor-coated screen that glows when struck by the charged particles. Variable electric fields at right angles to the electron beam are used to steer the electrons. If these fields are controlled by an oscillating or fluctuating voltage (such as the output from a microphone) the electron beam will "paint" lines on the screen that correspond to the shape of the incoming signal.

Electric Currents
Cells, meters, and resistances

Electric current heats wires in hair dryers, in TV electron guns, and in filament lamps and, with expensive loss of energy, it heats the cables that transmit power from electrical generating stations. But current doesn't just heat – every motor or transformer relies on the magnetic field that surrounds a conducting wire. Patterns in currents carry encoded information; current through ionic solutions can cause chemical separation of compounds. Current does not occur spontaneously – almost all materials have resistance to it (apart from superconductors). An energy supply, called a potential difference, is therefore needed to maintain a current, and this can be provided by a battery or by an electromagnetic generator.

Conductors and insulators
An electric current is a movement of charged particles. In the copper core of a cable, these particles are electrons, which are free to move from atom to atom in the lattice of the metal crystal. All metals possess this property, both as solids and as liquids. Some materials, however, do not normally conduct electricity. These are called *insulators* – an example is the plastic cladding around a cable.

A metal – like any substance – can be heated until it becomes a gas. This state of matter consists of separate individual atoms, each with a number of bound electrons. There are no free electrons, so no current can be transmitted. In very extreme conditions a gas can act as a conductor. Inside a fluorescent tube is mercury vapor at very low pressure. If a high voltage is placed across the tube, the gas molecules break down, freeing some electrons, which are able to conduct a small electric current.

Electrolysis
A current is not always made up of electrons. A solution of an *ionic* solid, such as sodium chloride (common salt), can carry an electric current. As it flows, the differently charged elements in the solution are pulled in opposite directions and separate out. This makes *electrolysis* a good technique for refining certain metals, such as sodium and potassium, the ores of which are otherwise very stable.

Sometimes, however, the effects can be unwanted. Most ships' hulls are made of steel, but other parts under water are also made of brass, an alloy of copper. The conducting properties of sea water – a salt solution – allow current to flow between the two dissimilar metals, leading to serious corrosion. This can be avoided by attaching a zinc plate, called a *sacrificial anode*, to the hull. The current now flows to the zinc, which is eventually eaten away instead of the important brass components.

The electrons surrounding an atom are only allowed in certain specific "orbits," each with a specific energy level [A]. In a metal crystal, there are many available levels. Some lower-energy electrons, in the innermost levels, are bound to individual atoms, but there are many with energy high enough to be free of a particular nucleus. They gather in bands of very closely spaced energy levels. These are either the fully occupied valence *bands (pink) which bind the crystal together, or the conduction* bands *(yellow) which enable electrons to travel over the entire crystal, carrying electric current.*

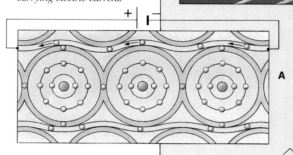

Air is normally a very good electrical insulator. But in a strong-enough electric field, its constituent molecules break down into a conducting plasma. The spectacular result is a flash of lightning (above), with a core at 54,000 °F. This flash was created to test the safety of an electricity pylon.

Most crystalline solids contain both valence bands and conduction bands [B]. The bands in a conductor (left) overlap, making electron jumps between them easy. In insulators (right) the energy gap is too wide for electrons to jump, and so they are unable to carry an electric current.

conduction band

electron

valence band

Molten sodium chloride – common salt – is an ionic liquid made up of charged particles able to conduct an electric current between two electrodes [C]. The sodium ions move to the positive anode, where they gain an electron and become atoms of sodium metal. At the negative cathode, chlorine ions lose an electron to form bubbles of chlorine gas.

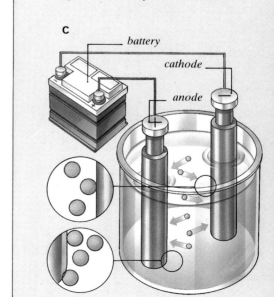

battery

cathode

anode

 sodium

 chlorine

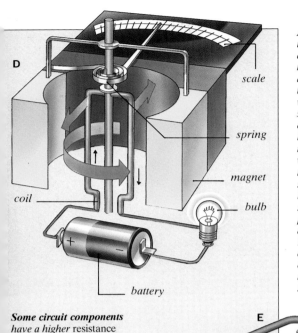

D

scale

spring

magnet

coil

bulb

battery

*A **moving coil ammeter*** *used to measure electric current resembles an electric motor [**D**]. As in a motor, an electric current flows through a coil of wire, which is mounted on a spindle between the poles of a magnet. The interaction between the current and the magnetic field produces a torque or turning force that moves the needle along the scale. The torque is opposed by a spring, attached to the spindle, which applies a progressively greater force as the pointer moves along. Because the force on the coil is proportional to the current in it, the distance the needle travels indicates the size of the current.*

Some circuit components *have a higher resistance than others – current flows through them with more difficulty and produces heat. This effect of the flow of electrons can be compared to the flow of water around a closed pipe-loop [**E**]. Performing the role of a battery is a pump, which*

E

battery

resistor

pump

constriction

provides the energy to push the water molecules – equivalent to electrons – around the circuit. A constriction in the pipe acts similarly to an electrical resistor, making the flow more difficult. So more energy is dissipated in forcing the water through the narrower pipe.

Electric batteries or, more *properly, cells, convert latent chemical energy into electric power. One type is the simple copper/zinc acid cell [**F**]. Unfortunately, the hydrogen gas produced as it works soon masks off the copper anode from the acid, making the cell less effective. Dry cells [**G**] work in a similar way, but use a paste instead of liquid.*

⊖ *electron*

⊖ *sulfate ion*

F

bulb   cathode

anode

G

sulfuric acid

● *zinc atom*

⊕ *zinc ion*

● *hydrogen atom*

⊕ *hydrogen ion*

Current and electrons in wires

An electron carries a tiny quantity of electric charge, but there are huge numbers of them even in the smallest electrical components. About six million billion electrons, together, are said to have a charge of 1 *coulomb*. A typical iron nail contains 10^{25} – ten followed by twenty-four zeroes – electrons, their total negative charge of a few million coulombs being balanced by the positive charge of the nuclei of the iron atoms. Only a small proportion of these electrons are free to flow within the metal.

Current is rate of flow of charge – a current of one ampere being a flow of one coulomb per second. Thus in an automobile headlight circuit, several million billion electrons drift pass any point in the circuit every second.

It is not practical to measure current in terms of the numbers of electrons flowing. So, instead, practical measurement relies on the measurement of a current's magnetic or similar effects.

Resistance and potential difference

It takes energy to drive a current through a conductor, because even the best conductors have some *resistance*. This is measured in units called ohms (Ω): the filament of a flashlight bulb, for example, has a resistance of around 7Ω.

A resistance dissipates electrical energy, producing heat. In most circuits, this heat is not desired and must be vented safely. In other appliances, such as hair dryers, the resistance is deliberately large for maximum heat output. A circuit with electrical resistance needs a constant input of energy to keep the current circulating. This is provided by a battery or generator, which sets up a *potential difference* – measured in volts (V) – across the *resistor*.

In a direct current circuit the relationship between resistance, potential difference, and current is a simple one. A potential difference of 1V across a resistance of 1Ω causes a current of one amp to flow.

Cells and electromotive force

Resistance in a circuit dissipates energy, so some source has to replace it to keep a current flowing. *Cells* (usually inaccurately called batteries) are sources of *electromotive force* (emf), which "pushes" electrons around a circuit. Most harness a chemical reaction between two electrodes, so that a positive charge builds up on one and a negative charge on the other. When a circuit is connected between the terminals electrons are taken up at the positive and given off at the negative, driving the current.

A simple cell can be made from copper and zinc electrodes in dilute sulfuric acid. Under acid attack, zinc enters the solution as positive ions, giving up electrons to leave the plate with a negative charge. Positive hydrogen ions from the acid take up electrons from the copper plate to emerge as bubbles of gas. The copper anode becomes positively charged, so that a lamp lights when connected to the electrodes.

Magnetism and Electromagnetism

The invisible forces guiding compasses and producing electricity

Magnetic fields are all around, produced naturally as well as by electrical cables and machinery. Yet humans cannot directly sense magnetism. It therefore appears to be a curious force, acting across apparently empty space. Magnetic effects are a manifestation, as is electrical behavior, of a fundamental force between charged particles. A moving charged particle creates a magnetic field – atoms are surrounded by a cloud of negatively charged orbiting electrons, and so each atom has its own magnetic field. Currents flowing in a wire are streams of fast-moving electrons, and so produce an associated magnetic field. The combined fields from several wires provide the forces inside electric motors and other machines.

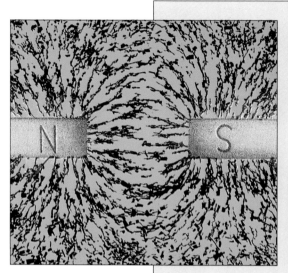

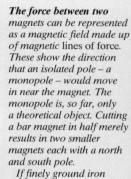

The force between two magnets can be represented as a magnetic field made up of magnetic lines of force. These show the direction that an isolated pole – a monopole – would move in near the magnet. The monopole is, so far, only a theoretical object. Cutting a bar magnet in half merely results in two smaller magnets each with a north and south pole.

If finely ground iron powder (left) is sprinkled over a bar magnet, the particles line up along the lines of force, clearly showing the pattern of the magnetic field.

Compasses and the earth's magnetic field

All magnets consist of a north and a south pole – no isolated north or south (or *monopole*) has yet be found, even in the tiniest of fundamental particles. Like poles always repel each other, whereas dissimilar ones attract. Magnets do not have to be in contact for these forces to be felt: the region surrounding a magnet is called its *magnetic field* – represented by a series of lines starting at the north pole and finishing at the south. At any point the line shows the direction in which a compass placed there would point. Right at the poles, where the lines are closest together, the field is strongest. Its strength diminishes with distance squared – so at twice the distance, the field is four times as weak.

The earth itself has a magnetic field, which makes the north poles of compasses point roughly northward. However, because like poles attract, this must mean that the earth's north pole is in fact a magnetic south pole!

Magnetic materials and domains

Materials can be categorized according to the effect they have on a magnetic field. *Diamagnetic* materials reduce the intensity of a field. This is because the electrons within atoms, each of which behaves like a tiny magnet, arrange themselves so as to oppose the direction of the field. In *paramagnetic* materials the opposite happens – the spinning electrons line up to reinforce the field. Iron, nickel, and cobalt – materials traditionally called "magnetic" – are strictly *ferromagnetic*. They are strongly paramagnetic, retaining their magnetism after the external field is removed.

Steel stays magnetized for much longer than pure iron when an external field is removed. This makes it a good material for permanent magnets, such as compass needles. Pure iron is used in the core of electromagnets to concentrate the field which they produce, but enables them to be switched on and off rapidly.

Pushing the like poles of two permanent bar magnets together produces a repulsive force that increases rapidly with proximity [A]. If one of the magnets is flipped over, an attractive force results between the two opposite poles [B] pulling the magnets toward one another.

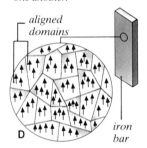

A magnetic compass is a bar magnet that lines up with the field of a much larger magnet – the earth itself [E]. Huge electric currents within the molten iron core of the earth are thought to create this immense magnetic field. The north-south axis of this field is roughly, but not exactly, aligned with the axis round which the earth spins. The actual direction of this magnetic field is not constant: every year the magnetic north pole wanders by a few degrees. Geological records show that several times in the past the field has reversed completely, the poles swapping places.

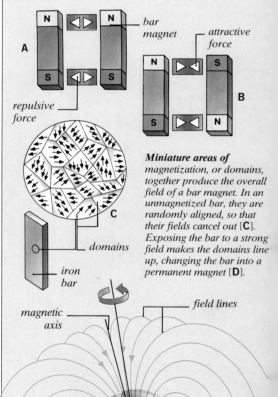

Miniature areas of magnetization, or domains, together produce the overall field of a bar magnet. In an unmagnetized bar, they are randomly aligned, so that their fields cancel out [C]. Exposing the bar to a strong field makes the domains line up, changing the bar into a permanent magnet [D].

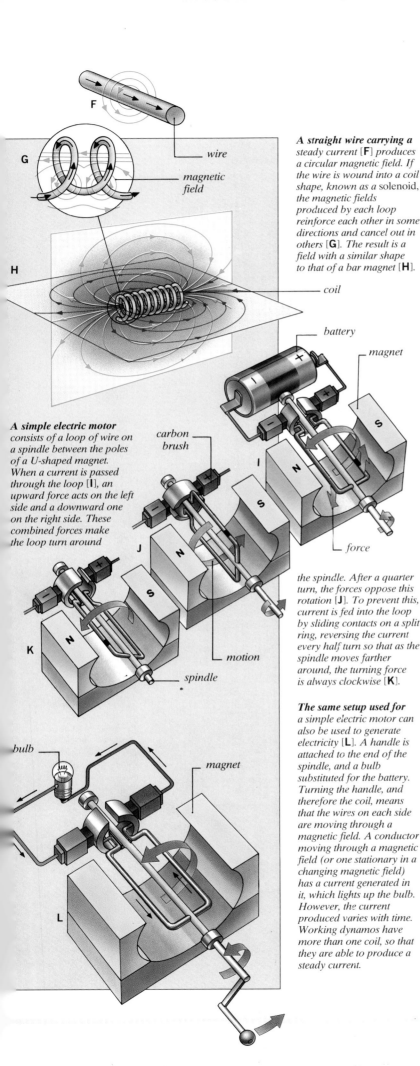

A straight wire carrying a steady current [**F**] produces a circular magnetic field. If the wire is wound into a coil shape, known as a solenoid, the magnetic fields produced by each loop reinforce each other in some directions and cancel out in others [**G**]. The result is a field with a similar shape to that of a bar magnet [**H**].

A simple electric motor consists of a loop of wire on a spindle between the poles of a U-shaped magnet. When a current is passed through the loop [**I**], an upward force acts on the left side and a downward one on the right side. These combined forces make the loop turn around

the spindle. After a quarter turn, the forces oppose this rotation [**J**]. To prevent this, current is fed into the loop by sliding contacts on a split ring, reversing the current every half turn so that as the spindle moves farther around, the turning force is always clockwise [**K**].

The same setup used for a simple electric motor can also be used to generate electricity [**L**]. A handle is attached to the end of the spindle, and a bulb substituted for the battery. Turning the handle, and therefore the coil, means that the wires on each side are moving through a magnetic field. A conductor moving through a magnetic field (or one stationary in a changing magnetic field) has a current generated in it, which lights up the bulb. However, the current produced varies with time. Working dynamos have more than one coil, so that they are able to produce a steady current.

Electromagnetism

Electrons do not only create magnetic fields by spinning in atoms. A flowing stream of billions of electrons – otherwise known as an electric current – also creates a field. First noticed by Michael Faraday almost 200 years ago, this phenomenon is known as *electromagnetism*. The magnetic field around a straight wire is made up of concentric lines of force, all centered on the wire. This shape of field is not particularly useful: however, if the wire is wound into a coil the interaction between the fields produced by each loop produces a field with the same shape as that of a bar magnet.

The important difference is that this is a magnet that can be switched on and off at will. For this reason they are used as hooks in scrapyards to move steel and iron around. Electromagnetic forces are the basis of electric motors, generators, and lifesaving machines such as magnetic resonance scanners.

Electric motors

Just as like poles of two magnets repel, the magnetic field produced by a current-carrying wire and that of a magnet can repel each other. This effect is harnessed in an electric motor, where a loop of wire, through which a steady current is running, is mounted on a spindle inside a magnetic field. The resulting magnetic fields, and therefore forces, on the opposite sides of the coil cause it to spin.

Practical motors are a good deal more complicated. They use multiple coils to ensure that the *torque*, or turning force, produced is approximately constant by only energizing the coil that is perpendicular to the magnetic field. Some are designed to run on constant voltage direct current (d.c.) electricity, others on varying voltage alternating current (a.c.). Compared to gasoline engines, electric motors are very efficient, converting more than 70 percent of energy input into turning power.

Induction, generators and transformers

An electric motor produces spinning motion when a current is fed into it. A generator uses the same components as a motor, but the coil is turned instead, and an electric current is *induced* in it. As the coil revolves, the magnetic field it experiences changes. Any changing magnetic field produces a current in a conductor – for this reason generators work equally well with stationary coils of wire, inside which turns a bar magnet or electromagnet.

If a coil of wire is wound around a ring of iron and an alternating current fed through it, a fluctuating magnetic field is induced in the iron. A second coil wound around the iron core will have a current induced in it due to this changing magnetic field. The size of the induced current is proportional to the relative number of turns in the coils, enabling the principle to be used in *transformers* to step up or step down voltages in power supplies.

Electrical Circuits

How simple circuits are put to use

Throwing a light switch completes an electrical circuit that allows current to flow through the wires connecting a power supply to the filament of a lightbulb. The nature of the components along the length of a circuit, and the way in which they are linked together determines the function that the circuit performs. Real circuits often resemble tangles of wire, and their anonymous components make it difficult to determine function. But by using circuit diagrams, in which standard symbols represent components and their connections, the path of current around a circuit can be readily understood. Circuit diagrams also use symbols to represent the type of power supply used to drive the current through the components.

Voltage and potential difference

An object lifted from the ground to a height of one meter gains energy or *gravitational potential*. If it is released, this energy is given out as it falls from the higher potential to the lower (the ground). The function of a battery in a circuit is to raise the *electrical potential* at one of its terminals relative to the other. Electrical current (a stream of electrons) moves from areas of high electrical potential to areas of low potential, just as an object falls to the ground under the influence of gravity. Therefore if the two terminals of the battery are connected by a length of wire or another conductor, current flows from one terminal to the other. The difference in potential between the battery's two terminals is expressed as its voltage. The current that flows through a circuit when such a potential difference is applied depends on the resistance of the circuit: a large resistance allows a small current to flow through, and vice versa.

Alternating and direct current

The battery in a handheld torch provides a steady potential difference that drives a current through the wire filament of the bulb. This current flows continuously in one direction and is known as direct current (d.c.). The domestic electricity supply, in contrast, is alternating current (a.c.). Here, an electrical potential changing from positive to negative more than 100 times a second causes a current to oscillate back and forth in a circuit at the same rate. These different types of current have different uses: d.c. is needed by most sensitive electronic devices, whereas a.c. is used to transmit power over long distances.

The size of a direct current can be measured by using an *ammeter*, in which the steady current produces a corresponding magnetic field that in turn displaces a pointer needle. Alternating current, which rapidly changes direction, cannot be measured in this way.

Decorative lights link up in a series circuit. Each bulb is a link in a chain between the live and neutral terminals in a plug [**A**]. The current through each bulb is the same, but unfortunately is cut off completely if just a single bulb burns out.

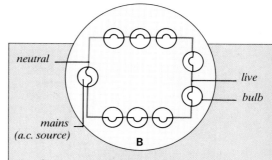

The apparent complexity of an electrical circuit is simplified when expressed in the form of a wiring diagram. This uses standard symbols for items such as batteries, switches and bulbs. The diagram for a set of decorative lights [**B**] shows that the bulbs are wired in series between the terminals of an a.c. supply.

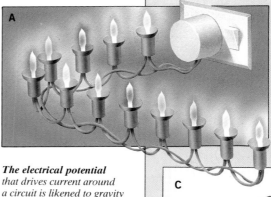

The electrical potential that drives current around a circuit is likened to gravity in this analogy of a series circuit [**C**]. The battery is represented by a lifting device, and resistances (for example, lights in a circuit) by slopes. Balls, equivalent to current, travel around the circuit. Their energy (potential) is raised by a lift. A portion of this energy is lost every time a ball rolls down a slope, until it all has been used up. The value of the voltage is equivalent to the height above ground level. The steeper the incline of the slope, the greater the energy, or potential, that is "dropped" across it.

The fuel gauge in a car is a simple ammeter [**D**]. The deflection of its needle depends on the current flowing through it. The current in the meter circuit is controlled by a variable resistor made of a carbon track along which a metal contact can slide. The position of the contact on the track determines how much current is "tapped off" and allowed through to the meter. The metal contact is in turn linked to a float, the position of which depends on the amount of fuel in the tank. The diagram for the circuit [**E**] shows more clearly how these simple components are wired together.

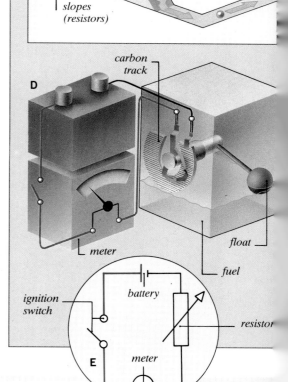

Diodes and a.c. rectification

Radios and other electronic devices are designed to run from batteries that produce a low voltage d.c. supply. If a radio is to be powered from the domestic a.c. supply, power must therefore be converted to lower voltage d.c. This conversion is carried out by diodes, semiconductor devices that allow current to flow through them in only one direction.

An alternating current regularly changes direction. If it is fed through a diode, only the forward-flowing parts of the current can get through. The resulting current flows always in one direction but for just half of the time: the other half is entirely blocked by the diode. This is known as *half-wave rectified direct current.*

A clever arrangement of four diodes allows full-wave rectification – the output of the system is not only always in the same direction, but flows during both halves of the a.c. cycle, whatever the polarity of the supply.

Parallel connection

The domestic electricity supply is at a voltage of either 110 (US) or 230 (Europe) volts. All appliances powered by this domestic supply are designed to work at one of these two voltages.

If domestic appliances were connected to the supply in series – like the bulbs in a set of fairy lights – the voltage across each one would only be a fraction of the supply voltage, and the power supplied to each appliance would be reduced. To avoid this problem, wall and light sockets are wired in *parallel.* This means that the circuit branches into separate lines for each appliance. Each of them, when switched on, is thus connected to the full potential difference, or voltage, of the power supply no matter whether the other appliances are on or off. The current flowing through each of the branches of a parallel circuit is *inversely* proportional to the resistance in that branch – the higher the resistance, the lower the current in the branch.

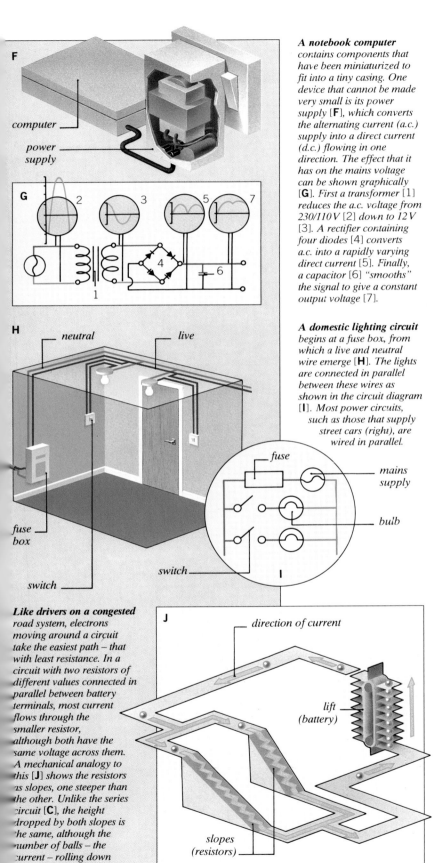

F

computer

power supply

G

2   3   5   7

4

6

1

A notebook computer contains components that have been miniaturized to fit into a tiny casing. One device that cannot be made very small is its power supply [F], which converts the alternating current (a.c.) supply into a direct current (d.c.) flowing in one direction. The effect that it has on the mains voltage can be shown graphically [G]. First a transformer [1] reduces the a.c. voltage from 230/110 V [2] down to 12 V [3]. A rectifier containing four diodes [4] converts a.c. into a rapidly varying direct current [5]. Finally, a capacitor [6] "smooths" the signal to give a constant output voltage [7].

H

neutral   *live*

fuse box

switch

A domestic lighting circuit begins at a fuse box, from which a live and neutral wire emerge [H]. The lights are connected in parallel between these wires as shown in the circuit diagram [I]. Most power circuits, such as those that supply street cars (right), are wired in parallel.

fuse

mains supply

bulb

switch

I

Like drivers on a congested road system, electrons moving around a circuit take the easiest path – that with least resistance. In a circuit with two resistors of different values connected in parallel between battery terminals, most current flows through the smaller resistor, although both have the same voltage across them. A mechanical analogy to this [J] shows the resistors as slopes, one steeper than the other. Unlike the series circuit [C], the height dropped by both slopes is the same, although the number of balls – the current – rolling down each slope is different.

J

direction of current

lift (battery)

slopes (resistors)

Semiconductors
Materials with special electrical properties

Some substances, most notably metals, are good conductors of electricity. Others, such as rubber, will not carry an electrical current, and thus make ideal insulators. A small number of elements and compounds – the *semiconductors* – have properties somewhere between these two extremes. A semiconductor, such as silicon, has current-carrying abilities that are profoundly altered by a change in its temperature or in the amount of light falling on it. These properties can be tailored more by diffusing minute quantities of other elements into their crystal structure, to build up devices such as diodes, transistors, and capacitors – the building blocks of computers and all other electronic devices.

Energy bands
Silicon, germanium, and crystalline compounds like cadmium sulfide are semiconducting materials. The nuclei of their atoms are surrounded by electrons that can occupy only certain energy levels. At low temperatures the electrons gather in the *valence band* – a set of energy levels that are associated with individual atoms. Electrons in this band are unable to move from atom to atom, the process by which an electrical current is carried, so semiconductors, when cold, act as electrical insulators.

But the special property of semiconductors is that they have only a small energy gap between the valence band and the *conduction band* – another set of energy levels in which electrons can move freely. As a semiconductor is heated, increasing numbers of electrons have enough energy to jump into the conduction band. The semiconductor starts to conduct electricity, and its conductivity depends on its temperature.

Moving holes
When an electron in a heated semiconductor jumps to the conduction band, it leaves a vacant space, or *hole*, behind in the valence band. An electron from a neighboring atom's valence band can move in to fill this hole, but will, in turn, leave another hole behind. An electrical current flowing through a semiconductor can therefore be thought of either as the flow of negative electrons in one direction, or the flow of positive holes in the other.

In a cold semiconductor, there are no free electrons or holes available to carry current. However, this situation changes if the semiconductor is *doped* – infused with a few atoms of another element. If this element (the dopant) has "extra" electrons in its valence band, these electrons become free to move throughout the semiconductor. Conversely, if the dopant has a deficit of valence electrons, charge-carrying holes are introduced into the material.

Electrons in the valence band of an atom are tightly bound to that atom [A]. But once they pass into the conduction band, they are free to move from atom to atom and carry an electric current. In metals [1], the two bands overlap, making these materials good conductors. In insulators [2], the two bands are separated by a wide energy gap. In semiconductors [3], the gap is smaller. At low temperatures [4] electrons remain in the valence band. But if heated, they jump to the conduction band, and a current can flow [5]. An electronic thermometer (below) uses this property to measure temperatures.

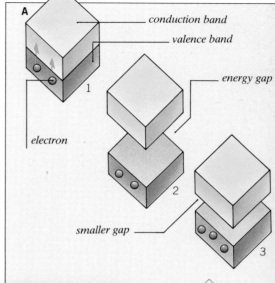

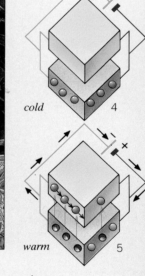

An atom of silicon has four electrons in its valence band [B]. A crystal of pure silicon is a regular array, or lattice, of such atoms [1]. When a crystal is doped with arsenic, some silicon atoms are replaced by arsenic atoms, which have five valence electrons. The "extra" electron moves into the conduction band and can carry current. Such doped semiconductors are called n-type [2]. If a crystal is doped with boron, which has three valence electrons, holes are introduced into the crystal. These can "jump" from atom to atom carrying a current. Semiconductors doped in this way are called p-type [3].

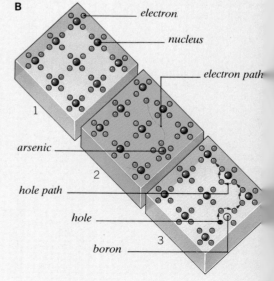

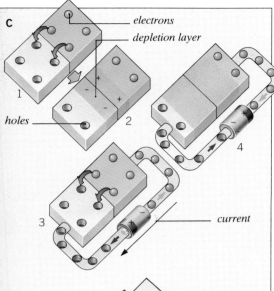

C electrons
depletion layer
1
holes
2
3
4
current

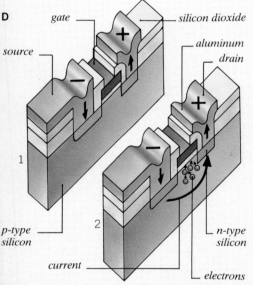

D gate   silicon dioxide
source   aluminum
drain
1
p-type silicon
2
current
n-type silicon
electrons

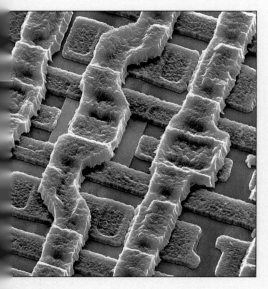

When blocks of p- and n-type semiconductor come into contact [C], free electrons in the n-type jump across to "fill" the positive holes in the p-type [1]. A depletion layer results [2], which blocks the passage of current. Attaching the positive end of a battery – in effect a source of holes – to the p-type allows holes to be replenished as soon as they are filled by electrons [3]. At the same time electrons from the negative terminal take the place of electrons lost from the n-type, and a current flows. Turning the battery around reverses the flow of current carriers and extends the depletion zone, so that no current flows [4].

A silicon chip contains many thousands of field-effect transistors [D], which act like tiny electronic switches. Each has three aluminum terminals. Two connect with the source and drain, islands of n-type silicon on a p-type substrate. Insulating silicon dioxide separates them from the third connection, the gate. In effect the path from source to drain goes via two diodes that stop a current from flowing [1]. But a positive voltage on the gate [2] attracts electrons to the region between the source and drain Because the path is now effectively n-type along its whole length, a current is able to flow.

A much enlarged view of a microprocessor (left) reveals the individual transistors on its surface. In a few years, chips have evolved from units containing hundreds of devices to circuits made up of millions of transistors. Conventionally, ultraviolet light is used to etch components onto the surface of a chip. But the details can be no smaller than the wavelength of the light, setting a limit on the size and density of the components. Using shorter-wavelength X-rays to create the features allows even more components to be crammed in, allowing faster chips, and therefore computers, to be built.

Diodes

In isolation, samples of both p-type and n-type materials can carry electric currents. However, when p- and n-types are brought into contact, interesting and useful properties emerge. The simplest device made in this way is the *diode*, which is a single n-p junction. Connecting a battery one way around to this junction results in the current-carriers' – holes (in the p-type) and electrons (in the n-type) – being sucked away from the junction area. This now acts as an insulator and stops any current from flowing.

When the battery terminals are swapped around, a current can flow because the current-carriers are constantly replaced at the ends as they cancel each other out at the junction. This ability to conduct a current in only one direction is used in a *rectifier*, in which alternating current (with a constantly changing direction) passes through a ring of diodes. The output is direct current that flows in one direction only.

Transistors

A diode is made up of two layers of differently-doped semiconductor, each connected to a terminal, or external electrical contact. A transistor has a third semiconductor layer, and correspondingly, three terminals. The path between two of the terminals can be thought of as a kind of electrical resistance. The value of this resistance, and hence of the current traveling through it, is controlled by a voltage fed to the third, central terminal.

This behavior makes it possible for a transistor to be used in two important ways. A voltage fed to the central terminal can switch on or off a current between the other two. This switching effect is the basis of all digital computers. Larger transistors can be used as amplifiers. The small output from a microphone can be fed to the central electrode. Its fluctuations produce corresponding changes in a much larger current passing between the other two terminals.

Miniaturization

The use of semiconductors has revolutionized modern life. The first transistor was built in 1948. Within a few years, transistors had taken over the role of vacuum tubes, earlier devices that were bulky and produced considerable waste heat. Transistors were first manufactured individually, but were soon produced in hundreds on *integrated circuits*, better known as chips. Other components, such as resistors and capacitors, could also be fashioned on the chip to produce a *microprocessor* containing all the circuitry for a complete computer.

Newer chip designs cram more and more components into ever smaller dimensions. The great numbers – more than a million individual transistors – mean that chips now consume reasonably large amounts of electricity. This electricity is inevitably converted into heat. For this reason, computers contain cooling fans to remove the heat and thus protect the chips.

Electronic Systems
Some basic electrical devices

As dusk falls, the street lights in a city switch on. Despite the changes in weather and length of day that the seasons bring, they seem to "know" just the right time to light up. This is because each is equipped with a light sensor on its upper surface. It is made from a piece of semiconductor – the same material from which transistors and microchips are built. The sensor exploits a very useful property of the semiconductor: it loses resistance as the intensity of light falling on it increases. Combined with a few basic building blocks – transistors, resistors, capacitors – the sensor becomes one of the many electronic devices that make life more convenient, but which are also often taken for granted.

Seeing the light
An automatic camera depends on light sensors to measure the brightness of a scene and set the appropriate aperture and shutter speed. These sensors are made from a *semiconducting* material called gallium arsenide (GaAs). When illuminated, GaAs changes its electrical resistance, and therefore regulates the amount of current flowing in a control circuit. Such a light sensor is called a *light-dependent resistor*.

A second type of semiconductor sensor is the *photodiode*. Like a normal diode this is a junction of two different types of semiconducting material, called p-type and n-type. In the dark the photodiode conducts electricity only in one direction. But when light falls on the junction between the two semiconductors, its behaviour changes so that it can carry electric current in both directions. Light can thus be sensed as a sudden increase in current flowing "the wrong way" through a photodiode in a control circuit.

Simple amplifiers
Transistors have many useful applications. They can be used as electronic switches and also as amplifiers, converting tiny signals into much higher currents. But when used as an amplifier, a transistor is not always accurate. In technical terms, its response is not *linear*. If the size of the input voltage is plotted on a graph against the size of the output voltage, the result is a curve, not a straight line, which tells us that large input voltages are boosted more than weaker ones. This leads to *distortion*.

Distortion is avoided by using a simple integrated circuit called an *operational amplifier* (or op-amp). Although this common component has eight external connecting pins, only five are actually used. Two are for inputs, one for the output, and the remaining two are for its power supply. For inputs of between –0.15 and +0.15 mV, it amplifies linearly, its output being a faithful, but magnified, copy of its input.

A light-dependent resistor or LDR [A] on a street light is a slice of gallium arsenide semiconductor, the resistance of which depends on the amount of incident light. When struck by light, some of the electrons in the semiconductor are promoted to a higher energy (called the conduction band) in which they are free to move and carry current. The changing resistance of the LDR can be shown in a simple circuit. When illuminated [1] the resistance of the LDR is low and a large current flows through the circuit. In the dark [2], its resistance increases and a far smaller current flows.

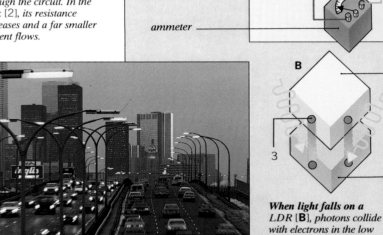

When light falls on a LDR [B], photons collide with electrons in the low energy valence band of the semiconductor [1]. The electrons [3] gain enough energy to jump to the higher-energy conduction band [2]. LDRs are at the heart of sensors which turn street lights on at dusk (left).

The black plastic casing of an operational amplifier [C] conceals a circuit that can convert a tiny input into a copy 100,000 times larger. It only uses 5 of its 8 legs. These are the positive and negative power supplies [1,2], the two inputs [3,4] and the output [5]. The component is represented by a standard symbol [D].

The graph [E] shows that for input values of between –0.15 and +0.15 mV the op-amp behaves in a linear fashion. However, above and below these values the amplifier saturates – it cannot amplify the signal to beyond the maximum possible output voltages of +15 and –15 V.

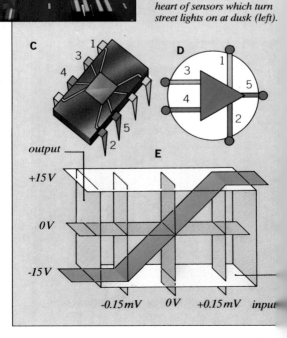

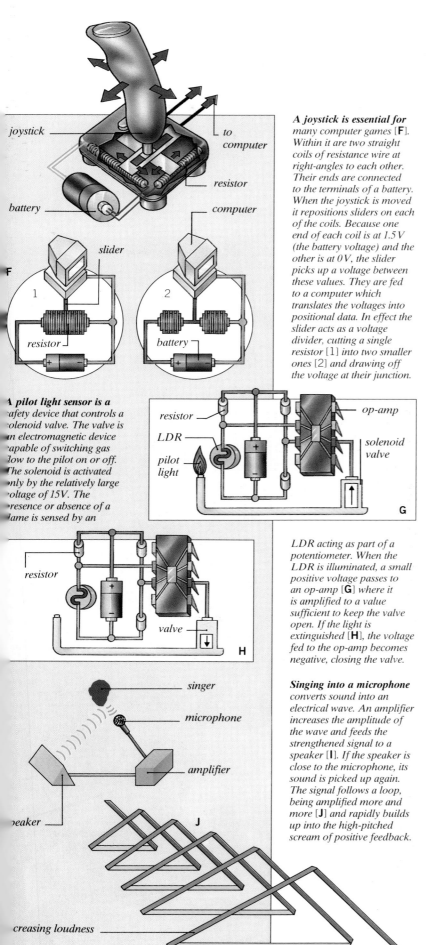

joystick

to computer

resistor

battery

computer

slider

F

1

resistor

2

battery

A joystick is essential for many computer games [F]. Within it are two straight coils of resistance wire at right-angles to each other. Their ends are connected to the terminals of a battery. When the joystick is moved it repositions sliders on each of the coils. Because one end of each coil is at 1.5 V (the battery voltage) and the other is at 0 V, the slider picks up a voltage between these values. They are fed to a computer which translates the voltages into positional data. In effect the slider acts as a voltage divider, cutting a single resistor [1] into two smaller ones [2] and drawing off the voltage at their junction.

A pilot light sensor is a safety device that controls a solenoid valve. The valve is an electromagnetic device capable of switching gas flow to the pilot on or off. The solenoid is activated only by the relatively large voltage of 15 V. The presence or absence of a flame is sensed by an

resistor

LDR

pilot light

op-amp

solenoid valve

G

resistor

valve

H

singer

microphone

amplifier

speaker

J

creasing loudness

LDR acting as part of a potentiometer. When the LDR is illuminated, a small positive voltage passes to an op-amp [G] where it is amplified to a value sufficient to keep the valve open. If the light is extinguished [H], the voltage fed to the op-amp becomes negative, closing the valve.

Singing into a microphone converts sound into an electrical wave. An amplifier increases the amplitude of the wave and feeds the strengthened signal to a speaker [I]. If the speaker is close to the microphone, its sound is picked up again. The signal follows a loop, being amplified more and more [J] and rapidly builds up into the high-pitched scream of positive feedback.

Dropping volts

The volume control on a radio is a variable resistance wired to act as a *potentiometer*. It contains a carbon track, which has high electrical resistance. A sliding metal contact can be moved to any point along the track by turning the volume control knob. If each end of the track is connected to the terminals of a 1.5 V battery, one end is at an electrical potential of 1.5 V, and the other is at a potential of 0 V. Points in between are at intermediate potentials, or voltages. Moving the sliding contact along the track allows different voltages to be "tapped off" and fed into the radio's amplifier circuits, determining the degree of amplification, and therefore volume.

Potentiometers (or "pots" as they are commonly known) are found in many different devices. A mixing desk in a sound-recording studio can contain almost one thousand pots, each adjusting a different aspect of the sound.

Safety circuits

The components described above – light dependent resistor (LDR), potentiometer and operational amplifier (op-amp) – can be combined to make a simple control circuit used to monitor the pilot light in a gas boiler. It functions by sensing the light emitted by the burning gas. If the light is inadvertently extinguished, a solenoid valve is automatically closed, shutting off the gas supply and so preventing a potentially dangerous leak.

In this circuit, the LDR is wired in series with a fixed resistor of high value and a battery to create a potentiometer. When light falls on the LDR, its resistance is low. The electrical potential at the junction between the LDR and resistor is therefore also low. But if no light falls on the LDR, its resistance increases, and a higher voltage is tapped off at the junction. This voltage difference is amplified by the op-amp and used to close the solenoid valve.

Automatic control devices

An electric heater converts an input of electricity into an output of heat. It contains a thermostat – a temperature sensitive switch. As heat is produced by the device the temperature rises to a threshold value at which the thermostat switches off the power supply. When temperature drops once again below this value, the power supply is switched on. In effect, the thermostat uses the output of the heater to control the electrical input. This type of control system, the purpose of which is to maintain a stable set of conditions (here a steady temperature), is called a *negative feedback loop*. Many biological systems, such as human temperature, are also regulated by negative feedback.

If the output of a system serves to reinforce the input, the system becomes unstable. This is called *positive feedback*, and is responsible for the whistle heard when a microphone picks up its own amplified signal from a loudspeaker.

Logic and Digital Devices

1s, 0s, and the circuits that add them together

A computer is, in essence, a machine that adds and subtracts numbers. The numbers represent a bewildering variety of different things, from points of color on a screen to data locations in the memory, and at any one time many millions of different numbers are being processed. The computer is able to do this because the numbers are binary, made up of only 1s and 0s, and very simply represented by "on" and "off" voltages.

The basic circuits that manipulate these voltages are called logic gates and can be as simple as a single transistor. A combination of only a few logic gates forms a circuit that can add two binary numbers – the process that is the basis of the way a computer works.

Number systems

The figures that we use every day are decimal numbers, a term derived from *decimus*, the Latin for "in tenths." Each number in itself is a kind of sum – the right-hand figure gives the amount of single units, the figure next to it the number of tens (10^1), followed by the hundreds (10^2), and so on, each further column representing the next greater power of 10. A number system needs as many different figures as the number it is based upon, so the decimal system has 10 digits, 0 to 9.

A computer, however, deals with voltages, and its circuits have only to distinguish between on and off states. For this reason computers use *binary* or *base 2* numbers, in which each successive column represents a greater power of two – 1, 2, 4, 8, etc. Only two figures are needed – 1, or "on," and 0, or "off." Every input to the computer has to be converted into a binary number before being stored and manipulated.

On in, off out

On a single silicon chip there are millions of *field-effect transistors* (FETs). These have three connections: a current can flow between two of them, the *source* and the *drain*. A voltage on the third connection, the *gate*, controls the flow of current between the other two. One type of FET will only pass a current if the gate voltage is zero or negative, so if a positive pulse, representing a 1, is applied to the gate, the current output from the drain is a 0. If there is no gate voltage – a 0 – a current is allowed to flow, representing a 1.

Binary digits can be compared and manipulated using the rules of *Boolean algebra*, named after the 19th-century mathematician, George Boole. In Boolean algebra, certain words, such as NOT, AND, and NOR describe given logical operations. For example, the FET described above, which always gives an output opposite to its input, is (logically) called a NOT gate.

Counting in the decimal system can be represented by placing blocks in shelves [B]. The first 9 blocks fit into the right-hand shelf, but to count 10 a block has to be placed in the next shelf along. The right-hand shelf is then refilled one block at a time as counting continues. The contents of the shelves represent a number just as a decimal figure gives the number of units, tens, hundreds, etc. in the number.

A similar system can show binary numbers [A]. It has room for only one block in each shelf, the contents of which would represent a 1 or a 0 in a binary number. Successive shelves give the number of each successive power of 2 present in the number – 2s, 4s, 8s, etc.

An abacus (right) is an ancient adding machine that uses beads to represent numbers. Instead of being based on the number 10, like decimal numbers, or the base 2 of the binary system, it counts in powers of 5.

Computers use binary numbers to communicate [C]. Each letter in the alphabet has an 8 digit ASCII code (American Standard Code for Information Interchange). There are 256 possible combinations of 8 1s and 0s, leaving many "spare" codes to represent special symbols and useful characters, such as accented letters.

A field-effect transistor can act as a NOT gate, taking a binary input (0 or 1) and outputting the opposite value (respectively 1 or 0) [D]. It has 3 terminals: between two of these, the source *and the* drain, *a current can flow. A voltage on the third terminal, the* gate, *governs the size of the current that passes. Current is fed to the source as a series of pulses. A positive voltage on the gate [1] (representing "on," or binary 1) stops the current passing to the drain, to give a 0. A negative gate voltage, standing for a 0, allows the pulse to pass through and out of the drain, to represent a 1 [2].*

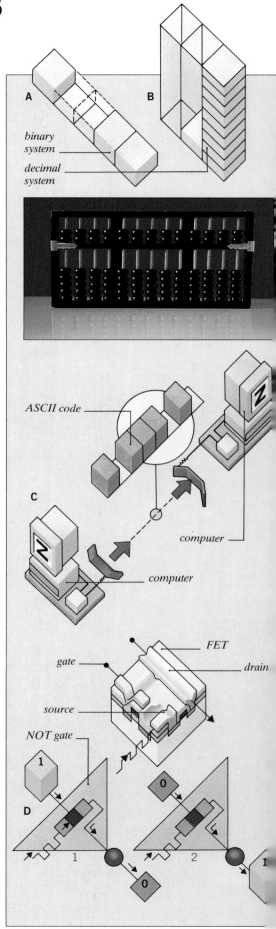

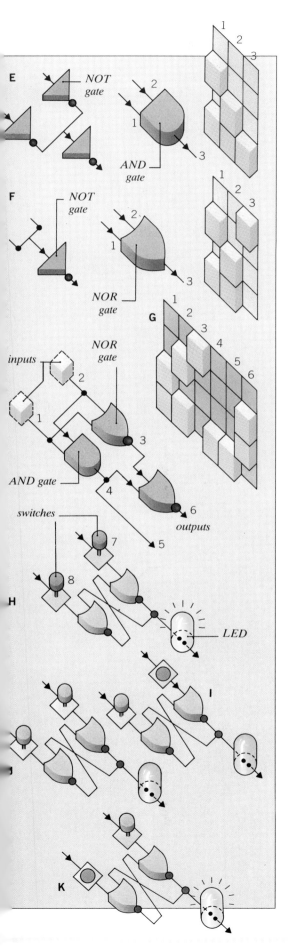

Three NOT gates can be linked up to form an AND gate [E]. Its output is only a 1 if both of its inputs are also 1s. Any other combination of inputs yields an output of 0.

It is easiest to visualize the different possible inputs and outputs for any logic gate as a truth table. In this, a protruding "tile" stands for a 1 and a flush area for a 0. The columns [1] and [2] give the possible inputs, while [3] shows the output. The truth table for a NOR gate [F] shows that its output is the inverse of that of an AND gate. This logical device gives an output of 1 only when both of its inputs are 0s.

A half adder is a simple combination of one AND and two NOR gates [G]. It is a useful circuit because it can add together two inputs (1s or 0s) and give the answer as two-digit binary code. For example, if inputs [1] and [2] are both 1, the output of the AND gate [4] will be a 1, while the output from the first NOR gate [3] is a 0. Both outputs are then fed to the second NOR gate, which gives an output [6] of 0 – the units figure of the binary code. The output of the AND gate gives the second digit of the binary code – a 1. So in binary code, adding 1 and 1 gives the answer 10 – equivalent to 2 in the decimal system.

A bistable, more popularly called a flip-flop, is a useful circuit made of two linked NOR gates [H]. Each gate's output is fed to one of the other gate's two inputs. The remaining two inputs link to two switches [7] and [8]. The act of pressing down and releasing a switch produces a pulse of current representing a 1. An illuminated light-emitting diode (LED) shows when the output of the bistable is a 1. At first the LED is on. Pressing switch [7] turns it off [I]. When the switch is released the LED remains unlit. [J]. But pressing down switch [8] "flip-flops" the circuit, turning the LED on once again [K].

Logical output

Transistors working as logic gates according to the rules of Boolean algebra can be used in different ways and combinations to perform different logical operations. A simple NOT gate, for example, gives an output of 1 for an input of 0 (and vice versa). But if two inputs are combined and fed into a single NOT gate, it becomes a NOR gate, in which the output is 1 only when *both* inputs are 0. Similarly, combining three NOT gates in a simple circuit gives an AND gate. Here, the output is a 1 only when *both* inputs are 1s. If either input is a 0, the output becomes 0. Each of these logic gates is represented by its own geometric symbol (see illustration). The way that they function can be expressed in a *truth table*, which lists all the possible combinations of inputs and the outputs that they yield. These outputs can be combined and processed through other logic gates so that a complex table of possibilities builds up.

Adding up

A computer works by manipulating numbers, which can represent many things, from points of light on a screen to the position of a robot arm. The most simple calculation that a computer needs to perform is adding two single-digit binary numbers. Three logic gates, one AND and two NORs, can be linked together to perform this calculation. The whole device gives two outputs – a *unit* figure and a *carry* figure. For example, if two 1s are fed into the adder, the yield is a "0" unit figure and a "1" for the carry digit. Together these make up 10, the binary representation of the decimal number 2. In a computer, much larger numbers can be added by feeding the unit and carry outputs to thousands of other adder circuits.

Similar circuits can be built that subtract one input from the other. With these simple building blocks much more complicated calculations can be performed.

Electronic flip-flops

Magnetic media such as disks and tapes can provide long-term storage of digital information in a computer. But the computer also needs a fast-acting *memory* to which it can rapidly refer. This type of memory is called *random access memory*, or RAM, and is based on electronic rather than magnetic storage of information. RAM exists in two different forms.

One type, called *dynamic* RAM, uses capacitors to store tiny electrical charges corresponding to digital 1s and 0s. Despite its name, this type of memory is relatively slow. *Static* RAM is a far faster, though more expensive, type of memory. It contains *bistable* circuits (or "flip-flops"), which make use of clever arrangement of two NOR gates. These create a circuit with a constant output that does not change until triggered to do so. Thus a 1 or a 0 pulse can be stored in the bistable and read off almost at the instant it is needed.

The Nature of a Wave

Carrying energy from place to place

Light waves from the sun transport energy across empty space to drive many of the processes on the earth. And water waves created by a storm make boats in a distant harbor bob up and down. Although water and light waves might seem to be very different phenomena, they actually have much in common. Both carry energy from place to place, but do so without transferring any material.

Different types of wave are characterized by their origins, the way in which they travel, and the speeds at which they move. But despite the diversity of wave types, all behave in basically similar ways. They can be reflected and bent (refracted), or otherwise manipulated for use in a variety of instruments from lenses to radar guns.

Waves and earthquakes

As an earthquake occurs, a huge amount of energy is released. This energy is transported through the earth by waves, causing detectable shocks at distant locations. The waves are oscillations in the medium (rock) that are passed from particle to particle. Although the particles oscillate, carrying the energy of the earthquake, the rock itself is not carried from place to place.

The energy of the earthquake is carried by two distinct types of wave. *Primary*, or P, waves are transmitted by the alternate compression and stretching of the rock (the medium). These waves are called *longitudinal*, because the oscillation of the rock is in the same direction as the wave's progress. *Secondary*, or S, waves move in a different way – in the form of side-to-side oscillations of the rock. These are called *transverse* waves. The longitudinal P waves move at a speed of 3 miles/s, while the transverse S waves move more slowly, at 1.8 miles/s.

Reflection and refraction

When a ray of light strikes a silvered surface, it is reflected according to a simple geometric rule: the angle at which the ray strikes the surface is exactly matched by the angle at which it rebounds. Not just *electromagnetic* waves (ones made up of oscillating electric and magnetic fields), like light and radio, are reflected: *mechanical* waves (ones composed of oscillating particles), such as sound, water, and earthquake waves behave in a similar way. This is what allows echoes to be heard.

Another property common to all waves is that they can be bent, or refracted. It is clearly visible when a ray of light passes through a lens, and is deflected from its straight path. This happens when any wave passes from a medium of one density into a medium of different density. Sound can be refracted in the same way as light: sound (acoustic) lenses can even be made from large blocks of wax.

Transverse waves are produced when an oscillating plank dips into and out of a water tank [A]. The water molecules move vertically, at right-angles to the direction of wave travel. The distance between successive peaks is called the wavelength, *and the height of each the* amplitude.

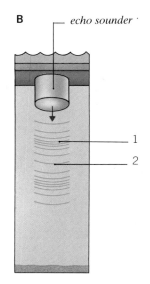

An echo-sounder produces longitudinal pressure waves [B] by moving water molecules back and forth. This creates compressions [1] and rarefactions [2], areas of high and low pressure. Measuring the time the wave takes to reflect from the sea bed indicates the depth of the water [C].

Waves are bent or refracted when they pass from one medium into another. The effect can clearly be seen when waves travel from deep water over an oblique step into a shallower area [D]. The waves move more quickly in the shallower water: this means that the crests of the waves tend to bunch up more closely together, and the wavelength decreases. Because the waves hit the deep/shallow boundary at an angle, wavelength is shortened at one end of the wavefront before the other. This has the effect of changing the direction in which the waves move. This process is known as refraction.

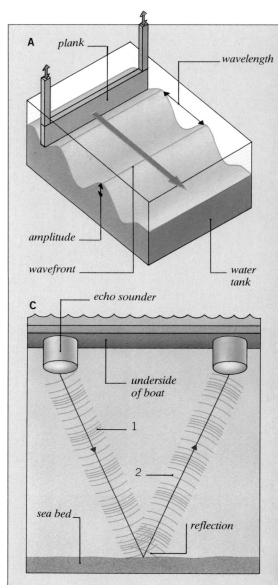

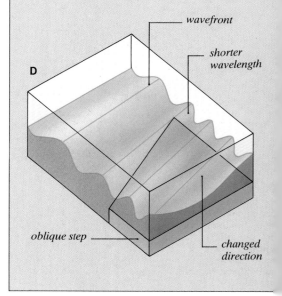

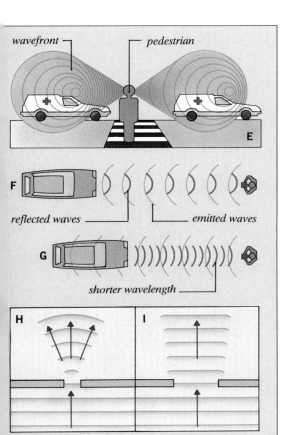

wavefront — pedestrian

E

F

reflected waves — emitted waves

G

shorter wavelength

H   I

J

As an ambulance passes, a pedestrian hears the Doppler effect *as a sudden drop in the pitch of its siren note* [**E**]. *As the ambulance approaches, sound waves given out by its siren bunch up, diminishing in wavelength and raising the frequency, or pitch, of the sound. As soon as the ambulance has passed the observer, the wavelength is lengthened and the pitch of the siren suddenly drops.*

The same principle is used in police radar guns. Radar waves reflected from a stationary vehicle [**F**] *have the same frequency as the emitted waves. If the vehicle is moving* [**G**] *their frequency becomes higher.*

When parallel wavefronts *pass through an opening of about the same size as their wavelength* [**H**], *they are diffracted – spreading out into a series of concentric waves. The effect is less marked when the opening is much wider* [**I**]. *Thus sound can be heard around a corner, even though there is no direct path between source and listener. Similar effects are observed with water waves. The tides generate waves in the oceans that may be hundreds of yards long. If these pass through a narrow land constriction, a diffraction pattern may be detectable from the air or from a satellite (left).*

The principle of *superposition is used to tune a guitar* [**J**]. *One string is plucked to give a known note (wave 1). The guitarist then tries to reproduce the same note on a different string* (wave 2). *If the frequency of the two waves differs slightly, their peaks and troughs coincide only periodically, producing a louder, combined wave. This is audible as a regular series of "beats." The guitar is correctly tuned only when the beats can no longer be heard.*

wave 1

wave 2

combined wave

----- envelope

The Doppler effect

Every child that mimics the sound of a passing police car is unknowingly imitating the Doppler effect. When the police car is not moving, its siren has a steady pitch because it emits sound as a series of equally-spaced wavefronts. When the car moves forward, each successive wavefront "catches up" slightly with the previous one, so that the waves are more closely spaced when they reach an observer. This shorter wavelength translates into a higher frequency, which is heard as a higher pitch than when the car was at rest.

The Doppler effect does not only apply to sound waves. Electromagnetic waves, such as light, show similar effects. Gases within stars such as the sun emit light at certain well-defined wavelengths. If a star is traveling away from us very quickly, its light appears to be *red-shifted* because the Doppler effect lends blue light the longer wavelengths of red light.

Diffraction and the limits of magnification

The effects of diffraction – when a wave spreads out after passing a gap or obstacle of a similar size to its wavelength – are usually not very noticeable. But diffraction is the mechanism by which sound be heard around corners, and it also provides confirmation of the theory that light is made up of waves.

Light waves have wavelengths of less than a millionth of a meter, and so create few everyday diffraction effects. But to a scientist using a high power optical microscope, diffraction is an intrusive phenomenon. Objects that are of a similar size to the wavelength of light produce strong diffraction effects, blurring edges and producing visible bright and dark "fringes." For the microscopist, these effects can be a barrier to observation. For this reason, very small objects are often examined in electron microscopes, which use ultra-high frequency electron waves, rather than light to "see" with.

Superposition, interference and radio

If two waves arrive at the same place at the same time, their *amplitudes* – or heights – are added together. This is the principle of *superposition*. On water, two peaks together produce a higher peak, two troughs produce a deeper trough, and if a peak and a trough of the same size coincide they cancel one another out.

Radio waves behave in a similar way. Two independent FM transmitters emitting waves of the same frequency and strength produce fixed patterns of reinforcement and cancellation across space. When such a pattern is set up, signal strength can vary from very strong to inaudible over a small area, making for highly patchy reception, particularly for users of car radios. For this reason, a single radio station is often broadcast on slightly different frequencies in different areas. Interference is avoided because a radio receiver can pick up only one of the signals.

Sound and Hearing
Audible energy

The noise of a space shuttle launch can shatter windows in buildings more than one mile away in all directions. This tremendous energy is transmitted as sound waves – minute vibrations of the molecules in the air and the earth. When carefully controlled, sound can carry detailed information, allowing a bat to locate a flying insect in pitch darkness, and conveying all the nuances of human language. The energy emitted when a word is spoken is very small, and spreads out as it moves away from its source. The human voice travels only a few hundred feet before its energy is dissipated: but its importance in communication has led to the manipulation of sound by electronics and skillful architecture.

The nature of sound waves

Any vibrating object – such as a loudspeaker diaphragm or human vocal cord – produces sound. As the vibrating membrane moves outward it compresses the air directly in front of it; and as it moves back inward it *rarefies* the air, creating a partial vacuum. The compressions and rarefactions are transmitted in a line from one air molecule to the next, carrying the sound energy away from its source. The distance between adjacent compressions gives the *wavelength* of the sound emitted, and the number of compressions in a second gives its *frequency*, measured in hertz (Hz).

Sound is transmitted by the vibration of any medium – solid, liquid, or gas – so it cannot pass through a vacuum. It travels more quickly through liquids and solids because their particles are closer together and more tightly bound. The speed of sound in air is about 745 mph; in water 3350 mph; in steel 11,180 mph.

Human hearing

As sound waves enter the ear, pressure differences between successive ridges of compression and rarefaction set the eardrum – a taut "drumskin" with an area of around 50 mm² – vibrating. Attached to the inner side of the eardrum is a linked group of three tiny bones – the auditory ossicles – that amplify the vibrations and transmit them to the fluid-filled inner ear or *cochlea*. The cochlea is essentially a tapering, fluid-filled tube rolled up upon itself to resemble a spiral. Incoming sound causes changes in the pressure of the fluid within the tube: as pressure increases, tiny hairs on the wall of the tube are pressed against an adjacent membrane. Receptor cells attached to these fine sensory hairs detect these contacts. For each contact a cell sends a nervous impulse to the brain. The brain registers these impulses as sound: the more impulses arriving at a time, the louder the sound heard.

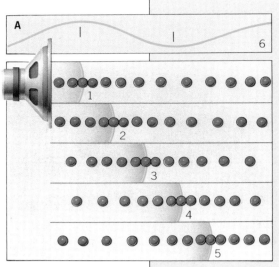

A *vibrating loudspeaker* [**A**] *compresses and rarefies air molecules. As these knock into their neighbors they pass on their vibrational energy. In this way sound is transmitted through the air* [1–5]. *The louder the sound, the more each molecule is displaced: but even the loudest noises displace a molecule by less than 0.5 mm from its normal resting position.*

Sound is transmitted by longitudinal *waves (where particles vibrate parallel to the direction of sound movement), but is often represented as a* transverse *wave, like a water wave* [6].

Just as light is bent or refracted *when it passes into a dense medium, such as glass or water, sound is bent when the medium it moves through changes in density. A frozen lake* [**B**], *for example, has cold, dense air immediately above it, while the air a little higher is warmer and less dense. Sound is refracted downward toward the ice, and then reflected from its surface. Under such conditions, sound can "bounce" across a lake, making even faint speech clearly audible over long distances. The degree of sound refraction depends on the difference in density between the layers of air.*

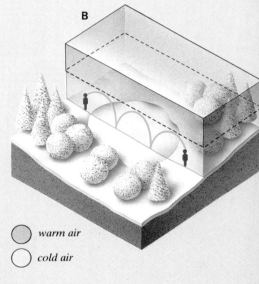

○ warm air

○ cold air

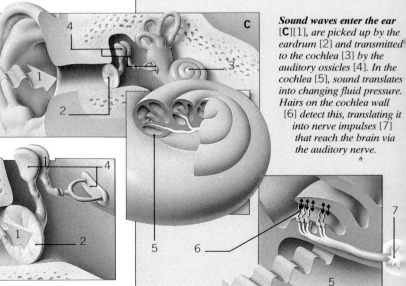

Sound waves enter the ear [**C**][1], *are picked up by the eardrum* [2] *and transmitted to the cochlea* [3] *by the auditory ossicles* [4]. *In the cochlea* [5], *sound translates into changing fluid pressure. Hairs on the cochlea wall* [6] *detect this, translating it into nerve impulses* [7] *that reach the brain via the auditory nerve.*

Hearing range

The human ear is incredibly sensitive, capable of picking up a wide range of frequencies, from low notes of 20 Hz up to high pitch sounds of 20,000 Hz – a range of more than ten octaves. But the range of audible *intensities* is even bigger. Intensity is a measure of amount of energy carried by a sound wave. A very loud sound, such as a rocket launch, carries more than a million million times more energy than the sound of a mosquito buzzing. The decibel (dB) scale of loudness is a convenient way of expressing this huge range of intensities. Zero dB is the threshold of hearing, and every *tenfold* increase in sound intensity is equivalent to an extra 10 dB. The buzzing mosquito produces sound with an intensity of 50 dB; the rocket launch is measured at around 1170 dB. While intensity level can be objectively measured in decibels, *loudness* is a subjective term that depends on the senses of the listener.

The sound of music

When plucked, a stretched string vibrates at a frequency determined by its tension, its length and its mass. The vibrating string, however, produces little sound. This is because the string is narrow, and can only pass on its vibration to a small volume of air. In stringed instruments, such as the violin and piano, the sound is therefore amplified by passing the string over a support connected to a wide wooden plate, or *soundboard*, which is forced to vibrate, and which disturbs a far larger mass of air.

The sound produced by a violin is easily distinguished from that of a piano, even when the two instruments play the same note. This is because the vibration of the violin string is far from simple: the pattern of vibration can be broken down into a number of components or *harmonics* particular to the instrument. It is the combination of harmonics that helps give each instrument its distinctive timbre.

Acoustics – designing for sound

In a modern concert hall, only about a twelfth of the sound that reaches the ears of the audience travels directly from the orchestra. The rest comes by way of reflections from the floors, walls, ceilings, and so on. Low-frequency sounds reflect up to 50 times before the surfaces absorb the last of their energy, whereas high-frequency sounds are damped more quickly.

The time taken for reflected sound to fade away – the *reverberation time* – is the critical factor in determining the quality of sound in a room or concert hall. Organ music requires a long reverberation time, and is suited to large churches where hard surfaces keep the sound reflecting. For speech, reverberation time must be kept short or one word will merge into the next – a common problem for train station announcers. In the past, concert halls were built without much consideration for acoustics, but today sound quality is a primary consideration.

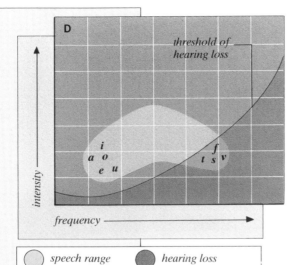

Most people's hearing deteriorates with age because the sensory hairs in the cochlea lose some of their flexibility. Such hearing loss affects high frequencies more than lower pitch sounds. The graph [**D**] shows the normal frequency and intensity range of human hearing. The central "bubble" represents the range of human speech. Many consonants in human speech are relatively high-frequency sounds (about 4000 Hz). If these cannot be heard, then even though the deeper vowel sounds remain audible, speech can be hard to understand.

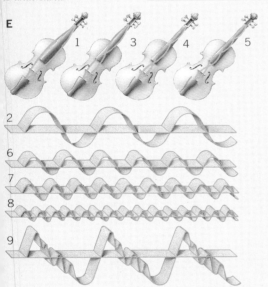

When plucked, a string of a violin [**E**] is made to vibrate and produce sound. Because it is fixed at both ends, the simplest mode of vibration involves the string moving most at its midpoint [1]. Thus simple form of vibration produces a sound at the string's fundamental frequency [2]. At the same time, the string vibrates in more complex ways [3,4,5]. These modes of vibration, produce discrete higher-frequency sounds known as harmonics [6,7,8]. The fundamental frequency is heard together with all the string's harmonics [9], giving the violin its particular, distinctive sound quality.

The acoustics of a modern concert hall (right) can be adapted to suit a particular type of performance. The sound quality of the room is altered by moving banners that reflect sound and by raising or lowering an acoustic canopy above the stage. Reverberation time – the time taken for a sound to die down – can be adjusted by opening doors to reverberation chambers. For the full sound of an orchestra the acoustic canopy is raised and reverberation chambers opened. For a solo performance the banners are lowered to give sound a more intimate quality.

Electromagnetic Waves

A spectrum of radiation

Radio, microwaves, light, and X-rays are all forms of electromagnetic radiation – fast-moving waves that transfer energy from one place to another. Some forms of this radiation are fundamental to life – for example, light and infrared waves from the sun bathe our planet in light and warmth – while others, such as gamma rays, are extremely hazardous.

Electromagnetic waves have much in common with sound waves and ocean waves: all are traveling disturbances that can be reflected and refracted (bent). But unlike these mechanical waves, which only travel through solids, liquids, or gases, electromagnetic waves can move through a vacuum, although they are also slowed down when passing through a medium.

The nature of light

The sense of sight provides us with plenty of information about what is happening in the outside world. Yet the human eye is sensitive only to light, which is just a small fraction of the full range of radiation given out and reflected by objects. The existence of radiations similar in nature to light, but undetectable by the human senses, was predicted by the Scottish physicist James Clerk Maxwell in the mid 19th century.

Maxwell also deduced that light (and other forms of radiation) travel in the form of *electromagnetic waves*. Such waves are created by a moving electric charge when it changes its speed or direction: this is true of any type of charge – a bolt of lightning, for example, produces electromagnetic waves, as does an electric current surging up and down in the antenna of a radio transmitter. The wave is formed because the moving charge produces a changing electric field. The changing electric field in turn produces a changing magnetic field. And as the magnetic field changes, it produces a changing electric field. The mutually sustaining fields make up the electromagnetic wave, which travels away from its source in a straight line at a speed of 300,000,000 m/s – equivalent to more than 1 billion km/h.

Electromagnetic radiation can be *refracted* and *diffracted*, behaving in a way similar to sound and water waves. But some aspects of its behavior can only be explained by thinking of it as a stream of discrete particles or *quanta* (a quantum of light is called a *photon*) rather than continuous waves. These ideas about the nature of electromagnetic radiation cannot easily be reconciled, and it is perhaps most useful to say that although this radiation travels as a wave, its *energy* travels in discrete "packets."

This idea applies not only to waves, which are found to have particle properties, but also to particles, such as electrons, that under certain conditions behave like waves.

Electromagnetic waves move in straight lines unless reflected or refracted. As distance from the source increases, the intensity of the radiation falls according to an inverse square relationship [A] – every doubling of the distance from the source results in a fourfold drop in intensity of the radiation. This applies equally to all types of electromagnetic waves.

The mutually sustaining electric (blue) and magnetic (red) fields of an electromagnetic wave [B] have equal energies. The fields fluctuate at right-angles to one another, and also at right angles to the direction of wave travel.

Radio waves, X-rays, and light are, despite appearances, very similar phenomena. All are produced when charged particles accelerate or decelerate. In a radio transmitter, negatively charged electrons oscillate up and down an antenna [C]. As each electron moves

[D] it generates a varying electric and magnetic field, which moves away from the antenna [1–5]. In a vacuum-filled X-ray tube [E], electrons [6] are accelerated to high speed away from a negatively charged electrode [7], and then "crash" into a positively charged metal target [8]. Their rapid deceleration produces highly energetic X-rays.

Light is given out [F] when electrons within atoms are excited to high energy states [9] by, for example, applying heat. When they fall back to their original energy level [10] they accelerate, emitting their excess energy as light [11].

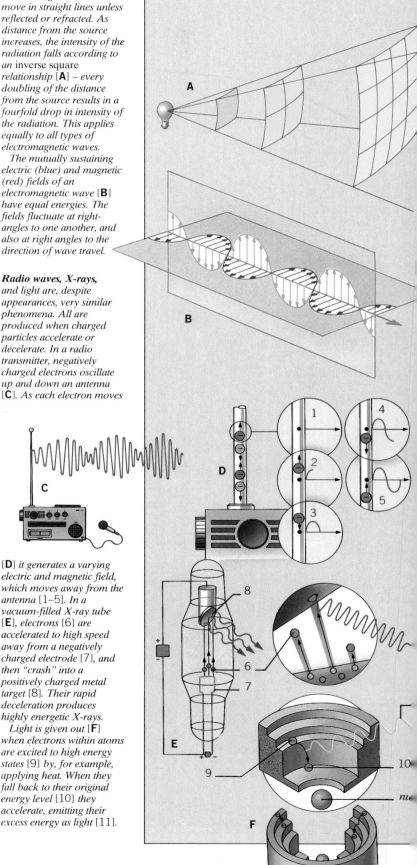

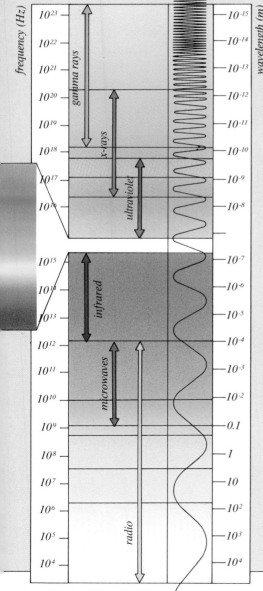

A triangular prism splits white light (left) into its component colors because the shorter-wavelength components (blue and violet) are bent or refracted more than the longer wavelengths (corresponding to red and orange).

Different materials refract light (and other electromagnetic radiations) by differing amounts. For example, light is bent through a greater angle when it enters a diamond than when it enters glass. The degree to which a material bends light is called its refractive index, and it is the high refractive index of a diamond that gives a cut stone its sparkle.

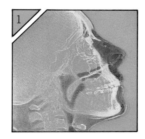

High-speed waves

All types of electromagnetic waves travel at the speed of light – 300,000,000 m/s, far faster than any type of mechanical wave. However, they only move at this speed when traveling through a vacuum: when they move through a *medium* – a solid, liquid, or gas – they are slowed down. A result of this change in speed is that electromagnetic rays are bent or *refracted* when they pass from one medium to another. It is this property that allows light to be focused by a glass lens.

Some media absorb selected types of radiation while allowing other types to pass. Glass, for example, transmits light but absorbs both ultraviolet and infrared radiation. This is why it is impossible to get a suntan from ultraviolet rays behind glass. A greenhouse heats up because light can enter and warm up the plants within, but the infrared radiation emitted by warm plants is blocked by the glass.

The electromagnetic spectrum

Electromagnetic waves can be described using the same concepts that apply to other waves – frequency, wavelength, amplitude, and speed. Just as many different frequencies of mechanical waves can be produced (a piano, for example, produces sound ranging in frequency from 30 to 3000 cycles per second or *hertz* (Hz) so there is a range of frequencies for electromagnetic waves. This extends from radio waves with frequencies of 1000 Hz to gamma rays with frequencies in excess of one billion billion Hz. Each frequency has a corresponding wavelength, the two always obeying the relationship, frequency x wavelength = the speed of light.

The range of frequencies and corresponding wavelengths is shown here in the form of an electromagnetic spectrum. The radio waves at one extreme have wavelengths of more than 1 km, while gamma rays at the other extreme have wavelengths as low as 10^{-16} m.

The energy carried by an electromagnetic wave is directly related to its frequency. The higher the frequency, the greater the energy. X-rays, for example, have enough energy to pass through the soft tissue of the human body, and can therefore be used in medical imaging [1]. Invisible ultraviolet (UV) radiation carries less energy than X-rays, but still enough to kill microorganisms (making it useful as a sterilizing agent) and damage human cells. When used in the laboratory, a visor must be worn to protect against damage of the retina [2].

Passing a current through a lightbulb filament makes it hot, causing it to glow with a white light. Cooler objects also give out radiation, but because they have less energy, they emit at lower frequencies. The human body, for example, emits energy as infrared radiation. Special detectors can pick up these emissions and convert them into a visible image [3]. Radio waves have less energy still: large transmitters are needed to send them over long distances [4].

Mirrors and Lenses

The principles at work in optical instruments

The first lenses were made in medieval times, and their crude abilities to magnify objects must have awed their makers. As techniques for grinding glass improved, and lenses began to be used in combinations, the range of human observation was extended to take in very small and very distant objects.

Modern science is rooted in the development of the microscope and the telescope. At their most basic, these instruments are combinations of two simple convex lenses: the different arrangement of the lenses accounts for their different function. But lenses do not have to be convex or even concave in shape. Fresnel lenses use many concentric prisms to concentrate light.

Rays and waves

Light is an electromagnetic wave, as are radio waves and X-rays. Like other waves, light behaves in a particular manner, changing speed and direction as it travels through different media. These properties have been exploited to produce the huge variety of optical instruments available today.

The speed of light in a vacuum – it travels 186,282 miles every second – is one of the fundamental constants of physics. But when light propagates through other materials it has a lower speed, governed by a property called the material's *optical density*. A beam of light that travels between materials with different optical densities, such as air and water, changes direction at the interface between the two in a process called *refraction*. This is the result of the change in the gap between successive wavefronts – the wavelength – as the light's speed changes on meeting a different material.

Powers of concentration

The effects that certain glass shapes have on light have been known since ancient times. A crystal ball can bring the rays of the sun to a focus – although not a very sharp one – by virtue of its spherical shape. The bending occurs at the boundaries of the glass and air, not in the glass bulk of the ball. A thin piece of glass with two spherical surfaces – a lens – thus achieves the same effect while minimizing weight. Additionally, the lens reproduces the shape of only the middle sections of the surface of the ball, where a sharp focus is produced, and so gives a more precise image.

The power of a lens can be described in terms of its *focal length*. This is the distance from the lens to the point at which parallel rays of light striking the lens are brought to a focus. A fat, highly curved lens has a higher power, and a shorter focal length than a thinner lens made from glass with the same optical density.

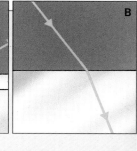

*When a light beam hits the surface of a mirror [**A**], it reflects back at an angle that is exactly equal to its incident angle. Because light travels more slowly through glass than through air, a ray is refracted when it enters a glass block [**B**]. The ray's direction of travel veers away from the surface.*

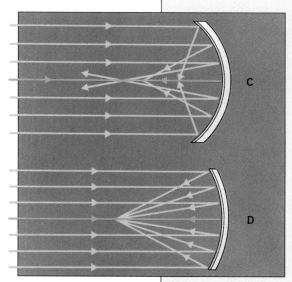

*Parallel rays of light falling on a concave mirror reflect and meet at one point – the rays are brought to a focus. Astronomical telescopes use this ability to gather light; some of them contain precisely ground mirrors more than 30 feet across. However, if the mirror is regularly concave (like a section cut from a ball), the focus is not sharp for rays reflected from its outer edges [**C**]. The glass reflector in a telescope is therefore ground to a parabolic shape. It focuses parallel rays hitting its surface to a sharp point [**D**]. The glass is silvered on its front face to avoid unwanted refraction effects.*

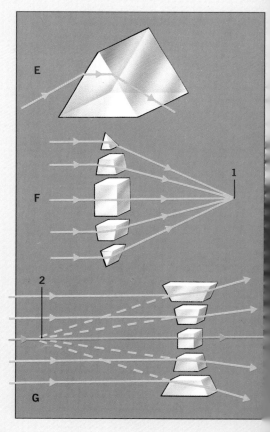

*A ray of light passing through a triangular glass prism is refracted at two surfaces [**E**]. At the first it travels from air to glass, and is bent away from the surface. The ray carries on in a straight line through the glass until it emerges back into the air. At this interface, the light bends toward the surface. The triangular shape thus has a double-bending effect on a ray.*

A lens can be thought of as a series of prisms, each with a slightly different angle between its faces. Each prism bends light to a different degree, and so grouped together, they bring parallel rays to a focus.

*In the case of a convex lens [**F**], this is a real focus [1], a point that the rays do all actually pass through. But for a concave lens [**G**] the focus is virtual [2]. The concave shape makes parallel rays spread out in such a way that they appear to be coming from a point to the left of the lens, although this is not their actual point of origin.*

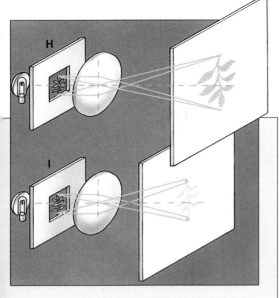

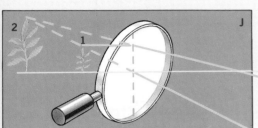

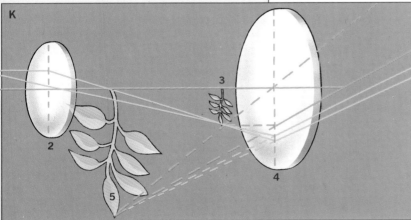

Construction rays from one point on an object show how a simple lens produces a focused real image in a projector [H]. If the screen is moved slightly [I], the rays of light no longer meet and the image blurs.

A convex lens acts as a magnifying glass [J] when an object (here a leaf) [1] is within one focal length of the lens. Rays from one point on the leaf are made to diverge, appearing to come from a larger virtual image [2] behind the lens. There are two lenses in a simple microscope [K]. The first [2] acts like a projector, making a magnified real image [3] of the object [1]. The second [4], acts like a magnifying glass, creating a still more enlarged virtual [5] version of the intermediate image.

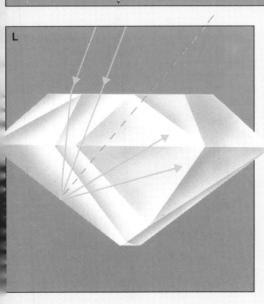

The optical density of a substance is a quantity related to the speed at which light travels through it. When light travels from one substance to another, the angle at which it strikes the interface determines whether refraction or reflection takes place. If the difference in optical density between the two substances is very great, rays of light striking a surface at large angles will still tend to be reflected back. Diamond has a very high optical density, and so a ray of light trying to escape from a diamond is reflected from very many facets before emerging [L]. This is seen as the inimitable sparkle of a true diamond.

Image makers

A movie projected on a movie screen is a rapid succession of still pictures, or frames, that momentarily interrupt an intense beam of light. A convex, or converging, lens *focuses* the light after it passes through the film, so that rays coming from one point on the film arrive at just one corresponding point on the screen. Light paths from the top and bottom of each frame cross over, with the result that the final image is inverted. The film in the projector must therefore be loaded upside down.

The image on the movie screen is known as a *real* image because it is formed where different rays of light from a single point on the frame actually meet up again on the screen. But not all images are formed in this way. When a convex lens is used as a magnifying glass, rays of light from a single point on an object placed on a table are made to diverge by the lens. The divergence is such that, to an observer, the rays appear to originate from a distant point somewhere below the level of the table. This is called a *virtual* image.

Concave lenses make parallel rays of light diverge in a similar way and also produce virtual images. They are used in conjunction with convex lenses to correct color aberrations in multielement compound lenses. Concave lenses are also used to correct short sight in humans, while convex lenses correct long sight.

The path that light takes through a lens can be calculated using two simple rules, if the focal length of the lens is known. Any ray of light that is initially parallel to the *principal axis* (an imaginary line through the center of the lens and perpendicular to it) is redirected through the *focal point*, a focal length away from the lens on the principal axis. Similarly, a ray that travels through the center of the lens will not be deviated and will continue in the same direction. These two construction rays are sufficient to fix the position, size, and nature of an image.

Total internal reflection

A ray of light is refracted whenever it crosses a surface between two substances with different optical densities, such as air and glass. Traveling from air to glass, a ray will always be bent away from the surface; going from glass to air it bends *toward* the interface. But if the angle at which a ray emerging from glass strikes the surface is shallow enough, it will not be refracted at all, but will reflect back just as if it had struck a mirror. This is *total internal reflection*, and is the reason that fiber-optic cables act as "light pipes." Rays are prevented from escaping by repeated reflections at the boundary between a core and cladding, made from glasses with different optical densities.

Reflecting prisms rely on total internal reflection. In a pair of binoculars they are used to reflect light traveling between the lenses, allowing a long optical path while keeping the binoculars compact.

Light, Color, and Vision
Different wavelengths and how they are detected

Light is a wave, but seldom is this nature apparent. Interference effects, like those that produce the swirling iridescent patches on a bubble's surface, are one of the few noticeable consequences of light-wave interaction. What we call light is in fact only a tiny portion of the entire electromagnetic spectrum, but of most interest to us because it contains the wavelengths our eyes are sensitive to. Most of the light that enters our eyes comes indirectly from the sun. Sunlight contains a huge range of different wavelengths, but different colors are picked out by substances that selectively reflect and absorb various wavelengths. For instance, the sky is blue because that color is dispersed much more than any other.

Color dispersion by interference

Light has a wavelength of slightly less than one thousandth of a millimeter. If a light wave is reflected off two surfaces separated by a distance comparable to this wavelength, *interference* occurs. Depending on the exact distance between the surfaces, the peaks and troughs of the two reflected waves can coincide, reinforcing one another to give a bright reflection. Alternatively, they may be out of step, canceling each other out to give no reflection. These phenomena – *constructive* and *destructive interference* – cause the shifting pattern of colors reflected by a layer of oil on water. The top and bottom surfaces of the oil are separated by a tiny but variable distance. Because the wavelength of light corresponds to its color (blue has a shorter wavelenth than red) some colors interfere constructively in certain areas of the oil patch, while other colors are reinforced in other parts of the layer.

Spectral visions

A rainbow can seem to be a mystical sign that a storm has passed. But, in fact, it is a result of the reflective and refractive properties of raindrops. Sunlight contains a range of different wavelengths – and hence colors. Raindrops reflect and refract sunlight, splitting it up into these colors and producing a rainbow.

Depending on where an observer is, a rainbow can take different shapes. The most familiar is the semicircular arc stretching from horizon to horizon. But a pilot at high altitude can sometimes see the complete circle of which the arc seen at ground level is only a part. Sometimes two rainbows are visible, the main arc concentric with a second one seen higher up in the sky. The lower arc comes from sunlight that has been reflected once in each raindrop before emerging, whereas the second, fainter rainbow is formed from light that has undergone two reflections in each droplet.

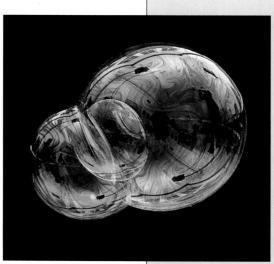

The skin of a soap bubble is a very thin layer of water (left). It is covered with a series of constantly shifting colored bands. These are a form of interference pattern, a consequence of the wave nature of light. Waves reflecting from the inner surface of the skin travel a slightly longer distance than reflections from the outer surface. The two reflections interfere: for a particular thickness of bubble-skin, some wavelengths (and hence colors) will cancel out while others will be reinforced: so the color of the skin depends on its thickness at that point.

Camera lenses are bloomed – given an ultrathin coating of resin to prevent unwanted reflections [A]. Light reflecting from the lower surface of the coating travels slightly farther than reflections from the top surface: the two waves emerge with a phase difference. For most light wavelengths, the waves interfere destructively [2], canceling out any reflection that might spoil the image. But shorter wavelengths, for example, blue and violet, interfere constructively [1] and reinforce each other. They reflect to and fro within the lens system producing its characteristic blue-violet color.

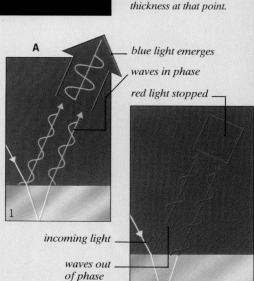

blue light emerges

waves in phase

red light stopped

incoming light

waves out of phase

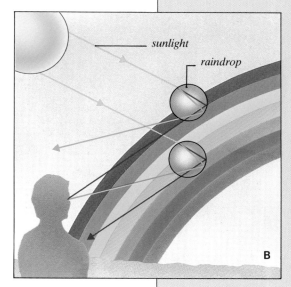

sunlight

raindrop

A rainbow forms when rays of sunlight reflect from a rain shower [B]. A ray enters a raindrop and is refracted at the front surface. It reflects back from the rear of the drop and is refracted again as it exits through the front. Each raindrop therefore acts as a prism, splitting the light into its component colors. Blue light emerges from each drop at a shallower angle to the ground than red light. Yet in a rainbow, red is seen at the top of the arch, and blue farther down. This is because the red light entering the eye comes from drops higher up in the sky than the ones from which it receives blue light.

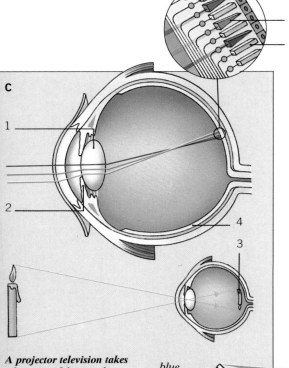

*A human eye brings light to a sharp focus [**C**] by changing the shape of its lens [1]. The amount of light admitted is controlled by opening and closing the iris [2]. The focused image is inverted [3] and forms on the retina [4], a tissue layer containing two types of light-sensitive cell. Rod cells [5] work at low light levels but give no color information. Cone cells [6] come in three varieties, sensitive to the red, green, or blue parts of the spectrum. The brain combines the signals they send out to form other colors. White light triggers all three types of cell, whereas yellow light triggers red and green cones only.*

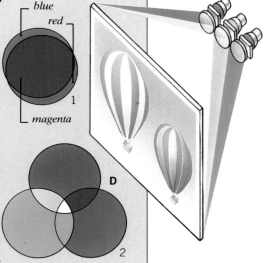

blue

red

magenta

1

D

2

*A projector television takes advantage of the way that our eyes sense color by forming full-color images from only red, green, and blue light [**D**]. These are called the additive primary colors because they can be combined in different strengths to form any shade in the spectrum. The television contains three tubes that emit narrow beams of light, which are focused by lenses on to a screen. Where red and blue beams overlap, magenta (a shade of pink) is seen [1]. Blue and green combine to give cyan, while red and green give yellow. Areas where all three colors overlap appear white [2].*

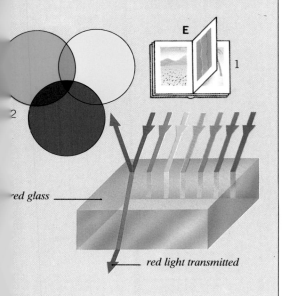

E

1

2

red glass

red light transmitted

*A piece of red glass has color [**E**] because it reflects or transmits only the red wavelengths contained in white light, and absorbs all other colors. The inks used to print a book [1] only absorb small parts of the spectrum: cyan ink absorbs red but reflects blue and green, magenta absorbs green while reflecting blue and red, and yellow ink absorbs blue but reflects green and red. So where cyan and magenta inks mix, only blue light is reflected. Magenta on top of yellow ink looks red, and cyan on yellow looks green. A mixture of inks of all three process colors absorbs all color and appears black [2].*

Vision and color

There are striking similarities between the ways that an eye and a camera bring light to a focus. Both have lenses and also light-sensitive surfaces – the film in the camera and the retina at the back of an eye. Focusing is accomplished in different ways, however. The lens in a camera simply slides forward and backward to produce focused images from objects at different distances.

In contrast, the human eye has a lens that is fixed in position. It is not the only refracting surface – the transparent cornea and the jelly-like humors that fill the eye achieve much of the task of light concentration. The lens takes care of the fine-tuning or *accommodation* – the ability to adjust for subjects at different distances. This is achieved by a ring of muscles that compresses or stretches the lens, changing its focal length so that both distant and close objects can be brought to a sharp focus.

Correcting far- and nearsightedness

Not everybody's eyes can achieve full accommodation. In some cases, the eyeball is too short to be able to bend light enough to produce sharp images of nearby objects. Such people are farsighted. In other cases, the eyeball is too long or the cornea is unusually concave. This means that light from distant objects is refracted too much to be brought to a sharp focus on the retina. Such people are near-sighted. Both of these problems can be remedied with the aid of external glass lenses.

Other problems occur if cataracts form – if the lenses become clouded, resulting in blindness. The cure is either to replace each cloudy lens with an artificial lens or with one from a donor, or, more simply and cheaply, to remove the lens altogether. This means that the eye bends light less than it should – it becomes farsighted. Again, this can be compensated for by wearing powerful spectacles.

Reproducing color

The color sensitive cells on the human retina are called cones. They respond to blue, green or red light by sending nerve impulses to the brain, which are interpreted as color. Colors other than red, green or blue are sensed as different combinations of these three *primary* colors.

To reproduce any color, it is therefore sufficient to mix together different proportions of light of the primary colors. This fact is used in the design of color televisions, the screens of which are covered with phosphor dots that emit only red, green, or blue light. Color built up in this way is termed *additive*. Color on a printed page is built up in a different way. This is because printed inks absorb different components of the white light (a mixture of all colors) that illuminates the page. Different inks are therefore used to absorb different parts of the spectrum, reflecting back only the desired color. This is known as *subtractive* color.

Radioactivity
Energy from unstable nuclei

Radioactivity is silent, invisible, and feared by many as a mysterious force. But it is a natural process that occurs all around us – in the air, in the rocks, and even in our food. In fact, it is radioactivity that heats the core of our planet, driving the earth's geological processes.

Radioactivity is a property of substances called *radioisotopes,* some of which are natural, others made by humans. Atoms of these substances have unstable nuclei, which become stable by expelling high-energy particles or by giving out electromagnetic radiation. This process is known as radioactive decay: it is independent of temperature, pressure, or chemical bonding and is a property of the radioisotope itself.

Types of radiation

The identity of an atom is given by the number of positively charged protons in its nucleus. Atoms of carbon always have six protons, nitrogen seven, and so on. The nucleus makes up a tiny fraction of the volume of an atom, and one would expect it to be torn apart by the mutual repulsion between the protons it contains. This, however, does not happen because the charge of the protons is "diluted" by neutrons, and because all the nuclear particles are held together by the so-called *strong nuclear force.*

The balance between protons, neutrons, and the nuclear force is a delicate one. In the nuclei of some atoms, the balance is maintained, while in others there are too many or too few neutrons to maintain stability. To become stable, these nuclei change their configuration by emitting radiation: this process is known as *decay.*

For example, most carbon atoms are stable, possessing 6 protons and 6 neutrons in the nucleus. This type, or *isotope,* of carbon is stable and is known as carbon-12. Carbon-14, on the other hand, is unstable, possessing 6 protons, and 8 neutrons. This isotope decays in a process by which one of its neutrons emits an energetic electron, or *beta particle.* In doing so, the neutron becomes a proton, and the nucleus now contains 7 protons and 7 neutrons. Because the atom gains a proton, it actually changes identity, becoming an atom of nitrogen. Beta particles are the type of radiation most commonly emitted by atoms with a surfeit of neutrons.

Alpha particles, on the other hand, are most often emitted by giant nuclei as they transmute toward smaller, more stable structures. Each alpha particle comprises 2 protons and 2 neutrons. The third main type of radiation is gamma rays. These are a form of electromagnetic radiation – like light but with short wavelengths (10^{-14}m). Gamma rays are given out when a nucleus undergoes alpha or beta decay, but still has excess energy to get rid of.

The nucleus of carbon's stable isotope (carbon-12) [**A**] *is made of six protons and six neutrons. In contrast, the unstable isotope carbon-14* [**B**] *has an additional two neutrons. It becomes stable when its complement of neutrons is reduced by beta decay, in which a neutron changes into a lighter proton, the excess mass being expelled in the form of a fast-moving electron, or beta particle. Beta particles* [**C**] *have low mass and can penetrate up to 1.5mm of aluminum.*

Some isotopes with an excess of protons achieve stability by an opposite process called electron capture, *in which a proton captures a nearby electron and becomes transformed into a neutron.*

In alpha emission, a cluster of two protons and two neutrons (an alpha particle) carries away the energy of a large unstable nucleus. Alpha particles are large, with poor penetrating power: a sheet of paper is enough to stop them [**D**].

Alpha and beta emission cause a change in identity of the element. For example, uranium decays to thorium with the loss of an alpha particle. In contrast, gamma emission rids the nucleus of excess energy but does not change its identity. Gamma radiation can penetrate up to 5cm of aluminum [**E**].

In general, stable nuclei have as many or slightly more neutrons than protons. The relationship between number of protons and neutrons, and the type of radioactive emission is shown in this graph [**F**]. *Alpha emission is typical of elements heavier than bismuth; beta of those with an excess of neutrons; electron capture of small nuclei with excess protons.*

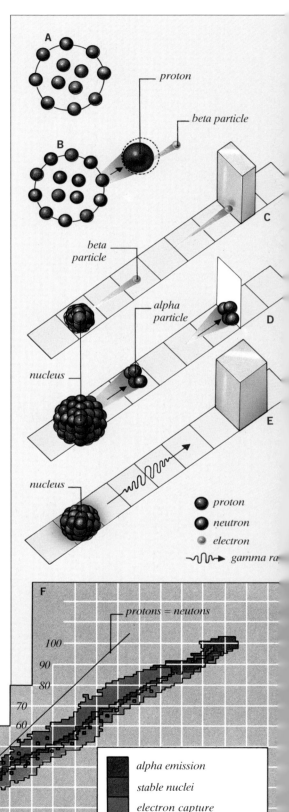

proton
beta particle
beta particle
alpha particle
nucleus
nucleus

● proton
● neutron
● electron
~⌁⤳ gamma ra[y]

F
protons = neutrons
100
90
80
70
60
50
40
30
20
10

0 10 20 30 40 50 60 70 80 90 100 110 120 130 140 150
number of neutrons

number of protons

alpha emission
stable nuclei
electron capture
beta emission

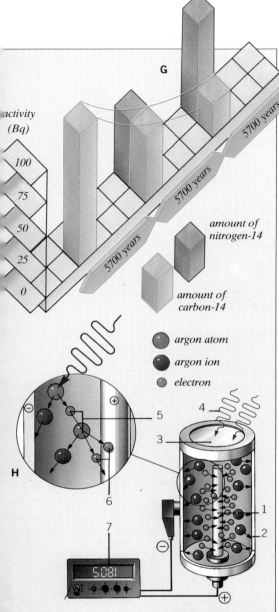

For any radioisotope, the half-life is constant regardless of the sample size [**G**]. Carbon-14, for example, has a half-life of 5700 years. This means that a sample of the element with an activity of 100 emissions per second (or 100 bequerels, Bq) will have an activity of 50 Bq after 5700 years. In this time, half of the sample will have decayed into nitrogen. And after another 5700 years, its activity will have fallen to 25 Bq, and the sample size will have halved again. The amount of the decay product (nitrogen) increases in parallel with the decrease in activity of carbon-14, doubling every 5700 years.

A Geiger counter comprises a metal tube filled with argon gas [**H**]. A wire electrode in the center of the tube [1] is charged to a positive voltage of 3000 V relative to the casing [2].

A mica window [3] allows radiation to enter. The radiation [4] ionizes argon atoms, producing argon ions and electrons. The electrons accelerate toward the positive electrode. As they do so, they collide with other argon atoms, causing further ionization [5]. This creates a "cascade" of electrons [6] – a pulse of current that hits the central electrode. These pulses are counted electronically [7] and displayed on an LCD.

Many techniques are used to give a visual representation of the path of particles emitted by unstable nuclei. In a cloud chamber (right), a gas is cooled to the point where it will condense into visible droplets if it is "seeded" with particles. Alpha and beta particles entering the chamber act as the necessary "seeds" and the trail of condensation betrays their paths.

Bubble chambers work in a similar way, but instead of gas, they a contain liquid that is on the point of boiling. Minute bubbles of gas form along the path of a particle passing through the chamber, forming a visible, tell-tale trace.

Half-lives

Radioactive decay is a random process. This means that it is impossible to state precisely at what instant the nucleus of any one radioactive atom will decay. Instead, the time taken for atoms of an isotope to decay is described statistically in terms of the *half-life*. This timespan is a measure of how long it takes for *half* of the nuclei in any large sample to break down.

Half-lives vary enormously from isotope to isotope: for example, it takes 4.5 billion years for half of a sample of uranium to decay to thorium; iodine-131 has a half-life of 8 days; while the synthetic element polonium has a half-life of one ten-thousandth of a second. Carbon-14 decays, turning into nitrogen-14, with a half-life of 5700 years. This period corresponds roughly to the history of human civilization: this is significant because the decay of carbon-14 into nitrogen can be used to date archeological relics.

The effects of radiation

The phenomenon of radioactivity was discovered in 1896, when the French physicist Henri Becquerel left some uranium salts on top of a photographic plate. When developed, the plate was found to be "fogged," just as though it had been exposed to light. This occurred because radiation – alpha and beta particles, and gamma rays – carries energy. When it hits a photographic emulsion, it collides with its atoms, dislodging some of their electrons. This process is called *ionization*. Instruments for detecting radioactive particles – for example, the Geiger-Müller counter – actually detect ionization caused by the particles, rather than the particles themselves. Units of activity are the *becquerel* (one becquerel is one emission per second) and the *sievert*, which is a measure of the effect that the radiation dose has on living cells.

All types of radioactivity affect living cells because they cause ionization. The ionized molecules are chemically unstable and take part in abnormal reactions, which can destroy structures within the cell, disrupt cell division, or change the way that the cell functions.

Though many fear the emissions of the nuclear power industry, by far the largest source of radiation to which people are regularly exposed is radon gas, which is of natural origin. This gas seeps out of the ground – particularly where the underlying rocks are rich in radioactive granite – and becomes trapped in the still air of houses. Moreover, one-fifth of a person's radioactive exposure comes from potassium present within the body. Potassium is an essential mineral. It is present in many foods, and makes its way into every cell of the body, including the cells from which sperm and eggs are derived. About 0.1 percent of all potassium is in the form of the radioactive isotope potassium-40, and it is thought that the mineral makes a significant contribution to the incidence of mutation in humans.

The Forces of Nature

The four fundamental interactions and their effects

Scientists and mathematicians are forever seeking simple patterns in the world around them. This has resulted in the revelation that every type of physical behavior can be explained in terms of four fundamental forces – gravity, electromagnetism, and two types of force (termed strong and weak) that are only felt within ranges equal to the tiny dimensions of atomic nuclei.

One of the goals of modern physics is to combine all four of these forces in a single mathematical theory. The electromagnetic, weak, and strong nuclear forces have already been synthesized into a single theoretical framework, but attempts to include gravity in this grand unified theory have so far met with little success.

Action at a distance

Anyone who has tried to push together the like poles of two magnets is familiar with the concept of "action at a distance." The mutual repulsion of the magnets can be detected over relatively large distances, even if materials are placed between them. Gravitational force is also transmitted over empty space, with no apparent connection between the interacting bodies, as are the forces of electrostatic attraction and repulsion. These puzzling interactions are described by physicists in terms of *fields* – regions of space around the bodies concerned, in which a force can be detected. Fields are commonly represented as lines of force emerging from the object. The force between two objects decreases with distance: this is represented graphically by a decrease in the density of the field lines. The field concept is a useful way of describing how these forces act, but it does not address the question of why they exist.

Strong nuclear force and particle exchange

According to the laws known to 19th-century scientists, atomic nuclei should not exist. They are incredibly dense collections of electrically neutral *neutrons* and positively charged *protons*. Even at the comparatively large distance of 1 mm, the repulsive force between two protons is more than 10,000 times greater than the earth's gravitational pull on them. As protons approach one another the repulsive force increases rapidly, and on the scale of a nucleus (1 femtometer, a million-billionth of a meter) the repulsion is so great that the nucleus ought to blow apart. Only the presence of an even stronger attractive force – known as strong nuclear force – holds the nucleus together.

This force acts only over tiny distances. No stable nuclei exist that have many more than 80 protons, because those on the opposite sides of such a nucleus are so far apart that the strong force is overcome by electrical repulsion.

Matter interacts via four fundamental forces. Each dominates the others on a particular scale. In the vast expanses of the Universe, gravity is the dominant force. It acts across huge distances to cluster the stars in a spiral galaxy (below). On a smaller scale, many interactions depend on

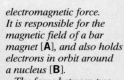

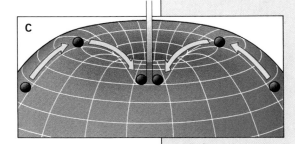

electromagnetic force. It is responsible for the magnetic field of a bar magnet [A], and also holds electrons in orbit around a nucleus [B].

The force between two protons varies with distance. At larger separations electromagnetic force causes repulsion between the protons. But at small separations – the scale of an atomic nucleus – strong nuclear force draws the protons together.

Electromagnetic force can be represented as a "hill" that the two approaching protons must climb before descending into the "valley" of attraction caused by strong nuclear force [C].

Forces seem to be able to act across empty space, almost as if interacting particles were "aware" of each other. It is now thought that each of the four fundamental forces is transmitted by a messenger particle [D]. This flies between two particles, carrying the force which acts on the recipient particle, just as a catcher recoils with the momentum of a thrown ball [3]. Protons and neutrons influence each other via strong nuclear force, carried by gluons [1]. Two negatively charged electrons repel one another via the electromagnetic force, carried by short-lived photons [2].

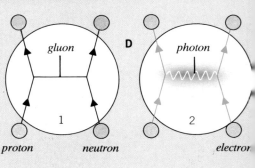

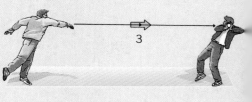

The world that we live in and the objects that we use are made of matter, which, at an atomic level, consists of electrons, protons, and neutrons. But all of these particles have almost identical antimatter counterparts, with the same mass and spin, but which carry an opposite charge. When an invisible gamma ray enters a bubble chamber and collides with an atom, it can produce an electron and its antimatter equivalent, a positron. The two particles shoot away from each other and their opposite charges make them spiral in opposite directions in a strong magnetic field inside the chamber (left).

Matter and antimatter

Quantum mechanics, the language that physicists use to describe the forces and particles that occur in atoms and nuclei, is a highly mathematical and abstract set of ideas. But as a theory it works, and its predictions are drawn upon in the design of electronic circuitry, in the manufacture of superconductors, and in many other important industries.

One of the more surprising predictions of the theory is that for every particle, such as a proton or an electron, there exists a particle with the same mass but an exactly opposite charge – an "electron" with a positive charge, for example. These are called *antiparticles*, or *antimatter*. Although the world is made almost entirely of matter, antimatter particles turn up often in detectors such as bubble chambers. The positive electron has its own name, the *positron*, whereas the "mirror-image" version of the proton is simply called the *antiproton*.

Particles and quarks

Electrons were the first subatomic particles to be identified, followed by the nucleus and the nuclear components – the proton and neutron. In time, ever more sophisticated detectors revealed more and more new particles. One group with properties between those of electrons and protons were named mesons. Four types were found – pi, mu, eta, and K. This growing number of new particles threatened the underlying simplicity of physics.

Rescue came in the form of a bold idea – a hypothetical particle named the quark. Most of the hundreds of particles found could be related together in simple patterns if they were thought of as being made up of different combinations of just six types of quark and their antimatter counterparts. Only one group of particles and antiparticles similar to electrons could not be explained by the quark combinations. The basic simplicity of particle physics was thus saved.

The particles inside a nucleus are composed of even smaller particles called quarks. Protons and neutrons are two members of a "family" of particles said to have a spin of 1/2 [E]. It contains all of the possible combinations of three of the six types of quark, called up, down and strange. The proton and neutron contain only up and down quarks. More exotic particles contain strange quarks. Σ, or sigma, particles come in three varieties with different charges, Σ⁻, Σ⁰ and Σ⁺. A further two positive and negative members of the family are called xi or Ξ particles, Ξ⁻ and Ξ⁺.

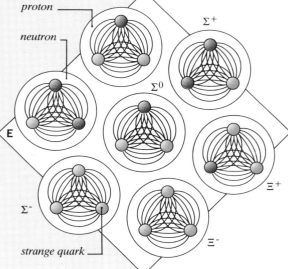

Neutron decay and the weak interaction

Most atomic nuclei are combinations of protons and neutrons. But when neutrons are isolated, they become unstable. After a few minutes of solitary existence, they change into protons, creating electrons and particles called antineutrinos in a process known as beta decay.

Although some of the particles involved carry charge, beta decay is not caused by the particles' electrical properties. The strong nuclear force cannot be responsible because it is experienced by neutrons and protons but not by electrons or antineutrinos. Instead, another interaction between particles, the weak nuclear force, is responsible. Like the electromagnetic force, the weak interaction is transmitted through particle exchange. In 1983 an experiment found evidence of the existence of these carriers, which were named W- and Z-bosons. The discovery enabled the two theories to be linked and named the *electroweak* interaction.

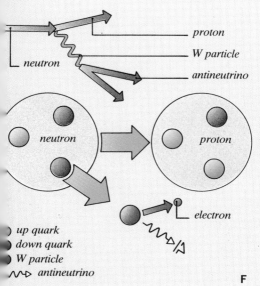

Neutrons are not stable particles in isolation. They eventually decay into a proton as one of their two down quarks changes to an up quark [F]. The decay is the result of the weak nuclear force which is transmitted by a W-particle. This has a mass a hundred times greater than the neutron that emitted it, an anomaly only allowed by the uncertainty principle that is part of the theory of quantum mechanics. This permits the excess mass to appear but only for a short time, too brief to be detected. However, the W-particle itself decays into an antineutrino and an easily detectable electron.

proton — W particle — antineutrino — neutron — proton — electron — neutron

up quark · down quark · W particle · antineutrino

F

Units of measurement

The SI System

The language of science is built on the seven fundamental physical quantities of length, mass. time, temperature, electric current, light intensity, and amount of matter. The units used by scientists around the world to measure these quantities are given by the International System of Units (or SI). All other scientific units are expressed in terms of these seven base units.

Physical quantity	Name of unit	Symbol
Length	Meter	m
Mass	Kilogram	kg
Time	Second	s
Temperature	Kelvin	K
Electric current	Ampere	A
Light intensity	Candela	cd
Amount of matter	Mole	mol

Exponential notation

The science of measurement has become so refined that it is possible accurately to measure distances between galaxies, as well as diameters of individual atoms. The huge and minute numbers involved in such measurement are conveniently expressed in terms of exponential notation, using powers of ten to simplify unwieldy numbers. These exponential numbers are given standard prefixes.

Number	Exponential	Prefix	Symbol
1 quintillion	10^{18}	exa-	E
1 quadrillion	10^{15}	peta-	P
1 trillion	10^{12}	tera-	T
1 billion	10^{9}	giga-	G
1 million	10^{6}	mega-	M
1 thousand	10^{3}	kilo-	k
1 hundred	10^{2}	hecto-	h
Ten	10^{1}	deka-	da
1 tenth	10^{-1}	deci-	d
1 hundredth	10^{-2}	centi-	c
1 thousandth	10^{-3}	milli-	m
1 millionth	10^{-6}	micro-	μ
1 billionth	10^{-9}	nano-	n
1 trillionth	10^{-12}	pico-	p

Conversions

Although SI units are used for most scientific measurements, units derived from SI, or based on older systems of measurement are sometimes encountered. Some of the most useful conversions are given below.

Length
1 in. = 2.54 cm
1 ft = 12 in. = 0.3048 m
1 yd = 3 ft = 0.9144 m
1 mile = 5280 ft = 1.61 km
1 light year = 9.46×10^{12} km

Area
1 in.^{2} = 6.452 cm^2
1 ft^2 = 144 in^2 = 0.0929 m^2
1 square mile = 2.59 km^2
1 acre = 4840 yd^2 = 0.4047 hectares (ha)

Volume
1 in.^{3} = 16.39 cm^3
1 ft^3 = 0.028 m^3
1 yd^3 = 0.764 m^3
1 pint (US) = 0.473 liters
1 gallon (US) = 3.79 liters

Mass
1 oz = 28.35 g
1 lb = 0.45 kg
1 ton (imperial) = 1.02 tonnes

Density
1 lb/in.^{3} = 27.68 g/cm^3
1 lb/ft^3 = 16.019 kg/m^3
1 lb/gallon (US) = 0.12 kg/liter

Speed
1 ft/s = 0.3048 m/s = 1.097 km/h
1 mile/h = 1.61 km/h
1 knot = 1.852 km/h

Force or weight
1 lbf = 4.45 N (newtons)
1 kgf = 9.81 N
1 tonf = 9.96 kN

Temperature
°F = 9/5 °C + 32

Pressure
1 lbf/in.^{2} = 6.89 kN/m^2
1 mmHg = 133.32 N/m^2
1 atmosphere = 760 mmHg
1 atmosphere = 14.7 lb/in.^{2} = 101,300 N/m^2

Glossary

Absolute zero (-459.67°F or -273.15°C) lowest temperature theoretically possible; achieved when all heat has been removed from a body.

Acceleration the rate of change of **velocity** of an object.

Achromatic lens combination of lenses of different shape or density designed to reduce **chromatic aberration**.

Acid compound that releases hydrogen ions when dissolved in water.

Acoustics the science of sound. In particular, the propagation of sound around a room.

Activation energy energy needed to start a chemical reaction.

Adhesion attractive force between atoms or molecules of different substances.

Air resistance frictional force that resists the passage of an object through air.

Alcohol organic compound characterized by the presence of an OH (hydroxyl) group.

Aliphatic compound any organic compound in which the carbon atoms are joined in straight chains.

Alkali a **base** that is soluble in water.

Allotropy ability of an element to exist in two or more distinct physical forms in the same state.

Alloy metal mixed with other metals or nonmetals to give it special qualities.

Alpha particle positively charged particle made up of two **protons** and two **neutrons** emitted in the process of radioactive decay of certain heavy radioisotopes.

Alternating current an **electric current** the direction of flow of which around a **circuit** changes at regular intervals.

Alternator electrical **generator** that produces an **alternating current**.

Altimeter instrument used in an aircraft for reading height above the ground.

Amine organic compound formed when a hydrogen atom of ammonia (NH_3) is replaced by a carbon group.

Amino acids organic compounds that are the constituents of **proteins**.

Ammeter an instrument that measures electric current.

Amplitude the maximum size of an **oscillation**, such as the height of a wave.

Amplitude modulation adapting a signal for transmission by modifying its **amplitude** in sympathy with the signal to be sent.

Analog representing a value by an infinitely varying signal.

Analog to digital converter electric circuit that converts an analog signal to a digital one.

Anion negative **ion**, that is, an atom with one or more extra electrons.

Anode positive **electrode**, which attracts negative particles (**anions**).

Antenna (aerial), device for transmitting and/or receiving electromagnetic waves.

Antibody protein produced by white blood cells as part of the body's immune response to foreign substances.

Aromatic compound organic compound containing a group of carbon atoms linked together in a ring.

ASCII acronym for *American Standard Code for Information Interchange*. Binary code for text and digits.

Atmosphere mixture of gases surrounding a planet.

Atom the smallest unit of an element that can take part in a **chemical reaction**. Each one has a nucleus of **protons** and **neutrons,** surrounded by **electrons.**

Atomic energy level region around the **nucleus** of an atom where an **electron** is most likely to be found.

Atomic force microscope instrument that moves a diamond-tipped probe across the surface of a specimen to produce a highly magnified image.

Atomic mass mass of an atom, measured in units one-twelfth of the mass of a carbon-12 atom.

Atomic number number of protons in an atom.

Background radiation steady low-level radiation emitted by radioactive substances in the environment

Ballistics study of the motion of objects in a gravitational field

Bandwidth spread of frequencies

Base a compound that forms hydroxyl (OH⁻) ions when dissolved in water. Bases react with **acids** to give **salts**.

Battery an energy storage device made up of two or more electric **cells**, and designed to release electricity on demand.

Baud measurement of the number of **bits** per second sent by a **modem**.

Beats regular variation in loudness arising from the interference of sound waves of nearly equal pitch.

Beta particle high energy electron emitted by decaying radioactive atom.

Binary system system of numbering that uses only the digits 0 and 1 to represent all numbers.

Binding energy energy holding together the protons and neutrons in a nucleus.

Biotechnology the use of living organisms and their components in industry and commerce.

Bit abbreviation of *binary digit*. The smallest unit of information that can be stored or processed by a computer. Represented by a 0 or 1 or on/off electrical pulse.

Blackbody perfect absorber and emitter of heat radiation.

Boiling point temperature at which a liquid becomes a vapor.

Brownian motion random motions of particles in a fluid resulting from collisions with molecules.

Buoyancy lifting force of a fluid on a body immersed in it.

Byte eight **bits** of information, usually representing a single character.

Cache memory extremely fast part of the main computer memory.

Capacitor electrical device for storing **electric charge**.

Capillary action spontaneous movement of a liquid up or down a narrow tube, due to **adhesive** attraction between the molecules or atoms of the liquid and those of the tube.

Carbon dating archaeological dating method, which compares the relative number of carbon isotopes present in a sample of organic material.

Carboxylic acid organic acid containing the carboxyl group –COOH.

Catalyst substance that increases the rate of a **chemical reaction** while itself remaining unchanged.

Catalytic converter device fitted to a motor vehicle that converts harmful exhaust products to less toxic ones.

Cathode negative **electrode** that attracts positive particles (**cations**)

Cathode ray beam of high-energy, high-speed electrons.

Cathode-ray tube (CRT) evacuated glass tube that produces a beam of electrons that can be steered and focused on to a fluorescent screen.

Cation positive **ion**, that is, an atom that has lost one or more electrons.

CD-ROM acronym for *Compact-Disc Read-Only Memory*. A **compact disc** used to store computer data, especially words and images, in digital form.

Center of mass or center of gravity, the point in an object, about which its weight is evenly balanced.

Central Processing Unit the main chip in a computer that interprets and executes instructions, and controls the operation of other components.

Centrifugal force apparent outward (center fleeing) force on an object moving in a circular path.

Centrifuge apparatus used to separate substances, especially biological molecules, according to their **densities** by spinning them at high speeds.

Centripetal force force needed to keep an object on a circular path.

Chain reaction self-sustaining reaction in which the products of one step form the reactants of the next.

Charge-coupled device (CCD) electronic imaging device containing layers of silicon that release electrons when struck by light.

Chemical bond the force that holds two atoms together in a **molecule**.

Chemical equilibrium condition in a reversible reaction in which the rates of the forward and reverse reactions are exactly equal and so the amounts of reactant and product remain unchanged.

Chip popular name for an **integrated circuit**, a complete electronic circuit etched onto a piece of semiconductor material, commonly silicon.

Chromatic aberration smearing of an image as different colors are brought to a focus at different points by a lens.

Chromatography method of separating the components of a mixture according to their speed of movement through a stationary medium, usually a liquid held on paper or on solid beads.

Cloud chamber device for tracking the paths of charged particles through a gas.

Coherent light two or more waves of light with the same **frequency** and a constant **phase** difference.

Cohesion attractive force between atoms or molecules of the same substance.

Colloid mixture containing groups of molecules of one substance evenly dispersed in another substance.

Compact disk data storage device that permanently records digital information as a series of pits and flats etched into the surface of a solid material by a **laser**.

Compound substance made up of atoms of two or more elements joined together by chemical bonds.

Compression region of increased pressure along a longitudinal wave.

Conduction (electrical) flow of electric charge through a substance.

Conduction (of heat) transfer of heat energy through a material without the bulk movement of any part of the material itself.

Conduction band atomic energy level at which electrons can conduct electricity.

Conductor any material that can easily transmit heat or electricity.

Convection transfer of heat energy through a fluid by bulk movement of the fluid (liquid or gas).

Covalent bond chemical bond in which two electrons are shared by two atoms.

Crystal solid composed of a regular three-dimensional array, or lattice, of atoms or molecules.

Density the ratio of the **mass** of a substance to its **volume**.

Deuterium an **isotope** of hydrogen, mass number 2, consisting of one proton, one neutron, and one electron. Occurs naturally in sea water.

Diffraction the slight spreading of a wave as it passes around an object or through an opening.

Diffusion the gradual mixing of the molecules of two or more gases as a result of their motion.

Digital representing a value as a limited set of discrete signals, usually on/off pulses of electricity.

Diode device that allows an **electric current** to flow in one direction only.

Direct current an **electric current** with a constant direction of flow.

Displacement distance traveled in a specific direction.

Distillation separation of a mixture of liquids according to their boiling points.

DNA (*Deoxyribo Nucleic Acid*) complex giant molecule containing the genetic code for a living organism.

Doppler effect the apparent change in frequency of a wave (electromagnetic or mechanical) caused by relative motion between the source and the observer.

Drag force that opposes the motion of an object through a fluid.

Elasticity ability of a solid to regain its original size and shape after being compressed, stretched, or deformed.

Electric charge quantity of unbalanced electricity (a surfeit or deficit of electrons) in an object.

Electric field the region of space surrounding an **electrically charged** object in which an attractive or repulsive force is experienced by any other charged object or particle.

Electricity all phenomena caused by static or dynamic electrical charges.

Electrode conductor used to make electrical contact with a nonmetal.

Electrolysis chemical change produced by passing an electric current through a solution or molten compound.

Electromagnetic force fundamental force of nature, operating between electrically charged particles.

Electromagnetic wave a **wave** made up of self-sustaining electric and magnetic fields traveling at right angles to each other and to the direction of propagation.

Electromotive force (emf) the **potential difference** used to drive an electric current around a circuit.

Electron negatively charged fundamental particle. A basic constituent of the atom.

Electron microscope instrument that uses beams of electrons to produce highly magnified images.

Electron-volt (eV) a convenient measure of the kinetic energy possessed by moving charges, especially electrons; defined by the kinetic energy an electron gains as it moves through a potential of one volt.

Electronegativity the relative ability of a bonded atom to attract electrons.

Electrophoresis motion of charged particles in an electric field; a method for separating charged molecules in a fluid.

Electrostatics the study of stationary electric charges.

Element substance that cannot be split into simpler substances by conventional chemical means.

Elementary particle subatomic particle that cannot be further divided. The three groups of elementary particles are quarks, leptons, and gauge bosons.

Endothermic reaction chemical reaction that absorbs heat from the environment.

Energy the capacity to do work.

Energy level discrete amount of energy possessed by an atom.

Entropy an abstract measure of the disorder of a system.

Enzyme a biological catalyst.

Equilibrium (mechanical) state in which all the forces acting on a body are completely balanced.

Escape velocity the velocity needed by an object to escape the earth's gravitational field.

Evaporation the change state from a liquid to a vapor at temperatures below the boiling point of a substance.

Exothermic reaction chemical reaction that produces heat.

Expansion increase in volume of a constant mass of a substance.

Fermentation the conversion of sugar into ethanol (alcohol) by yeast.

Field region in space around a body in which a physical force can be detected.

Floppy disk computer storage device made of a flexible plastic disk with a magnetic coating.

Fluid any substance that can flow.

Focal length distance between the center of a lens or curved mirror and its principal focus.

Focus the meeting point, **real** or **virtual,** of parallel light rays passing through a lens or reflecting from a mirror.

Force any influence that accelerates or decelerates an object.

Fossil fuel fuel formed from the fossilized remains of plants and animals, e.g. coal, oil, and natural gas.

Freezing point temperature at which a liquid becomes a solid.

Frequency the number of oscillations in a unit of time.

Frequency modulation adapting a signal for transmission by modifying its **frequency** in sympathy with the signal to be sent.

Friction force that resists the relative motion of two bodies in contact.

Gamma radiation high-energy **electromagnetic** radiation emitted during nuclear reactions or decay.

Gas state of matter in which atoms or molecules move randomly to completely fill a container.

Gas turbine an internal combustion engine, which uses the products of combustion, a stream of hot gases, to drive a turbine.

Gene a sequence of DNA that contains the chemical code for a specific characteristic of a living organism.

Generator machine that converts mechanical energy into electrical energy.

Geosynchronous (geostationary) orbit satellite orbit, 22,000 miles above the earth's equator, with a period of exactly 24 hours, which makes a satellite appear stationary above one point on the earth's surface.

Geothermal energy energy harnessed for electricity or heating from hot rocks or steam inside the earth's crust.

Gluon elementary particle, carrier of the **strong nuclear force.**

Graviton elementary particle, yet to be experimentally detected, carrier of the **gravitational force.**

Gravitational force fundamental force of nature; the force of attraction between two masses.

Greenhouse effect warming of the atmosphere due to the entrapment of heat by gases in the air.

Ground state the lowest **energy state (level)** of an atom.

Half-life time taken for half of a sample of a radioisotope to decay.

Hard disk computer storage device made of a rigid metal disk with a magnetic coating.

Heat form of energy transferred between two bodies at different temperatures.

Hole vacancy in a crystal structure normally filled by an electron, so carrying a net positive charge.

Hydraulics study of the flow of fluids, especially how they transmit pressure.

Hydrocarbons organic compounds made up of carbon and hydrogen.

Hydrogen bond relatively weak electrostatic interaction between the hydrogen atom(s) bonded to an **electronegative** atom and another electronegative atom. Hydrogen carries a small positive charge, the electronegative atom a negative charge.

Hypersonic velocities of five times the speed of sound or more.

Ideal gas hypothetical gas with molecules of negligible size, which exert no forces on each other.

Image picture or appearance of a real object formed by a mirror or lens.

Induction change in the physical properties of a body by a field.

Inertia tendency of an object to resist a change in its motion by virtue of its mass.

Infrared radiation electromagnetic radiation with wavelengths in the range 0.75 to 10,000 μm.

Insulator any material that resists the flow of heat or electricity.

Integrated circuit electronic circuit with thousands of components printed on a small wafer of semiconducting material.

Intensity amount of power per unit area carried by a wave.

Interference the superposition of waves.

Internal combustion engine heat engine that burns fuel in a combustion chamber within an engine to give an output of mechanical energy.

Internal energy the energy stored in a material system.

Internet international computer network, linking university, commercial, private, and government databases.

Ion a charged particle created when a neutral atom or molecule gains (**anion**) or loses (**cation**) one or more electrons.

Ionic bond electrostatic force that holds **anions** and **cations** together in a ionic compound.

Ionization energy the energy required to remove an electron or electrons from an atom.

Isomers compounds with the same molecular formula and therefore mass but with different structural formulae, and therefore different physical and chemical properties.

Isotopes atoms with the same number of protons and electrons, but with different number of neutrons, and therefore differing atomic masses.

Jet engine form of **internal combustion engine** that propels a vehicle by the reactive force produced by the expulsion of hot gases.

Kinetic energy energy of an object arising from its motion.

Kinetic theory theory that explains the physical properties of gases by the movement of its molecules.

Laser acronym for *l*ight *a*mplification by the *s*timulated *e*mission of *r*adiation. An optical device that produces an intense beam of parallel, cohesive light of a single wavelength.

Latent heat heat energy absorbed or released by a substance as it changes state (e.g. from gas to liquid) without changing temperature or pressure.

Lens device for focusing light, or other forms of **electromagnetic** radiation.

Lift The upward force on an aircraft that keeps it in airborne.

Light electromagnetic radiation with a wavelength of between about 400 and 800 nm, that can be detected by the human eye.

Light dependent resistor electronic component, the electrical resistance of which changes with the amount of light striking its surface.

Light-year the distance traveled by light in a vacuum in one year (approximately 5.875×10^{12} miles / 9.46×10^{12} km).

Linear motor type of electric induction motor in which the stator and rotor are straight and parallel.

Liquid state of matter in which it takes the shape of its container. A liquid has a fixed volume at a fixed temperature.

Lithography printing method in which a plate is treated so that some areas attract ink, while other repel it.

Logic gate basic electronic component used in building integrated circuits.

Longitudinal wave wave in which the particle oscillations are parallel to the direction of propagation.

Loudness a subjective measure of the intensity of a sound.

Loudspeaker device that converts an electrical signal into sound by the oscillation of a diaphragm or horn.

Magnet any material capable of producing a magnetic field.

Magnetic field region of space around a magnet in which a force acts on any other magnet or electric charge present.

Magnetic tape ribbon with a magnetic coating on one side, which stores data as varying patterns of magnetization.

Magnetism all phenomena associated with magnets and magnetic fields.

Mass the quantity of matter in a body.

Mass number the number of protons and neutrons in the nucleus of an atom.

Mass spectrometer scientific instrument for analyzing chemical composition.

Micrograph photographic image taken with a microscope.

Microphone device that converts sound waves into an electrical signal.

Microscope instrument that magnifies the image of a small object through a combination of lenses.

Microwave electromagnetic radiation with a wavelength between 1 mm and 30 cm, used in radar, communications, and microwave ovens.

Mixture a substance that contains two or more elements or compounds that are not chemically combined.

Modulation adaptation of a signal for transmission by combining it with a (usually high frequency) carrier wave.

Molecule a unit of two or more atoms held together by chemical bonds.

Momentum the product of the mass and velocity of a moving object .

Monomer molecule that is the basic building block of a polymer.

Monopole hypothetical magnet with a single pole.

Motherboard main circuit board of a microcomputer.

Nanotechnology building of devices on a molecular scale.

Network the electronic interlinking of telephones, computers, or other communication devices.

Neutrino elementary particle with zero charge and zero mass.

Neutron basic constituent of the atom, located in the nucleus, with a **mass number** of 1 and no electric charge.

Newton's first law of motion an object will remain at rest or in constant motion unless acted on by a force.

Newton's second law of motion the acceleration of a body is directly proportional to the force applied to it and inversely proportional to its mass.

Newton's third law of motion every action (force) produces an equal and opposite reaction.

Noble gas unreactive element belonging to Group 8 of the **Periodic Table.**

Node point of minimum displacement in a wave. Points of maximum displacement are called antinodes.

Noise unwanted sound or signal.

Nuclear fission the splitting of an atomic nucleus into two smaller nuclei, with the release of energy and neutrons.

Nuclear fusion the fusion of light atomic nuclei into a heavier nucleus, with the release of energy.

Nucleon a **proton** or **neutron.**

Nucleus the positively charged central part of an atom, which consists of protons and neutrons.

Objective principal lens or mirror of an optical instrument such as a microscope or telescope.

Observatory building that houses an astronomical telescope.

Optical fiber thin, ultrapure glass fiber through which light travels by total internal reflection.

Optics the study of light and vision.

Orbit the circular or elliptical path of a body in space around another.

Orbital the region in space around an atomic nucleus occupied by an **electron.**

Ore a naturally occurring rock or sediment that contains valuable minerals.

Organic compound compound that contains carbon in combination with elements such as hydrogen, nitrogen, oxygen, and sulfur.

Oscillator any device that produces a mechanical or electrical oscillation.

Osmosis the movement of solvent molecules through a semipermeable membrane from a dilute solution to a concentrated one.

Oxidation the loss of electrons from a substance in a chemical reaction.

Ozone an **allotrope** of oxygen comprising three oxygen atoms.

Parallel circuit electric circuit in which the current is divided between two or more parallel conductors.

Particle accelerator scientific apparatus for accelerating charged particles to high speeds and energies.

Pasteurization the heat treatment of food at temperatures below the boiling point of water to kill harmful organisms.

Period the time taken for one complete cycle of an oscillation.

Periodic Table table of all the elements, arranged in order of atomic number, which groups together elements with similar chemical properties.

Petrochemical chemicals derived from crude oil or natural gas.

Phase (chemistry) one of the three states of matter: gas, liquid, or solid.

Phase (physics) a stage in the cycle of a wave or oscillation.

Photodiode semiconductor diode used to detect or measure the intensity of light.

Photoconductivity property of a material that conducts electricity only when struck by light.

Photoelectric effect the emission of an electron from the surface of a metal when struck by a **photon** of light.

Photon particle of light or other electromagnetic radiation.

Piezoelectric effect the property of certain crystals by which they produce electricity when under mechanical stress, and conversely change shape when a voltage is applied across them.

Pixel acronym of picture element, smallest unit of controllable color and brightness in an electronic image, such as a video or computer image.

Plasma electrically conductive gas composed of equal numbers of positive ions and free electrons.

Polarized light light that oscillates in just one direction at right angles to the direction of propagation.

Polymer giant organic molecule, comprising repeating sequences of smaller molecules.

Positron positively charged elementary particle with zero mass; the antiparticle of the **electron.**

Potential (electric) the potential energy of a point charge in an electric field.

Potential difference the work done in moving an electric charge between two points at different electric potentials.

Potential energy the energy a body has by virtue of its position or state.

Power the rate of change of energy, or the rate of doing work.

Pressure force applied per unit area.

Primary colors the three pure colors, or wavelengths, of visible light (red, blue, and green) that produce white light when in combination.

Prism A piece of glass or other transparent material used in optics to bend a beam of light or to separate light into its component colors.

Protein organic polymer made up of chains of amino acids.

Proton basic constituent of the atom, located in the nucleus, with a mass number of 1 and carrying a positive electric charge.

Quantum indivisible unit of any form of physical energy.

Quark elementary particle that is the fundamental constituent of all hadrons (protons, neutrons, and mesons).

Quasar a very distant starlike object.

Radar acronym for *radio detection and ranging*. Device for locating objects using the detected reflections of high frequency radio waves.

Radiation emission of heat, radioactive particles, or electromagnetic waves.

Radio telescope telescope that collects and detects radio waves.

Radio wave electromagnetic radiation with a wavelength between 1 mm and 1000 m, produced by the oscillation of an electric charge.

Radio dating archaeological dating technique in which, by knowing the **half-life** of a radioactive material and its abundance in a sample, the age of the sample can be determined.

Radioactivity the spontaneous disintegration of atomic nuclei, accompanied by the emission of particles and/or **electromagnetic** radiation.

Radiography the production of images using X-rays.

Random-Access Memory (RAM) computer memory that provides rapid access to temporarily stored information.

Rarefaction region of decreased pressure along a longitudinal wave.

Ray narrow beam of light.

Reactant substance that takes part in a chemical reaction.

Read only memory (ROM) computer memory that provides a permanent store of information.

Rectifier electrical device that converts alternating into direct current.

Redshift apparent change in the wavelength of light produced when the distance between the source and the observer is increasing.

Reduction the gain of electrons by a substance in a chemical reaction.

Reflection the deflection of waves when they strike a surface.

Refraction the bending of waves when they pass from a medium of one density to another.

Refractive index the ratio of the speed of light in one medium to its speed in another. A measure of the angle through which light is bent (the degree of refraction) as it passes from one medium to another.

Resistance a measure of the degree to which a substance restricts the flow of electric current through it when a **potential difference** is applied.

Resolution ability of an imaging system to differentiate between two closely spaced objects.

Resolving power a measure of the **resolution** of an optical system.

Resonance a strong vibration produced when a body is made to oscillate at its natural frequency.

RISC acronym for *Reduced Instruction Set Computing*, a design of microprocessor that carries out relatively few simple instructions, allowing faster processing and simpler design.

RNA (*RiboNucleic Acid*) giant organic molecule. Like **DNA**, a store of genetic information in living cells.

Salt an **ionic** compound formed by the reaction of an **acid** with a **base**.

Salt common name for sodium chloride.

Sampling the measurement of the magnitude of an analog signal in order to turn it into a digital form.

Satellite natural or artificial object orbiting a celestial body.

Scanning electron microscope (SEM) instrument that scans a focused beam of electrons across the surface of a specimen to produce a magnified three-dimensional image of the specimen.

Scanning tunneling microscope instrument that moves a charged tungsten probe across the surface of a specimen to produce a magnified image.

Semiconductor material with electrical conductivity somewhere between that of a metal and an insulator.

Series circuit electric circuit in which the current flows through each conducting section in turn.

Sidereal time measurement of time based on the rotation of the earth with respect to the stars.

SIMMS acronym for *Single In-line Memory Modules*, plug-in memory expansion cards widely used in today's personal computers.

Simple harmonic motion oscillatory movement in which a particle (or body) moves so that its acceleration is proportional to its distance from a fixed point and is directed toward that point.

Software general term for all computer programs or applications.

Solar wind a stream of charged particles from the sun.

Solenoid coil of wire that creates a magnetic field like that of a bar magnet when electric current flows through it.

Solid state of matter in which a substance has a definite shape. Its molecules do not move relative to each other, but vibrate about their fixed positions.

Solute the substance dissolved in a solvent to make a solution.

Solution a homogeneous mix of two or more nonreacting substances.

Solvent the medium in which a solute is dissolved to form a solution.

Sonar acronym for *sound navigation and ranging*, The location of underwater objects using sound waves.

Sound mechanical wave energy, in the form of compressions and rarefactions, detectable by the human ear.

Specific heat capacity amount of heat energy that needs to be absorbed by a unit mass of a body to raise its temperature by one Celsius degree.

Spectrometer scientific instrument used to measure the wavelength of electromagnetic radiation emitted by a source, such as a star, and hence to determine its composition.

Spectroscopy the study of **spectra**.

Spectrum the full range of wavelengths of electromagnetic radiation.

Speed the rate at which an object moves.

Speed of light speed at which light and other forms of electromagnetic radiation travel (speed of light in a vacuum is 3×10^8 m/s).

Spherical aberration loss of definition of an image as a result of different parts of a lens or mirror having different **focal lengths** and so bringing parallel light rays to a focus at different points.

Spontaneous emission process by which atoms naturally release a photon.

Standing wave a wave whose **node** and antinode (points of maximum and minimum amplitude) do not change position along the wave.

Static electricity stationary electric charge on an object.

Stimulated emission process by which a photon interacts with an atom, causing it to emit a second photon. Distinct from **spontaneous emission**.

Strain measure of the deformation of a material when a force is exerted on it.

Stress deforming force per unit area acting on a body.

Strong nuclear force fundamental force responsible for binding **protons** and **neutrons** together in atomic nuclei.

Sublimation the change of a substance from a solid into a vapor, without passing through the liquid phase.

Superconductivity property of certain materials that have zero electrical resistance at temperatures approaching **absolute zero**.

Supercooling the cooling of a substance below its normal boiling or freezing point without a change of phase.

Supersonic speeds greater than Mach 1, that is, exceeding the speed of sound.

Surface tension tension caused by attractive forces between the molecules or atoms of a liquid, which makes the surface of the liquid behave as if it were a thin elastic skin.

Suspension mixture in which large, visible particles are dispersed throughout a liquid.

Swash plate helicopter component that controls the angles of the rotor blades and so the direction of flight.

Telescope optical instrument that uses two or more sets of lenses or mirrors to magnify the distant objects.

Tension reactive force exerted by a body when it is subjected to stress.

Thermodynamics study of energy flow and conversion.

Torque external force that causes an object to turn.

Transducer device that transforms one form of energy into another.

Transformer electrical device that increases or decreases the voltage of an alternating current.

Transistor semiconductor device that regulates a current passing through it.

Transition metals elements of the Periodic Table that have incompletely filled d or f orbitals.

Transverse wave wave in which the oscillations are at right angles to the direction of propagation.

Turbine machine consisting of a set of blades mounted on a central shaft, that is made to rotate by a moving fluid, usually in order to drive a generator.

Ultrasound very high-frequency sound waves that lie outside the range of human hearing. Used in imaging.

Ultraviolet radiation electromagnetic radiation with wavelengths between 10 and 400nm.

Valence band energy level at which the electrons cannot take part in the conduction of electricity.

Velocity rate at which an object moves in a specified direction.

Viscosity property of a fluid that causes it to resist flow or the motion of an object through it.

Voltage another term for electrical **potential difference**.

Voltmeter instrument that measures potential difference.

Volume the physical space occupied by an object.

W particle elementary particle; a carrier of the weak force.

Wave oscillating movement of particles or fields that propagates through space, carrying energy.

Wave-particle duality the theory that matter and radiation exhibit both wave and particle properties.

Wavelength the distance between successive crests or troughs of a wave.

Weak nuclear force fundamental force responsible for radioactive decay and other subatomic reactions.

Weight the force of gravity exerted on an object.

Work a measure of the change in energy when a force causes an object to move.

X-ray electromagnetic radiation with wavelengths in the range 0.001 to 10nm, produced when high-speed electrons strike a solid target.

X-ray diffraction the scattering of X-rays by the atoms and molecules of a crystalline solid. The phenomenon is used to determine the structure of crystals.

Xerography dry photographic copying process that uses **electrostatic** attraction to produce an image.

Z particle elementary particle; a carrier of the weak force.

Acknowledgments

Picture Credits

Abbreviations
SPL: Science Photo Library
T: Top B: Bottom C: Center L:Left R:Right

Jacket L SPL/Pasieka Jacket TR Spectrum/Bavaria Verlag Jacket BR SPL/Syred

1 ZEFA/Kalt 2 Robert Harding Picture Library/Koch 10 SPL/Brake 16 Volvo UK Ltd 21 Fiat Ferroviairia 23 ZEFA/Streichan 25 Rex Features 28T Airbus Industrie 28/9 Quadrant Picture Library 29T Airbus Industrie 31 Kamen Aircraft 35 Alberto Incucci 37 Rex Features/Timbault 41 Cliff Bolton/Siemens Plessey Systems 42 SPL/Burgess 44 Fotocentrum Zimmerman GmbH 50 ZEFA/Streichan 51 Sony UK Ltd 53 Digidesign UK Ltd 54 SPL 58 Frank Spooner Pictures 61L/R SPL/ Plailly 62/63 Richard Clark 66 Michael Freeman 67TL/BL Sygma 67BC First Independent 67BR Rex Features 68 ILFORD Anitec UK 69 Richard Clark 71 Jeff Robb 77 SPL 79 SPL/Morgan 82 Impact 86 SPL/Taylor 98 ZEFA/Horowitz 100 ZEFA 102 SPL 104 SPL/Vick 106 ZEFA 112T SPL 112B SPL 113 ZEFA 114 ZEFA/Halin 117 The National Grid Company plc 118 Spectrum Colour Library 121 QA Photos 122 ZEFA/Raga 125 SPL 126 SPL/Shambroom 128 SPL/Menzel 129 ZEFA 134 SPL136 SPL/Plailly 137L SPL/Burgess 1237R JEOL (UK) Ltd 138T SPL/Yoshihaki 138B SPL/Gohiers 139 SPL/Greenwich Observatory 140 SPL/Miller 141 Rex Features/Sipa 142L SRS Daresbury 142R SPL/Menzel 145 CERN 146T SPL/Plailly 146B IBM 147 SPL/Menzel 151TL/TR Philips Medical Systems 151C Siemens plc 153 SPL/Kulyk 154 Acuson UK Ltd 155TL Philips Medical Systems 155TR Acuson UK Ltd 156L SPL 156R SPL/Mason 157 SPL/Fielding 159 KeyMed 160 SPL/ESC 163 Japan Society of Aeronautics and Space Sciences 165 Associated Press 166 SPL/NASA 167 TRH Pictures 169 SPL/Ressmeyer 170T SPL/NASA 170B NASA 173 SPL/NASA 174 SPL/Feldman/NASA 175 ESA 176 SPL/NASA 177 SPL/NASA 179 ZEFA 181T SPL/Ressmeyer 181B SPL/NASA 182 SPL/Paseika 184 SPL/Holmes 186 Robert Harding Picture Library/Tettoni 189 Rex Features/Gardner 190 ZEFA 195TL SPL/Livermore National Library 195TR SPL/US Department of Energy 195C SPL/Plailly 196 SPL/Phillips 197 ZEFA 198T Holt Studios 198B DBP 199TL DBP 199TR SPL/Scharf 199C Ian Howes/DBP 200L Biofotos 200C SPL 200R FLPA/Hosking 201 SPL 202 SPL 204 SPL/Burgess 205 SPL/Parker 207 SPL/CNRI 208 SPL/CERN 212 SPL/Plailly 213L SPL/Joyce 213C ZEFA 213R SPL/Burgess 214 SPL/Fielding 218 Hutchison Library 219 Hutchison Library/Smith 220T Robert Harding Library/Woolfitt 220C Impact Photographers/Visa/CEDRI 221 SPL/Freeman 222 SPL/Kage 223T SPL/Revy 223C Impact Photographers/Perri 224T SPL/Nuvidsavy & Perennou 224B SPL/Parker/ESA 225T ZEFA 225B Frank Spooner Pictures/Stoddart 226T SPL/Gugliemo 226B FLPA/Life Science 227 Hutchison Library/Taylor 228 SPL/Nuvidsavy & Perennou 229 FLPA 230 SPL/McLenaghan 231 ZEFA/Heilman 232 SPL/NASA 234 SPL/Ressmeyer 235T LAT 235B SPL/NASA 236T ZEFA/Walther 236B Arcaid/Bryant 239 ZEFA 241C SPL/McGrath 241B SPL/Takehara 243T Impact/Ernoult 243C LSO/Keith Glossop 246 SPL/Ressmeyer 248 SPL/Mequa 251 ZEFA 252 SPL/Seth Loel 253 SPL/Scharf 254 ZEFA 256 ZEFA 259 ESA/FRSI 261 Impact/Stephens 263T SPL/Parker 263C1 SPL/CNRI 263C2 SPL/Tompkinson 263C3 SPL/Pasieka 263B Zefa/Streichan 266 SPL/Sinclair Stammers 296 SPL/Lorenz/CERN 270 SPL 271 SPL/Berkeley Laboratory

The publishers gratefully acknowledge the generous assistance of the following companies and institutions:

Transport Airbus Industrie, Blagnac, France; British Aerospace Airbus Ltd, Filton, Bristol, UK; British Hovercraft Corporation, Isle of Wight, UK; Boeing, Seattle, USA; Brittany Ferries, London, UK; Ford Motor Co. Ltd, Brentwood, UK; GMC-CANDIVE Ltd, Aberdeen, UK; Hoverspeed Ltd, Dover, UK; GEC Alsthom, Paris, France; Kaman Corporation, Bloomfield, USA; Ministry of Defence, London, UK; Oceaneering International Services Ltd, Aberdeen, UK; P&O European Ferries, Dover, UK; Rodriquez Cantieri Navali SPA, Messina, Italy; Rolls Royce, Civil Engine Office, Derby, UK; Rolls Royce, Filton, Bristol, UK; Rolls-Royce Industrial & Marine Gas Turbines Limited, Coventry, UK; Singapore Mass Rapid Transit Corporation, Singapore; Today's Railways, Douai, France; Westinghouse Signals Limited, Chippenham, UK; Westland Group plc, Yoevil, Somerset, UK; Woods Hole Oceanographic Institution, Massachusetts, USA

Information and Entertainment AGFA-GEVAERT LTD, Middlesex, UK ; Apricot Computers Limited, Birmingham, UK ; ARRI (GB) Ltd, Middlesex, UK; BT Museum, London, UK; BT Headquarters, London, UK; Canon (UK) Ltd, Surrey, UK; Cinemeccanica, Milano, Italy; Graph Techniques, Maidenhead, UK; Hitachi (UK) Limited, Middlesex, UK; IBM United Kingdom Limited, Hampshire, UK; Ilford Photo Co., London, UK; JFPR Ltd, Egham, UK; JVC (UK) Limited, London, UK; KBA (UK) LTD, Watford, UK ; Kinoton GmbH, Industriestasse 20a, Germany; Komori U.K. LIMITED, Leeds, UK; National Museum of Photography Film & Television, Bradford, UK; Nikon UK Limited, Kingston upon Thames, UK; O'Brien Associates Ltd, London, UK; Panasonic (Matsushita), London, UK; Pentax UK Ltd, South Harrow, UK; Philips Communication Systems, Cambridge, UK; Sony Semiconductor Europe, Basingstoke, UK

Power and Manufacturing ABB Robotics Ltd, Milton Keynes, UK; Bonas Machine Company Limited, Gateshead, UK; British Coal Corporation, London, UK; British Nuclear Fuels plc, Warrington, UK; BP, UK; British Textile Machinery Association, Manchester, UK; Canary Wharf Ltd, London, UK; ETSU, Didcot, UK; Friends of the Earth, London, UK; Fanuc Robotics (UK) Ltd, Coventry, UK; GEC Alsthom, UK; James Howden and Co Ltd, Renfrew, Scotland, UK; JET Joint Undertaking, Abingdon, UK; The National Grid Company plc, Coventry, UK; National Power plc, London, UK; Nuclear Electric plc, Gloucester, UK; PowerGen plc, London, UK; Pulp and Paper Information Centre, Wiltshire, UK; Shell U.K., Aberdeen, UK; Siemens AG, Erlangen, Germany; Southern Water plc, Worthing, UK; Staubli Unimation Ltd, Telford, UK; Thames Water Utilities, Reading, UK; Mr J. Walisiewicz, London, UK; Yolles Partnership Limited, London, UK

Research and Medicine Department of Crystallography, Birkbeck College, University of London, UK; Carl Zeiss, Heidenheim, UK; European Laboratory for Particle Physics (CERN), Geneva 23, Switzerland; Fermi National Accelerator Laboratory, Batavia, Illinois, USA; JEOL (UK) Ltd, Welwyn Garden City, UK; KeyMed (Medical and Industrial Equipment) Ltd, Southend-on-Sea, UK; Kratos Analytical, Manchester, UK; Nikon UK Limited, Kingston upon Thames, UK; Omega Laser Systems Limited, London, UK; Philips Medical Systems Ltd, London, UK; Rover Group Ltd, Birmingham, UK; USAF Phillips Laboratory, New Mexico, USA; WM Keck Observatory, Hawaii, USA

Space British Aerospace (Space Systems) Ltd, Filton, Bristol, UK; European Space Agency, Paris, France; Langley Research Center, Virginia, USA; Lyndon B. Johnson Space Center, USA; Marshall Space Flight Center, Alabama, USA; McDonnell Douglas Aerospace, Hungtington Beach, USA; NASA Headquaters, Washington DC, USA

Molecules and Materials Acuson Limited, Uxbridge, UK; Aluminium Federation Ltd, Birmingham, UK; The British Plastics Federation, London, UK; BP Educational Service, Alton, UK; Brewers Society, London, UK; British Aerospace (Operations) Ltd, Filton, Bristol, UK; Esso UK plc, London, UK; Fuller Smith & Turner plc, London, UK; ICI plc, London, UK; ICI Katalco, Cleveland, UK; LINPAC Plastics International Limited, Knottingley, UK; Malaysian Rubber Producers' Research Association, Hertford, UK; MW Kellogg Ltd, Wembley, UK; Polymer Industry Education Centre, University of York, York, UK; Shell International, London, UK; Siemens Medical Engineering, Bracknell, UK; Zeneca, Cleveland, UK

Principles Bogod Machine Company; Ford Motor Co. Ltd, Brentwood, UK; Royal National Institute for Deaf People, London, UK; Norson Power Ltd; Glasgow, UK.